Introduction to Mobile Network Engineering

Introduction to Mobile Network Engineering

GSM, 3G-WCDMA, LTE and the Road to 5G

Alexander Kukushkin
PhD, Australia

Registered Offices
John Wiley & Sons, Inc., 111 River Street, Hoboken, NJ 07030, USA
John Wiley & Sons Ltd, The Atrium, Southern Gate, Chichester, West Sussex, PO19 8SQ, UK

Editorial Office
The Atrium, Southern Gate, Chichester, West Sussex, PO19 8SQ, UK

For details of our global editorial offices, customer services, and more information about Wiley products visit us at www.wiley.com.

Wiley also publishes its books in a variety of electronic formats and by print-on-demand. Some content that appears in standard print versions of this book may not be available in other formats.

Library of Congress Cataloging-in-Publication Data

Names: Kukushkin, Alexander, author.
Title: Introduction to mobile network engineering : GSM, 3G-WCDMA, LTE and
    the road to 5G / by Alexander Kukushkin.
Description: Hoboken, NJ : John Wiley & Sons, 2018. | Includes
    bibliographical references and index. |
Identifiers: LCCN 2018012499 (print) | LCCN 2018021194 (ebook) | ISBN
    9781119484103 (pdf) | ISBN 9781119484226 (epub) | ISBN 9781119484172
    (cloth)
Subjects: LCSH: Mobile communication systems. | Wireless metropolitan area
    networks.
Classification: LCC TK5103.2 (ebook) | LCC TK5103.2 .K85 2018 (print) | DDC
    621.3845/6–dc23
LC record available at https://lccn.loc.gov/2018012499

Cover design by Wiley
Cover image: © pluie_r/Shutterstock

Set in 10/12pt WarnockPro by SPi Global, Chennai, India

10 9 8 7 6 5 4 3 2 1

To my family

Contents

Foreword

From the 1990s to the present, three generations of mobile radio networks have been deployed in every country of the world. Those networks connect billions of customers and provide mobile communications services. Mobile radio communications have become ubiquitous throughout the world. People are getting used to the technology through commercial mobile phones. The mobile network infrastructure that enables communications has become a normal part of the urban environment in which people live. There is also a great number of other applications for mobile radio that are essential in the modern world and are used in navigation, transportation, machine-to-machine communications (M2M), robotics, emergency and low enforcement services, broadcasting, space exploration, the military, and so on. The mobile radio is, in fact, a part of a more widely defined wireless technology that, of course, includes wireless LANs (Wi-Fi) with fixed and nomadic access.

The content of this book is limited to three major mobile communication technologies: GSM, 3G-WCDMA and LTE with the major focus on Radio Access Network (RAN) technology. We introduce some basic concepts of mobile network engineering used in the design and rollout of mobile networks. Then we cover principles, design constraints and provide a more advanced insight into the radio interface protocol stack, operation and dimensioning for three major mobile network technologies; the Global System Mobile (GSM), third (3G-WCDMA) and fourth generation (4G-LTE) mobile technologies that have been recently deployed or are shortly to be deployed. Enhancements of fourth generation technology in LTE-Advanced (LTE-A) are described at the level of conceptual design.

The concluding sections of the book are concerned with further development towards the next generation of mobile networks (5G). The last section describes some key concepts that may bring significant enhancements in network operation efficiency and quality of services experienced by customers. A development of the fifth generation of mobile networks can be regarded as a mix of evolutionary advances in 4G LTE through LTE-A and new radio technology likely operating in newly allocated spectrum bands. This development covers a broad area of applications and many different topics that require specifically dedicated study. Therefore, many interesting and important topics such as the Internet of Things, massive MTC, developments in new technology for emergency services based on LTE, integration of the mobile radio access network and Wi-Fi are out of the scope of this book.

Since the standards for 5G are still in development, most of the features of the new radio technology are related to 3GPP Release 15. Some breakthrough technological advances are planned for further releases of 5G, such as a Full Duplex and self-backhauling and are described as concepts rather than commercially available technology.

While many excellent books on mobile radio networking are available, I think many more will be published in the near future since the subject is continuously evolving. This book is intended to provide a generalist and compressed description of major technologies utilized in the radio access part of modern mobile networks. I envisage readers are engineers in relatively early stages of their careers in the mobile wireless industry. Some of them may be taking a post-graduate course to enhance their knowledge. They may include operation support engineers, technical sale/presale engineers, technical and account managers who may need or wish to enhance or expand their knowledge of mobile network system engineering. Each major technology section of the book consists of introductory material, a more advanced part and a summary.

Alexander Kukushkin

Acknowledgements

I thank Professor Branka Vucetic, School of Electrical and Information Engineering, University of Sydney, for the invitation to teach at the University that led to the writing of this book. I wish to thank the reviewers of the book for their constructive comments that helped to improve and extend the content, especially on the 5G related topics.

Abbreviations

3G	Third Generation
3GPP	3rd Generation Partnership Project
5G	Fifth Generation of mobile networks
5GC	5G Core network
5G-S	5G System
TMSI	Temporary Mobile Subscription Identifier
AA	Antenna Array
AAA	Authentication, Authorization & Accounting
AAS	Active Antenna System
ACK	ACKnowledgement
ADC	Analogue to Digital Converter
AF	Application Function
AGCH	Access Grant CHannel
AICH	Acquisition Indicator CHannel
AKA	Authentication and Key Agreement
AM	Acknowledged Mode
AMC	Adaptive Modulation and Coding
AMF	Access and Mobility Management Function
AMR	Adaptive Multi-Rate (coding)
ARFCN	Absolute Radio Frequency Channel Number
ARQ	Automatic Repeat reQuest
ATCA	Advanced Telecommunications Computing Architecture
AUC	AUthentication Centre
AUSF	Authentication Server Function
BALUN	BALanced to UNbalanced conversion
BBU	Base Band Unit
BCCH	Broadcast Control CHannel
BCH	Broadcast CHannel
BLER	BLock Erasure Rate
BMC	Broadcast/Multicast Control
BS	Base Station
BSC	Base Station Controller
BSIC	Base Station Identity Code
BSS	Base Station Subsystem
CA	Carrier Aggregation

CAC	Call Admission Control
CC	Component Carrier
CCCH	Common Control Channel
CCE	Control Channel Element
CCPCH	Common Control Physical Channel
CCTrCH	Coded Composite Transport Channel
CDD	Cyclic Delay Diversity
CDM	Code Division Multiplexing
CDMA	Code Division Multiple Access
CIR	Carrier to Interference Ratio
COMP	COordinated MultiPoint transmission and reception
CP	Cyclic Prefix
CPCH	Common Packet Channel
CPICH	Common Pilot Channel
CP-OFDM	Cyclic Prefix-OFDM
CPRI	Common Public Radio Interface
CQI	Channel Quality Indicators
C-RAN	Centralized Radio Access Network
CRC	Cyclic Redundancy Check
CRNC	Controlling RNC
CRNTI	Cell Radio Network Temporary Identifier
CRS	Cell RS
CSCH	Compact Synchronization Channel
CSFB	Circuit Switched Fall Back
CSI	Channel State Information
CSI-RS	Channel State Information Reference Signal
CTCH	Common Traffic Channel
DAC	Digital-to-Analogue Convertor
DC	Dual Connectivity
DCCH	Dedicated Control Channel
DCH	Dedicated Transport Channel
DCI	Downlink Control Information
DeNB	Donor eNB
DFT	Discrete Fourier Transform
DFTS-OFDM	DFT Spread-OFDM
DL PCC	Downlink Primary Component Carrier
DL SCC	Downlink Secondary Component Carrier
DLL	Data Link Layer
DL-SCH	Downlink Shared CHannel
DMRS	DeModulation Reference Signal
DN	Data Network
DNN	Data Network Name
DPCCH	Dedicated Physical Control CHannel
DPCH	Dedicated Physical CHannel
DPD	Digital Pre-Distortion
DPDCH	Dedicated Physical Data Channel
DRNC	Drift RNC

DRX	Discontinuous Transmission and Reception
DSCH	Downlink Shared Channel
DTCH	Dedicated Traffic Channel
e2e	End to End
E-AGCH	E-DCH Absolute Grant CHannel
ECCE	Enhanced Control Channel Element
ECM	EPS Connection Management
E-DCH	Enhanced Dedicated Channel
EDGE	Enhanced Data rate for GSM Evolution
E-DPCCH	E-DCH Dedicated Physical Control CHannel
E-HICH	E-DCH Hybrid ARQ Indicator CHannel
EIR	Equipment Identity Register
eMBB	Enhanced Mobile Broadband
EN-DC	E-UTRA-NR Dual Connectivity
EPC	Evolved Packet Core
EPDCCH	Enhanced Physical Downlink Control CHannel
EPS	Evolved Packet System
EREG	Enhanced Resource Element Group
E-RGCH	E-DCH Relative Grant Channel
E-TFC	E-DCH Transport Format Combination
ETSI	European Telecommunications Standards Institute
E-UTRA	Evolved UMTS Radio Access
E-UTRAN	Evolved UTRAN
FACCH	Fast Associated Control Channel
FACH	Forward Access Channel
FBI	Feedback Information
FCCH	Frequency Correction Channel
FDD	Frequency Division Duplex
FDM	Frequency Division Multiplexing
FDMA	Frequency Division Multiple Access
F-DPCH	Fractional DPCH
FDPS	Frequency Domain Packet Scheduling
FEC	Forward Error Correction
FER	Frame-Error Rate
FFT	Fast Fourier Transform
FN	Frame Number
FR	Full Rate
GBR	Guaranteed Bit Rate
GGSN	Gateway GPRS Support Node
GMSC	Gateway MSC
GMSK	Gaussian Minimum Shift Keying modulation
GPRS	GSM Packet Radio Service
GSM	Global System Mobile
GTP	GPRS Tunnelling Protocol
HARQ	Hybrid ARQ
HLR	Home Location Register
HR	Half Rate

HSDPA	High Speed Downlink Packet Access
HS-DPCCH	High-Speed Dedicated Physical Control CHannel
HS-DSCH	High-Speed Downlink Shared CHannel
HSS	Home Subscriber Server
HS-SCCH	High-Speed Shared Control Channel
HSUPA	High Speed Uplink Packet Access
HW	Hardware
iFFT	inverse FFT
IMEI	International Mobile Station Equipment Identity
IMS	IP Multimedia Subsystem
IMSI	International Mobile Subscriber Identity
IPsec	IP Security protocol
ISHO	Inter-System Handover
ISI	Inter-Symbol Interference
IWF	Interworking Function
LA	Location Area
LAC	Location Area Code
LAI	Location Area Identifier
LAN	Local Area Network
LLC	Logical Link Control
LNA	Low Noise Amplifier
LOS	Line Of Sight
LPMA	Lattice Partition Multiple Access
LTE	Long Term Evolution
M2M	Machine to Machine communications
MAC	Medium Access Control
MAHO	Mobile Assisted HandOver
MAPL	Maximum Allowable Path Loss
MCC	Mobile Country Code
MCG	Master Cell Group
MCS	Modulation Coding Scheme
MeNB	Master eNB
MgNB	Master gNB
MHA	Mast Head Amplifier
MIB	Master Information Block
MIMO	Multiple Input Multiple Output
MME	Mobility Management Entity
MMI	Man-Machine Interface
MN	Master Node
MNC	Mobile Network Code
MRC	Maximum Ratio Combining
MR-DC	Multi-RAT Dual Connectivity
MS	Mobile Station (mobile phone)
MSC	Mobile Switching Centre
MSISDN	Mobile Subscriber ISDN Number
MSRN	Mobile Station Routing Number
MT	Mobile Termination

MTC	Machine Type Communications
MTCH	Multicast Traffic Channel
MU-MIMO	Multi-User MIMO
MUST	Multiuser Superposition Transmission
NACK	Negative ACKnowledgement
NAS	Non-Access Stratum
NB-IoT	Narrow-Band Internet of Things
NDC	National Destination Code
NE-DC	MR-DC with the 5GC
NEF	Network Exposure Function
NF	Network Functions
NFV	Network Function Virtualization
NGEN-DC	NG-RAN E-UTRA-NR Dual Connectivity
NGMN	Next Generation Mobile Network Alliance
NG-RAN	New Generation Radio Access Network
NOMA	Non-Orthogonal Multiple Access
NR	New Radio
NRF	NF Repository Function
NSS	Network Switching Subsystem
NSSAI	Network Slice Selection Assistance Information
NSSF	Network Slice Selection Function
OAM	Operation, Administration and Maintenance
OBSAI	Open Base Station Architecture Initiative
OFDMA	Orthogonal Frequency Division Multiple Access
OMC	Operation and Maintenance Center
OSI	Open System Interconnect
OSS	Operation Support Subsystem
OVP	Over Voltage Protection
OVSF	Orthogonal Variable Spreading Factor
PACCH	Packet Associated Control Channel
PAPR	Peak-to-Average Power Ratio
PCC	Primary Component Carrier
PCCCH	Packet Common Control Channel
P-CCPCH	Primary Common Control Physical Channel
PCell	Primary Cell
PCF	Policy Control Function
PCFICH	Physical Control Format Indicator Channel
PCH	Paging Channel
PCPCH	Physical Common Packet Channel
PCRF	Policy Charging and Rules Function
PCU	Packet Control Units
PDCH	Packet Data CHannel
PDCP	Packet Data Convergence Protocol
PDP	Packet Data Protocol
PDSCH	Physical Downlink Shared CHannel
PDTCH	Packet Data Traffic CHannel
PDU	Packet Data Unit

P-GW	Packet Data Network Gateway
PHICH	Physical Hybrid-ARQ Indicator Channel
PICH	Paging Indicator Channel
PIN	Personal Identification Number
PLMN	Public Land Mobile Networks
PMI	Precoder Matrix Indication
PRACH	Physical Random Access Channel
PRB	Power Resource Block
P-RNTI	Paging Group Identity
PSC	Primary Scrambling Code
P-SCH	Primary Synchronization Channel
PSS	Primary Synchronization Signal
PSTN	Public Switching Telephone Network
PTCCH	Packet Timing advance Control Channel
PTCH	Packet Traffic Channel
PT-RS	Phase-Tracking Reference Signals
PUCCH	Physical Uplink Control CHannel
QCI	QoS Class Indicator
QoE	Quality Of user Experience
QoS	Quality of Service
RAB	Radio Access Bearer
RACH	Random Access CHannel
RAN	Radio Access Network
RAT	Radio Access Technology
RAU	Routing Area Update
RB	Resource Block
RDN	Radio Distribution Network
REG	Resource Element Group
RF	Radio Frequency
RI	Rank Indication
RLC	Radio Link Control
RN	Relay Node
RNC	Radio Network Controller
RP	Reference Point
RRC	Radio Resource Control
RRH	Remote Radio Head
RRM	Radio Resource Management
RRU	Remote Radio Unit
RS	Reference Signals
SACCH	Slow Associated Control Channel
SAE	System Architecture Evolution
SAW	Stop-And-Wait
SCC	Secondary Component Carrier
SCell	Secondary Cell
SC-FDMA	Single Carrier FDMA
SCG	Secondary Cell Group
SCH	Synchronization Channel

S-CPICH	Secondary Common Pilot Channel
SDCCH	Standalone Dedicated Control Channel
SDN	Software Defined Networking
SDR	Software Designed Radio
SDU	Service Data Unit
SF	Spreading Factor
SFN	System Frame Number
SFP	Small Form factor Pluggable
SgNB	Secondary gNB
SGSN	Serving GPRS Support Node
S-GW	Serving Gateway
SIB	System Information Block
SIC	Successive Interference Cancellation
SIM	Subscriber Identity Module
SINR	Signal to Interference and Noise Ratio
SIP	Session Initiation Protocol
SIR	Signal to Interference Ratio
SM	System Module
SMF	Session Management Function
SMG	Special Mobile Group
SN	Subscriber Number
SecN	Secondary Node
SNDCP	Subnetwork Dependent Convergence Protocol
S-NSSAI	Single Network Slice Selection Assistance Information
SON	Self-Organizing Network
SRB	Signalling Radio Bearer
SRNC	Serving RNC
SRS	Sounding RS
S-SCH	Secondary Synchronization Channel
SSS	Secondary Synchronization Signal
STR	Simultaneous Transmission and Reception
SU-MIMO	Single User-MIMO
SVD	Singular-Value Decomposition
SW	Software
TA	Terminal Adapter
TAB	Transceiver Array Boundary
TAG	Timing Advance Group
TAU	Tracking Area Update
TB	Transport Block
TBF	Temporary Block Flow
TCH	Traffic Channel
TCP	Transmission Control Protocol
TDMA	Time Division Multiple Access
TE	Terminal Equipment
TF	Transport Format
TFC	Transport Format Combination
TFCS	Transport Format Combination Set

TFI	Temporary Flow Identifier
TM	Transparent Mode
TMA	Tower Mounted Amplifier
TMSI	Temporary Mobile Subscriber Identity
TPC	Transmit Power Control
TrCH	Transport Channel
TRXUA	Transceiver unit array
TS	Time Slot
TTI	Transmission Time Interval
UDM	Unified Data Management Function
UE	User Equipment
UL PCC	UpLink Primary Component Carrier
UL SCC	UpLink Secondary Component Carrier
UL-SCH	UpLink Shared CHannel
UM	Unacknowledged mode
UMTS	Universal Mobile Telecommunication System
UPF	User Plane Function
URLLC	Ultra-Reliable and Low Latency Critical Communications
USB	Universal Serial Bus
USF	Uplink State Flag
USIM	Universal Subscriber Identity Module
VAS	Value Added Services
VLR	Visited Location Centre
VoIP	Voice over Internet Protocol
WCDMA	Wideband Code Division Multiple Access
Wi-Fi	Wireless local area networking

1

Introduction

Over the last few decades, mobile radio communications have become ubiquitous throughout the world. People have become accustomed to the technology through commercial mobile phones. The mobile network infrastructure that enables communications has become a normal part of urban environment in which people live.

There is also great number of other mobile radio applications essential in the modern world that are used in navigation, transportation, machine-to-machine communications (M2M), robotics, emergency and low enforcement services, broadcasting, space exploration, the military and so on. Mobile radio is, in fact, a part of more a widely defined wireless technology that, of course, includes wireless LANs (WiFi) with fixed and nomadic access.

Each application was developed on the basis of specific needs and, in some aspects, the mobile radio networks for emergency services and commercial mobile services are different. Nonetheless, the underlying principles in mobile communications, such as radio link design given performance constraints, separation of control and traffic channels, mobility support, principles of the channel allocation in the cell, radio network management and so on, have lots in common in many applications. Moreover, some of the commercial technologies, such as LTE, now appeared to support land mobile radio applications for emergency and public safety services.

This book is written as a modified and expanded set of lectures on the wireless engineering course I had privilege to teach at the University of Sydney, Australia for a couple of years. Most of the concepts of these lectures were adopted from published standards and also based on personal experience in the field as well as from some works of other authors. The course was delivered as post-graduate study. The assumption was made that the fundamentals of digital communications were already known to attendees and the objective was to explain the subject using mathematical arguments as little as possible; that is, close to common practice in the commercial communications industry. The target audience are engineers who are involved in either network operations or technical pre-sale. The content is limited to major three mobile communication technologies: GSM, 3G-Wideband Code Division Multiple-Access (WCDMA) and LTE with the major focus on radio access network (RAN) technology. The core part of the network is a complex subject on its own and is described only to discuss its role in e2e procedures and interfaces with the radio network.

Introduction to Mobile Network Engineering: GSM, 3G-WCDMA, LTE and the Road to 5G, First Edition. Alexander Kukushkin.
© 2018 John Wiley & Sons Ltd. Published 2018 by John Wiley & Sons Ltd.

2

Types of Mobile Network by Multiple-Access Scheme

Mobile radio networks can be distinguished by operation modes, services and applications and multiple-access schemes. A major influence on the development of commercial radio communication systems is the scarcity of radio spectrum available for utilization. An apparent objective is to assign the maximum number of users to an available radio frequency segment. This objective is achieved by using various multiple-access schemes. Here, we list the four most common technologies:

1) frequency division multiple access (FDMA)
2) time-division multiple access (TDMA)
3) code division multiple access (CDMA)
4) orthogonal frequency division multiple access (OFDMA)

Figure 2.1 illustrates the principles of multiple-access schemes used in mobile communications.

In FDMA, each mobile user (or user group) is allocated a frequency channel for the duration of the call, while in the TDMA scheme a group of callers use the same frequency channel but during different time intervals. Most of the systems using TDMA do, in fact, combine both schemes: FDMA and TDMA. In this approach, the system allocates a set of frequency channels to several groups of users, one frequency channel per group. One user in each group accesses an allocated frequency channel during a system assigned time slot. We will have a detailed look at the frequency-time-domain channel structure when considering the Global System Mobile (GSM) based on combined FDMA/TDMA multiple-access technology.

In CDMA, all users occupy the same frequency channel and can transmit/receive at the same time. The information stream of each user is coded by a specific code ensuring orthogonality between users. It can be achieved by allocating additional frequency bandwidth to each user in excess of the bandwidth required for transmitting user source data. The third-generation mobile system, WCDMA, utilizes this technology. The WCDMA system will be considered in Chapter 9.

In OFDMA, a large spectrum segment is allocated as a channel pool available to one or many simultaneous users. As seen in Figure 2.1, user allocated channel bandwidth and duration can be varied according to user service requirements and instant availability of common resource/channel pool. User channels are mapped on the set of orthogonal narrowband carriers, thus excluding mutual interference. The details of the OFDMA scheme will be discussed in the Chapter 11 discussion about LTE technology.

Introduction to Mobile Network Engineering: GSM, 3G-WCDMA, LTE and the Road to 5G,
First Edition. Alexander Kukushkin.
© 2018 John Wiley & Sons Ltd. Published 2018 by John Wiley & Sons Ltd.

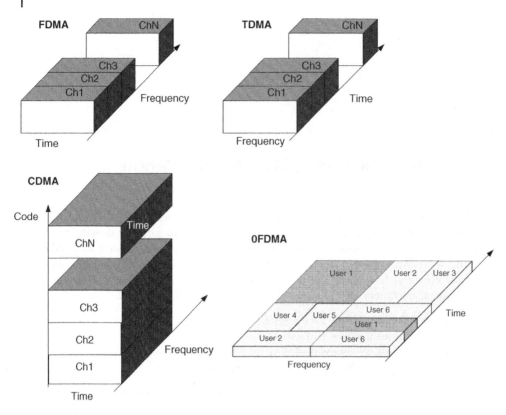

Figure 2.1 Common multiple-access schemes.

3

Cellular System

3.1 Historical Background

A scarcity of the available frequency spectrum is a major issue in the development of mobile networks. We consider a well quoted and quite convincing example of a GSM system. For example, only 25 MHz of the radio spectrum is available for the GSM system in the 900 MHz frequency range. That may allocate a maximum of 125 frequency channels each with a carrier bandwidth of 200 kHz. Within an eightfold time multiplex for each carrier, a maximum of 1000 channels can be realized. This number is further reduced by guard bands in the frequency spectrum and the overhead required for signalling.

Apparently, 1000 simultaneous users cannot produce sufficient revenue to justify the licence cost of 25 MHz of spectrum. In order to be able to serve several hundreds of thousands or millions of subscribers in spite of this limitation, frequencies must be spatially reused; that is, deployed repeatedly in a geographic area. In this way, services can be offered with a cost-effective subscriber density and acceptable blocking probability.

3.2 Cellular Concept

The spatial frequency reuse concept led to the development of the cellular principle, which allowed a significant improvement in the economic use of frequencies. The essential characteristics of the cellular network principle are as follows:

- The area to be covered is subdivided into cells (radio zones). These cells are often modelled in a simplified way as hexagons (Figure 3.1) with a base station located at the centre of each cell. Assume that the operator has a licence on a set of channels, called, for example, set S.
- To each cell i a subset of the frequencies S_i is assigned from the total set (bundle), which is assigned to the respective mobile radio network. In the GSM system, the set of frequencies assigned to a cell is called the Cell Allocation (CA). Under normal circumstances the number of channels in a subset S_i is driven by traffic capacity requirements.
- Neighbouring cells do not normally use the same frequencies since this would lead to severe co-channel interference from the adjacent cells.

Introduction to Mobile Network Engineering: GSM, 3G-WCDMA, LTE and the Road to 5G,
First Edition. Alexander Kukushkin.
© 2018 John Wiley & Sons Ltd. Published 2018 by John Wiley & Sons Ltd.

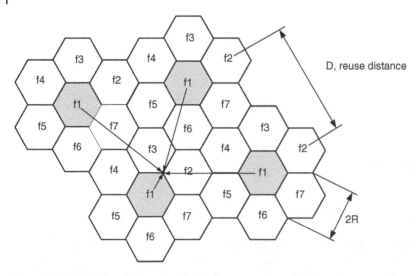

Figure 3.1 Model of a cellular network with frequency reuse. Shadowed hexagons represent cells with the same set of allocated frequencies.

- Only at distance D (the frequency reuse distance) can a frequency from the set S_i be reused (Figure 3.1); that is, cells with distance D to cell i can be assigned one or all of the frequencies from the set belonging to cell i. When designing a mobile radio network, D must be chosen to be sufficiently large, such that the co-channel interference remains small enough not to affect speech quality.
- When a mobile station moves from one cell to another during an ongoing conversation, an automatic channel/frequency change may occur (handover), which maintains an active speech connection over cell boundaries.

The spatial repetition of frequencies is done in a regular systematic way; that is, each cell with the cell allocation sees its neighbours with the same frequencies again at a distance D (Figures 3.1 and 3.2). Therefore, exactly six such neighbour cells exist. The first ring in the frequency set always contains six co-channel cells in frequency reuse system independent of the form and size of cells, not just in the hexagon model.

3.3 Carrier-to-Interference Ratio

The signal quality of a connection is measured as a function of received useful signal power and interference power received from co-channel cells and is given by the Carrier-to-Interference Ratio (CIR or C/I):

$$\text{CIR} = \frac{c}{I} = \frac{\text{Wanted signal power}}{\text{Interference power (from other cells)}} \tag{3.1}$$

The intensity of the interference is essentially a function of co-channel interference depending on the frequency reuse distance D. From the viewpoint of a mobile station, the co-channel interference is caused by base stations at a distance D from the current

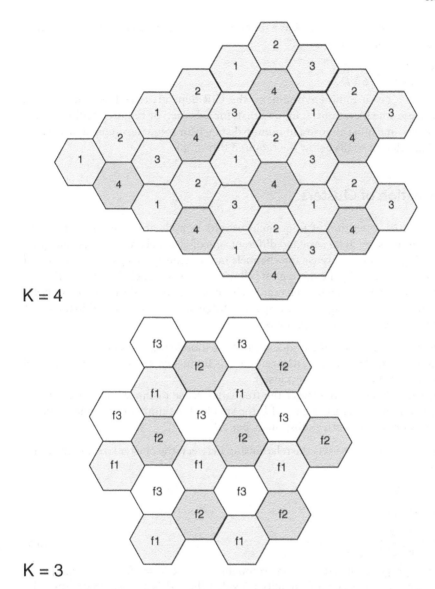

K = 4

K = 3

Figure 3.2 Frequency reuse and cluster formation.

base station, see Figure 3.1. A worst-case estimate for the CIR of a mobile station at the border of the covered area at distance $d = R$ from the base station can be obtained by assuming that all six neighbouring interfering transmitters operate at the same power and are approximately equally far apart (a distance D that is large compared with the cell radius R).

$$CIR = \frac{P_t R^{-\gamma}}{\sum_{i=1}^{6} P_t D^{-\gamma}} \qquad (3.2)$$

Finally, we find the worst-case CIR as a function of the cell radius R, the reuse distance D and the attenuation exponent γ as

$$CIR = \frac{R^{-\gamma}}{6D^{-\gamma}} = \frac{1}{6}\left(\frac{R}{D}\right)^{-\gamma} \tag{3.3}$$

Therefore, in a given radio environment, the CIR depends essentially on the ratio R/D. From these considerations, it follows that, for a desired or required CIR value at a given cell radius, one must choose a minimum distance for frequency reuse above which co-channel interference falls below the required threshold.

3.4 Formation of Clusters

The regular spatial repetition of frequencies results in a clustering of cells. The cells within a cluster must each be assigned different sets of channels, while cells belonging to neighbouring clusters can reuse the channels in the same spatial pattern. The size of a cluster is characterized by the number of cells per cluster k, which determines the frequency reuse distance D when the cell radius R is given. Figure 3.2 shows some examples of clusters. The numbers designate the respective frequency sets S_i used within the single cells. For each cluster, the following holds:

- A cluster can contain all of the frequencies of the mobile radio system.
- Within a cluster, no frequency can be reused. The frequencies of a set S_i may be reused at the earliest in the neighbouring cluster.
- The larger the cluster is, the larger the frequency reuse distance and the larger the CIR. However, the larger the values of k, the smaller the number of channels and the number of supportable active subscribers per cell.

The geometry of hexagons sets the relationship between the cluster size and the reuse distance as:

$$D = R\sqrt{3k} \tag{3.4}$$

The CIR is then given by

$$CIR = \frac{1}{6}\left(\frac{R}{D}\right)^{-\gamma} = \frac{1}{6}(3k)^{-\gamma/2} \tag{3.5}$$

assuming the propagation attenuation exponent $\gamma = 4$, $CIR = \frac{3}{2}k^2$. For example, if the system can achieve acceptable quality provided the C/I is at least 18 dB, then the required cluster size is 6.5. Hence, a cluster size of $k = 7$ would fit. Not all cluster sizes are possible due to the restrictions of the hexagonal geometry. The hexagon geometry results in following equation for cluster size

$$k = i^2 + ij + j^2, \tag{3.6}$$

where i, j are integers.

Possible values of k include 3, 4, 7, 12, 13, 19 and 27. The smaller the value of C/I, the smaller the allowed cluster size. Hence the available channels can be reused on a denser basis, serving more users and producing an increased capacity. In the example here, had the path loss dependence on radius been slower (i.e. the propagation exponent was less than 4), the required cluster size would have been greater than 7, so the path loss characteristics have a direct impact on the system capacity. Another constraint on the value of cluster size is that each base-station site often serves a cloverleaf of three cells. (This can be designated, for example, by specifying 21 cells as a 3 × 7 cluster.) Commonly used cluster sizes are multiples of three.

3.5   Sectorization

One way to reduce cluster size, and hence increase capacity, is to use sectorization. The group of channels available at each cell is split into three cells (sectors), each of which is confined in coverage to one-third of the cell area by the use of directional antennas, as shown in Figure 3.3.

Interference now comes from just two rather than six of the first-tier interfering sites, reducing interference by a factor of three and allowing cluster size to be increased by a factor of $\sqrt{} = 1.72$ in theory.

Sectorization has some disadvantages:

- Mobiles have to change channels more often, resulting in an increased signalling load on the system.
- The available pool of channels has to be reduced by a factor of 3 (in a three-sector site) for a mobile at any particular location; this reduces the trunking efficiency given same cell size.

Despite these issues, sectorization is used very widely in modern cellular systems, particularly in areas requiring high traffic density. More than three sectors can be used to further improve the interference reduction.

The effective radiated power and, consequently, CIR can be increased with directional antennas. In a three-sector site the radiation pattern of sector antenna spans 120° in the horizontal plane, as shown in Figure 3.4. In fact, the horizontal lobe of the sector antenna extends over 120° creating overlapping regions between site sectors where a mobile can receive a signal from both sectors. These regions facilitate an intra-sector handover; that is, they enable an MS travelling between sectors to be switched from one sector to another.

While sectorization does significantly increase the CIR, it often decreases the carried traffic in time-division multiple access (TDMA) and frequency division multiple access (FDMA) systems. For example, an omnidirectional site is allocated N frequency channels and carries a traffic *Aomni* Erlang with a defined probability of cell blockage. After sectorization, each sector may be allocated $N/3$ channels and may carry traffic of *As*

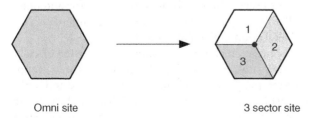

Omni site   3 sector site

Figure 3.3 Sectorization.

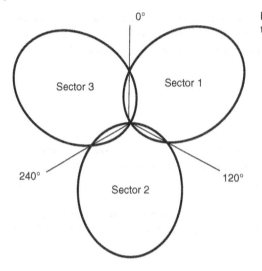

Figure 3.4 Antenna patterns for a cell site with three 120° sectors.

Erlang per sector with the same probability of cell blockage as the omnidirectional cell. One may observe that *Aomni* > 3*As*, where *Aomni* is the traffic carried by the omnidirectional site. The reason is that the traffic in Erlangs (see Section 3.7) is non-linearly related to the number of channels, and as each sector only has *N*/3 channels, then each sector carries less than a third of *Aomni*. This effect is known as *trunking efficiency*.

In CDMA systems, the situation is very different. Given the orthogonality of the cell codes, the same frequency channels can be reused in each sector without loss in trunking efficiency. In a system with perfect sectorization the increase in capacity at a cell site will be equal to the number of sectors; that is, a three-fold increase for three sectors. In practice, interference caused by overlapping antenna patterns and side and back lobes reduces this gain to around 80% of the ideal case.

3.6 Frequency Allocation

The reuse of frequencies in TDMA/FDMA systems may result in increasing co-channel and adjacent channel interference, especially with tight frequency reuse. If a large reuse distance is applied, the interference levels will be decreased, but the capacity is too. A short reuse distance is beneficial for the system capacity, but the interference will increase. The trade-off between capacity and quality is resolved in frequency planning. A better frequency plan will offer a higher capacity at maintained quality.

One base-station site is often used to serve three cells by means of sector antennas. For instance, a cluster of 7 × 3 cells implies seven sites each serving three cells (see Figure 3.5). The respective channel allocation is given in Table 3.1.

The shaded area inside the thick border in the figure comprises a cluster of cells. The cluster contains seven base-station sites, A–G, with each site having three groups of channels numbered 1–3. If, for example, 10 channels per cell are needed to handle the traffic, each base-station site must be allocated 30 channels. Adjacent clusters can use the same radio channels, as the reuse distance between nearby co-channel cells is such that co-channel interference causes only negligible degradation of the transmission

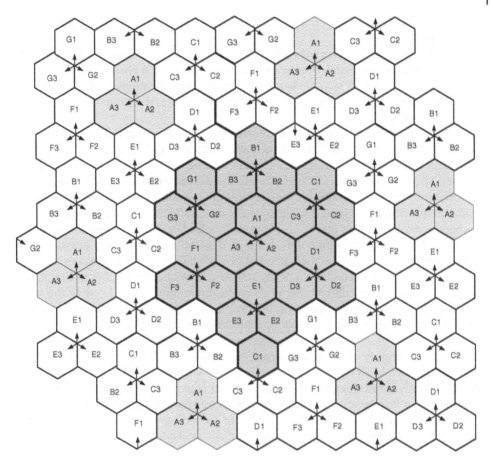

Figure 3.5 Illustration of 7 × 3 frequency reuse cluster.

Table 3.1 Channel allocation in a 7 × 3 cluster.

A1	B1	C1	D1	E1	F1	G1	A2	B2	C2	D2	E2	F2	G2	A3	B3	C3	D3	E3	F3	G3
1	2	3	4	5	6	7	8	9	10	11	12	13	14	15	16	17	18	19	20	21
22	23	24	25	26	27	28	29	30	31	32	33	34	35	36	37	.	.	.		

quality. This is known as geographical reuse of frequencies or channels. Thus, the system needs a total allocation of $30 \times 7 = 210$ channels irrespective of how many times the cluster pattern is repeated.

3.7 Trunking Effect

In a traditional public switched telephone network, each subscriber has a dedicated wire connection to the local switch, but the number of lines continuing from the local switch towards the next bigger switch is typically much smaller than the sum of subscribers served in that area.

The same applies to cellular networks as well, although the traditional subscriber line has been replaced with wireless access to the base station. This phenomenon is known as the trunking effect. In fact, the trunking effect reduces the number of lines at every network element concentrating traffic (merging several lines) if the number of incoming lines is big enough.

One can assume that calls take place during the busy hour and that the duration of each call is constant. Subscribers initiate calls randomly during the observation time. On one occasion, the traffic is increased by the number of one new call (top row in the Figure 3.6). Apparently, with a sufficient number of available lines (channels) there might be communication gaps available for placement of new calls as illustrated in Figure 3.6.

In example shown in Figure 3.6, allocation of new traffic was possible with the minimum of seven channels available. This became feasible because randomly arriving calls of a fixed duration create a randomly distributed 'silent' gaps of random duration; that is, random intervals when the channel is not occupied. During those gaps, the available channel can be assigned to carry a newly arrived call. While the number of simultaneous calls cannot exceed the number of channels, the number of users using the same pool of the channels over a period of time may exceed the number of lines due to the fixed duration of each call and random distribution of time of arrival. This is what is called a *trunked effect*, meaning that, given the random nature of call duration and time of arrival, the number of users may exceed the number of available lines. The trunking effect takes place in all traffic concentrating points when the pool of resources (number of channels or lines) is rather large. In a cellular system, the first such concentration point is the air interface. A fixed but sufficiently large number of traffic channels available in a transceiver can support large number of users camping on the cell; this number is much greater than the number of traffic channels available in the cell.

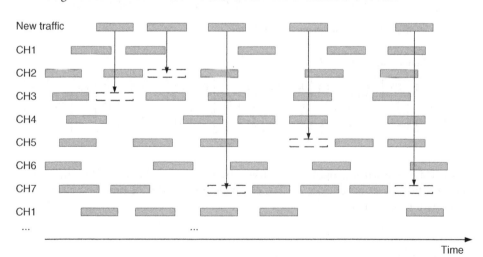

Figure 3.6 Illustration of the trunking effect.

3.8   Erlang Formulas

The trunking effect need to be estimated quantitatively in order to calculate the number of resources (channels, lines) to meet traffic demand from users of a communications system. The estimate of channel resources depends on many statistical factors related to traffic, such as call duration, time distributions of call arrivals and other statistical parameters. The unit of traffic is the *Erlang* (named after Agner Krarup Erlang (1878–1929) who invented it).

One Erlang equals the maximum traffic available on one line. The traffic is calculated using a simple formula:

$$x \text{ Erlangs} = \frac{(calls\ per\ hour) \times (average\ conversation\ time)}{3600\ \text{Seconds}} \tag{3.7}$$

It means that one call of a duration of 3600 seconds (i.e. 1 hour) produces 1 Erlang of traffic. Erlang derived two formulas for different systems:

- If all resources are used, additional calls are lost (Erlang B case). This is the case for voice calls in mobile cellular systems.
- If calls are put into a queue for certain time and will be served sequentially as resources become free again, the traffic capacity is described by Erlang C formulas. This is applicable to many trunked radio systems.

3.9   Erlang B Formula

The Erlang B formula determines the probability that a call is blocked. This probability defines a measure for the Grade of Service (GOS) for a trunked system that provides no queuing for blocked calls (i.e. blocked calls are instantly lost). The Erlang B formula uses the following assumptions:

- Call requests are memoryless. That is, all users, including blocked users, may request a channel at any time all free channels are fully available for calls until all channels are occupied.
- Probability of channel holding (i.e. usage) times is exponentially distributed. That is, longer calls are less likely to happen than short calls.
- A finite number of channels available in the resources pool time between channel requests follow a Poisson distribution (inter-arrival times).
- Inter-arrival times of call requests are independent of each other.
- The number of busy channels is equal to the number of busy users.

Offered traffic (in Erlangs) A is related to the call arrival rate, λ, and the average call-holding time, h, by

$$A = \lambda h \tag{3.8}$$

Let us define λ as a call arrival rate, h, mean holding time (duration of the call), then $\lambda h T$ is a mean operating time of a single user during period T, also called the time of

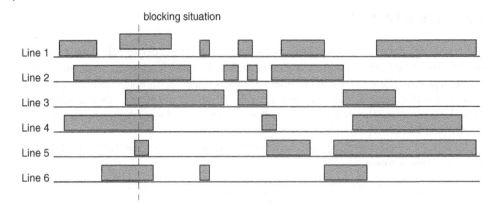

Figure 3.7 Blocking an incoming call.

occupancy of the channel. Relative operation time $\lambda hT/T = \lambda h = \alpha$ is a traffic load from a single user measured in Erlangs, $0 \leq \alpha \leq 1$. The traffic load from N users is then $A = N\alpha$, also called offered traffic: we have assumed that statistical characteristics of all calls by any user are the same.

Under all the assumptions here, the probability that in a system with n channels, k channels are occupied is given by the Erlang formula:

$$P_k = \frac{A^k/k!}{\sum_{m=1}^{n} A^m/m!} \tag{3.9}$$

The probability that all n channels are busy and, therefore, a new call is blocked is called the *blocking probability* and is given by the Erlang bocking formula (Erlang B):

$$P_n = \frac{A^n/n!}{\sum_{m=1}^{n} A^m/m!} \tag{3.10}$$

The Erlang B formula shows relations between offered load A and blocking probability with a total number n of available channels. Given a fixed amount of resources, the higher the acceptable blocking probability, the more traffic could be offered. Figure 3.7 illustrates, using a dashed line, the occurrence of blocking when at one instance all channels $n = 6$ are occupied.

Note that call arrival rate is often called BCHA (busy call hour attempt). Erlang values for a given set of resources are often tabulated in telecommunication engineering handbooks. The target blocking probability in a system is called Grade of Service, GOS, and is a percentage measure of service performance in mobile communication systems. For instance, GOS = 1% corresponds to an Erlang blocking probability of 0.01. A sample of an Erlang B table is presented in Table 3.2.

3.10 Worked Examples

3.10.1 Problem 1

Let us assume a traffic load per user of $\alpha = 25 \ mErl$ and GOS = 1%. Compare the number of supported subscribers in two trunked systems: the first comprised of four

Table 3.2 Erlang B table.

Erlang B Traffic Table

Maximum Offered Load Versus B and N

B is in %

N/B	0.01	0.05	0.1	0.5	1.0	2	5	10	15	20	30	40
1	.0001	.0005	.0010	.0050	.0101	.0204	.0526	.1111	.1765	.2500	.4286	.6667
2	.0142	.0321	.0458	.1054	.1526	.2235	.3813	.5954	.7962	1.000	1.449	2.000
3	.0868	.1517	.1938	.3490	.4555	.6022	.8994	1.271	1.603	1.930	2.633	3.480
4	.2347	.3624	.4393	.7012	.8694	1.092	1.525	2.045	2.501	2.945	3.891	5.021
5	.4520	.6486	.7621	1.132	1.361	1.657	2.219	2.881	3.454	4.010	5.189	6.596
6	.7282	.9957	1.146	1.622	1.909	2.276	2.960	3.758	4.445	5.109	6.514	8.191
7	1.054	1.392	1.579	2.158	2.501	2.935	3.738	4.666	5.461	6.230	7.856	9.800
8	1.422	1.830	2.051	2.730	3.128	3.627	4.543	5.597	6.498	7.369	9.213	11.42
9	1.826	2.302	2.558	3.333	3.783	4.345	5.370	6.546	7.551	8.522	10.58	13.05
10	2.260	2.803	3.092	3.961	4.461	5.084	6.216	7.511	8.616	9.685	11.95	14.68
11	2.722	3.329	3.651	4.610	5.160	5.842	7.076	8.487	9.691	10.86	13.33	16.31
12	3.207	3.878	4.231	5.279	5.876	6.615	7.950	9.474	10.78	12.04	14.72	17.95
13	3.713	4.447	4.831	5.964	6.607	7.402	8.835	10.47	11.87	13.22	16.11	19.60
14	4.239	5.032	5.446	6.663	7.352	8.200	9.730	11.47	12.97	14.41	17.50	21.24
15	4.781	5.634	6.077	7.376	8.108	9.010	10.63	12.48	14.07	15.61	18.90	22.89
16	5.339	6.250	6.722	8.100	8.875	9.828	11.54	13.50	15.18	16.81	20.30	24.54
17	5.911	6.878	7.378	8.834	9.652	10.66	12.46	14.52	16.29	18.01	21.70	26.19
18	6.496	7.519	8.046	9.578	10.44	11.49	13.39	15.55	17.41	19.22	23.10	27.84
19	7.093	8.170	8.724	10.33	11.23	12.33	14.32	16.58	18.53	20.42	24.51	29.50
20	7.701	8.831	9.412	11.09	12.03	13.18	15.25	17.61	19.65	21.64	25.92	31.15
21	8.319	9.501	10.11	11.86	12.84	14.04	16.19	18.65	20.77	22.85	27.33	32.81
22	8.946	10.18	10.81	12.64	13.65	14.90	17.13	19.69	21.90	24.06	28.74	34.46
23	9.583	10.87	11.52	13.42	14.47	15.76	18.08	20.74	23.03	25.28	30.15	36.12
24	10.23	11.56	12.24	14.20	15.30	16.63	19.03	21.78	24.16	26.50	31.56	37.78
25	10.88	12.26	12.97	15.00	16.13	17.51	19.99	22.83	25.30	27.72	32.97	39.44
26	11.54	12.97	13.70	15.80	16.96	18.38	20.94	23.89	26.43	28.94	34.39	41.10
27	12.21	13.69	14.44	16.60	17.80	19.27	21.90	24.94	27.57	30.16	35.80	42.76
28	12.88	14.41	15.18	17.41	18.64	20.15	22.87	26.00	28.71	31.39	37.21	44.41
29	13.56	15.13	15.93	18.22	19.49	21.04	23.83	27.05	29.85	32.61	38.63	46.07
30	14.25	15.86	16.68	19.03	20.34	21.93	24.80	28.11	31.00	33.84	40.05	47.74
31	14.94	16.60	17.44	19.85	21.19	22.83	25.77	29.17	32.14	35.07	41.46	49.40
32	15.63	17.34	18.21	20.68	22.05	23.73	26.75	30.24	33.28	36.30	42.88	51.06
33	16.34	18.09	18.97	21.51	22.91	24.63	27.72	31.30	34.43	37.52	44.30	52.72
34	17.04	18.84	19.74	22.34	23.77	25.53	28.70	32.37	35.58	38.75	45.72	54.38
35	17.75	19.59	20.52	23.17	24.64	26.44	29.68	33.43	36.72	39.99	47.14	56.04
36	18.47	20.35	21.30	24.01	25.51	27.34	30.66	34.50	37.87	41.22	48.56	57.70
37	19.19	21.11	22.08	24.85	26.38	28.25	31.64	35.57	39.02	42.45	49.98	59.37
38	19.91	21.87	22.86	25.69	27.25	29.17	32.62	36.64	40.17	43.68	51.40	61.03
39	20.64	22.64	23.65	26.53	28.13	30.08	33.61	37.72	41.32	44.91	52.82	62.69
40	21.37	23.41	24.44	27.38	29.01	31.00	34.60	38.79	42.48	46.15	54.24	64.35
41	22.11	24.19	25.24	28.23	29.89	31.92	35.58	39.86	43.63	47.38	55.66	66.02
42	22.85	24.97	26.04	29.09	30.77	32.84	36.57	40.94	44.78	48.62	57.08	67.68
43	23.59	25.75	26.84	29.94	31.66	33.76	37.57	42.01	45.94	49.85	58.50	69.34

switches, each of 10 channels in capacity and the second with one switch of 40 channels in capacity.

Solution:
From the Erlang B table, we find an offered load A by a switch with 10 channels is 3.09 Erlangs. The total load by four switches is $4 \times 3.09 = 12.36$ Erlangs. The number of supported subscribers is then $\frac{A}{\alpha} = 12.36/0.025 = 494$.

A second system with 40 channels offers a load of 24.44 Erlangs. The number of supported subscribers is 977. As observed, the trunking efficiency is significantly increased with an increase in channel pool size.

3.10.2 Problem 2

Consider a cellular system with a cluster size of $k = 7$ with a total of 395 channels. The call-holding time is 3 minutes, a user makes one call per hour and the blocking probability is 1%. Blocked calls are cleared, Erlang B distribution applies.
   Determine:

1) The average number of calls/hour in the case of an omnidirectional site antenna.
2) The average number of calls/hour in the case of a three-sector site configuration antenna. Calculate the decrease in trunking efficiency compared with an omni configuration.
3) The average number of calls/hour in the case of a six-sector site antenna configuration. Calculate the decrease in trunking efficiency compared with an omni configuration.

Solution:
Number of channel/cell N = total number channels/cells per cluster. For an omni site $N = 395/7 = 57$
   Call-holding time H = 3 min/60 min = 0.05 hour.

1) Omni configuration.
   Offered traffic load for a 0.01 blocking probability (1% GOS) is A = 44.2 Erlangs for 57 channels available.
   Number of calls A/H = 44.2/0.005 = 884 calls/hour
2) three-sector configuration.
   Number of channels per sector $N = 57/3 = 19$. Offered traffic load for a 0.01 blocking probability (1% GOS) is A = 11.2 Erlang per sector.
   Number of calls A/H = 11.2/0.005 = 224 calls/hour per sector.
   Number of calls per site 3 × 224 = 672 call/hour.
   Decrease in trunking efficiency is (884 − 672)/884 = 24%
3) six-sector configuration.
   Number of channels per sector N = 57/6 = 9.5 channel. Offered traffic load for a 0.01 blocking probability (1% GOS) is A = 4.1 Erlangs per sector.
   Number of calls A/H = 4.1/0.005 = 82 calls/hour per sector.
   Number of calls per site 6 × 82 = 492 call/hour.
   Decrease in trunking efficiency is (884 − 492)/884 = 44%
   It should be noted that there is a trade-off between reduction in co-channel interference versus loss of trunking efficiency and an increase of number of handovers between sectors.

3.10.3 Problem 3

A cellular operator is interested in providing GSM coverage at 900 MHz in a new entertainment centre. The marketing plan has targeted approximately 1 000 000 visitors attending shows every year, of which it is believed that around 90% are typical mobile phone users with an average traffic load $\alpha = 25$ mErl.

The following assumptions apply for the centre:

1) Busy-hour traffic takes about 30% of the total daily traffic.
2) The traffic in the entertainment centre is distributed as follows:
   40% is carried in the major exhibition pavilion, 50% in the foyer and 10% in the car park.
3) Each user makes an average of three phone calls of 2 min of duration during the busy hour.
4) Three cells are required for this system: one in the exhibition pavilion, one in the foyer and the other one in the car park.
5) Limit considerations to one of the mobile operators with a market penetration in the centre of 20%.

Determine the required number of channels per cell with a GOS $= 2\%$.

Solution:

The busy-hour number of mobile phone users needs to be calculated. If 1 000 000 passengers per year use the airport, then the average number of users per day is 1 000 000/365 $= 2740$.

As 90% of the passengers use a mobile phone, the cellular operator with 20% of market penetration carries

2740 × 0.9 × 0.2 = 494 mobile phone users/day

This number represents only the number of mobile phone users per day for this cellular operator. As capacity needs to be dimensioned on a per busy-hour basis:

Maximum number of user per hour $N = 494$ users/day × 0.3 $\sim$ 149 users/busy hour

Assume that the operator has installed its own infrastructure in the entertainment centre. A case where the infrastructure is shared between operators is not considered here.

We need to distribute the estimated number of users between three cells. As the foyer takes 50% of the traffic, the pavilion 40% and the car park 10%, we have:

In the foyer $N_{foyer} = 50\% \times N = 75$ user/hour
in the pavilion $N_{pavilion} = 40\% \times N = 59$ user/hour and
in the car park, $N_{car_{park}} = 10\% \times N = 15$ user/hour.

Given this number of users per cell, and the traffic per user, we can compute the total traffic per cell:

$A_{foyer} = 75$ users × 30 mE/user $= 2.25$ Erl
$A_{pavilion} = 59$ users × 30 mE/user $= 1.77$ Erl
$A_{car\_park} = 15$ users × 30 mE/user $= 0.45$ Erl

Given a 2% blocking probability, an estimate of the required number of channels per can be obtained from the Erlang B table.

$C_{foyer} = \sim 6$ channels
$C_{pavilion} = \sim 5$ channels
$C_{car_{park}} = \sim 3$ channels

4

Radio Propagation

The electromagnetic field radiated by an antenna propagates through the radio channels between transmitter Tx and receiver Rx. The mechanism of radio propagation can generally be described as reflection, diffraction and scattering.

Propagation models are traditionally focused on prediction of the average received signal strength used to define the coverage area in a mobile cellular environment. The statistics of the random component in the signal are used in RF planning and system deployment to evaluate the uncertainties in prediction and define a necessary margin to counteract signal fading. In addition, some statistical properties of the radio channel, such as fading parameters, multipath and associated intersymbol interference heavily impact the system design and various system parameters.

4.1 Propagation Mechanisms

4.1.1 Free-Space Propagation

The simplest possible scenario is that both transmit and receive antenna are placed in a free space. Energy conservation dictates that the integral of the power density S over any closed surface surrounding the transmit antenna must be equal to the transmitted power P_t. If the closed surface is a sphere of radius R, centred at the transmitter (Tx) antenna and if the Tx antenna radiates isotropically then the power density on the surface is

$$S = \frac{P_t}{4\pi R^2} \tag{4.1}$$

Assume that receiver (Rx) antenna has an 'effective area' or aperture A, then the received power is given by

$$P_r = S \cdot A = A \cdot \frac{P_t}{4\pi R^2} \tag{4.2}$$

If the transmit antenna is not isotropic, then the energy density has to be multiplied with a transmit antenna gain G_t in the direction of the receive antenna. The relationships between the aperture A and gain G is introduced by equation (4.3)

$$G = 4\pi A / \lambda^2 \tag{4.3}$$

Introduction to Mobile Network Engineering: GSM, 3G-WCDMA, LTE and the Road to 5G,
First Edition. Alexander Kukushkin.
© 2018 John Wiley & Sons Ltd. Published 2018 by John Wiley & Sons Ltd.

and the equation for receive power P_r with a directive transmit antenna is given by

$$P_r = A_r P_t G_t \cdot \frac{1}{4\pi R^2} \tag{4.4}$$

The product of transmit power and gain in the considered direction is also known as Effective Isotropic Radiated Power (EIRP).

$$EIRP = P_t G_t \tag{4.5}$$

The effective antenna area is proportional to the power that can be extracted from the antenna radiators with a given energy density. For example, for a parabolic antenna, the effective antenna area is roughly the geometrical area of the surface. However, antennas with a very small (in terms of wavelength) geometrical area – for example dipole antennas – can also have a considerable effective area. Isotropic antenna can be represented by a short dipole with a gain $G = 1$, so the effective aperture for isotropic antenna A_i is given by

$$A_i = \lambda^2 / 4\pi \tag{4.6}$$

Substitution of these formulas leads to the 'Friis law' or equation for a relation between received and transmitted power in a free space:

$$P_r = G_r P_t G_t \cdot \frac{\lambda^2}{(4\pi)^2 R^2} \tag{4.7}$$

The free-space equation is valid in a 'far-field' region, for example, at the distance R from transmitter not less than a Fraunhofer distance (also called Rayleigh distance):

$$d_f = 2D_a^2 / \lambda \tag{4.8}$$

where D_a is the largest dimension of the antenna. The far field requires that $R \gg D_a$ and $R \gg \lambda$.

It is important to note that the free-space propagation mechanism has to be independent of wavelength or frequency. This is due to the fact that the radiated energy is not lost, but rather redistributed over a sphere surface of area $4\pi R^2$, as observed from equations (4.1) and (4.2). On the other hand, Friis' law seems to indicate that the 'attenuation' in free space increases with frequency. This seeming contradiction is caused by the definition of antenna gain via effective aperture (4.3) and the following presumption that antenna gain G is independent of frequency.

In the case of transmit and receive isotropic antennas, the free-space equation takes the form

$$P_r = P_t \cdot \left(\frac{\lambda}{4\pi R} \right)^2 \tag{4.9}$$

Equation (4.9) can be re-written introducing a path loss L, which is measure of attenuation factor caused by propagation mechanism and propagation media in general case.

$$P_r = P_r / L \tag{4.10}$$

The free-space path loss factor is

$$L = \left(\frac{4\pi}{\lambda} \right)^2 R^2 \tag{4.11}$$

This is a well-known free-space equation for an isotropic antenna. When written in a logarithmic scale, it takes the form

$$L_0 = 33.44 + 20\log f(MHz) + 20\log R(km), \ \text{dB} \tag{4.12}$$

As follows from (4.12), the received power in a free space falls by 6 dB when the range is doubled; in other words, the path loss increases by 20 dB per decade. Similarly, path loss increases by 6 dB if the frequency is doubled.

In a general case, radio propagation path loss formulas are based on the free-space loss formula with additional empirical correction factors. The generic formula for path loss takes the form

$$L = L_0(r) + \alpha \ \log(R - r) \tag{4.13}$$

where L_0 is a loss at reference distance (can be 1 m or 1 km) and the second term states that the losses are exponential with distance. This is illustrated in Figure 4.1, where the second term of equation (4.13) determines a 'clutter loss factor'; that is, the slope of exponent in different environment related to land-usage. The slop is defined in dB/decade, and nominal values of slop are 20–25 dB/dec for free-space–open area loss, 30 dB/dec in a suburban area and 40 dB/dec in an urban area, respectively.

The radio signal attenuation depends on the environment the signal passes through. A rise in signal strength can be observed despite of increasing distance, when the receiver re-enters an open area after passing through a small urban cluster causing a higher attenuation exponent, see Figure 4.2. Since the received signal strength depends on the close environment of the receiver, there is no abrupt rise in signal strength but a gradual increase as the mobile enters an open area.

The total propagation path can be divided conditionally into long-distance and local propagation. Because the mobile terminal's antenna is often much lower than the height of surrounding objects, the direct line of sight to the horizon is obscured. Typically, the received signal will comprise a number of reflected waves; that is, there is distinct multipath propagation. The base-station antenna is often in a more open location, but reflections or shadow effects from nearby buildings and the terrain configuration can sometimes affect the propagation conditions.

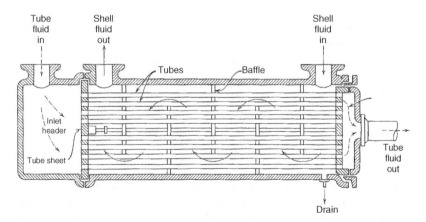

Figure 4.1 Path loss examples with different clutter loss factors.

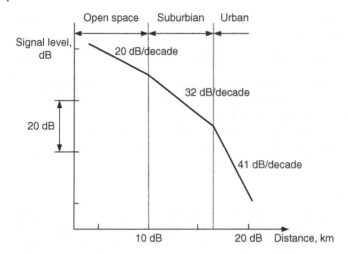

Figure 4.2 Path loss in a mixed environment.

Multipath propagation gives rise to a fine structure of signal fluctuations, which results in corresponding variations (fast fading) in the input signal to the mobile unit. Averaging with respect to the fast fading structure gives the local mean of the propagation path loss or received signal strength.

In a typical propagation situation there are considerable shadow effects. The propagation loss with respect to the local mean is generally much greater than when there is free-space propagation, owing to the strong influence of the terrain. This influence manifests itself partly as an increase in the average ('global') propagation loss and partly as a varying loss due to the irregularity of the terrain. The large-scale effects can be estimated from data on the terrain and built up areas. The result of the calculation is the global mean. However, the detailed structure causes random variations in the local mean of the signal received by the mobile unit (called shadow fading, slow fading or *lognormal fading*).

Following this discussion, it is useful to divide propagation loss into three components: the *global mean* of the propagation loss, that is the deterministic part, and two random elements resulting from shadow fading (*slow fading*) and multipath propagation (*fast fading*), see Figure 4.3. The global mean of the received signal power is typically inversely proportional to the fourth power of the propagation distance $P \sim R^{-n}$ ($n = 4$), where n is the propagation exponent.

4.1.2 Propagation Models for Path Loss (Global Mean) Prediction

The 'global mean' component together with the 'shadowing' (long term fading) component of the signal are used for RF coverage prediction in mobile cellular systems. Revising the path loss definition in the previous section, consider a general case of a propagation mechanism with losses in an antenna system, see Figure 4.4.

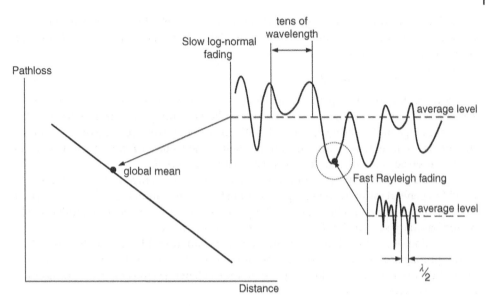

Figure 4.3 Three components in path loss.

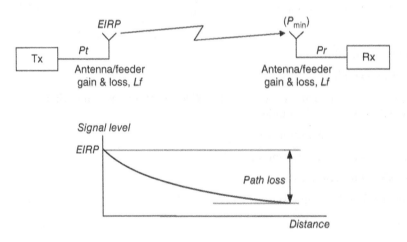

Figure 4.4 Basic components of link budget.

Introducing losses in the antenna-feeder system for transmitter and receiver, L_{ft} and L_{fr}, respectively, the output power at antenna connector P_t, one may obtain a more general equation for EIRP that includes feeder losses:

$$EIRP = P_t + G_t - L_{ft}, \text{ dBm} \tag{4.14}$$

in dB notation.

The received power P_r at the receiver input takes propagation path loss L into account and is given by

$$P_r = P_t + G_t - L_{ft} - L + G_r - L_{fr}, \text{ dBm} \tag{4.15}$$

Similar to transmitter EIRP, one can define a 'design level' of minimal received power at the receiving antenna input, P_{min}. Then the relation between acceptable path loss and design level is as follows:

$$L = MAPL = EIRP - P_{min} \tag{4.16}$$

The main goal of propagation modelling is to predict path loss L as accurately as possible, allowing the range of radio system to be determined before installation. The maximum rage of the system occurs when the received power drops below P_{min}, a level that provides acceptable communication quality. The value of L for which this power level is received is the maximum acceptable path loss, $MAPL$.

A general model for propagation loss is very often semi-empirical and intended to provide a general estimate of radio propagation based on nominal characteristics rather than specific path data. During the radio planning process, any generic model is calibrated based on specific land-usage data and measurements.

The tuning of propagation model involves determination different coefficient values in the propagation equation so that the residual RMS value of the error reaches the global minimum; that is, the lowest possible value. General semi-empirical propagation equation is as follows:

$$P_r = P_t + G_t + G_r - L = P_t + G_t + G_r + C_{CT} + C_d \log(R) + C_{dh} \log(R) \log(h_B)$$
$$+ C_h \log(h_B) + C_{dk} K_{dk} + C_{dr} K_{dr} + C_{Cl} K_{Cl} \tag{4.17}$$

where:

P_r = received power, dBm

P_t = transmit power, dBm

C_{CT} = the fixed correction term that accounts for effects of frequency and other non-specific factors of the model

R = the distance from transmitter, km;

h_B = the base-station antenna height, m;

K_{dk} = a knife-edge diffraction loss;

K_{dr} = a rounded-hill diffraction loss and

K_{Cl} = the clutter loss.

Table 4.1 Pre-set values for correction coefficients in semi-empirical equation for propagation loss.

Parameter	Initial Value
C_d	−44.9
C_{dh}	6.55
C_h	0
C_{dk}	−0.5
C_{Cl}	1

The parameters C_{CT} and K_{CI} combine to give a power level at the distance of 1 m. The initial value is normally zero and is then tuned to ensure it provides an overall RMS error of about 6–7 dB for the entire data set for each base station and the whole clutter.

The parameters in (4.17) are defined in Table 4.1.

Apparently, the optimal set of parameters and coefficients in equation (4.17) produce a global minimum of prediction errors only for specified clutter or land-usage type. The method based on the tuning of the equation (4.17) model involves dividing the prediction area into a series of clutter and terrain categories; namely, open, suburban and urban. These are summarized as follows:

- *Open area*: Open space, no tall trees or buildings in the path, a plot of land cleared for 300–400 m ahead; for example, farmland, open fields.
- *Suburban area*: Village or highway scattered with trees and houses, some obstacles near the mobile but not very congested.
- *Urban area*: Built up city or large town with large buildings and houses with two or more storeys, or larger villages with close houses and tall, thick tree cover.

5

Mobile Radio Channel

For wireless communications, the transmission medium is the radio channel between transmitter Tx and receiver Rx. The signal can get from the Tx to the Rx via a number of different propagation paths. In some cases, a Line-Of-Sight (LOS) connection might exist between Tx and Rx. In a typical mobile environment LOS is absent and the signal can get from the Tx to the Rx by means of being reflected at or diffracted by different objects in the environment: houses, mountains (in some hilly areas), windows, walls and so on, as shown in Figure 5.1. The number of these possible propagation paths is normally very large. Each of the paths has a distinct amplitude, delay (runtime of the signal), direction of departure from the Tx and direction of arrival; most importantly, the components have different *phase shifts* with respect to each other.

The envelope of the received signal is composed by many signal replicas arriving via different paths with different amplitudes and phases, respectively. When the phase shift between each pair of signal components is equal to 180°, these components tend to cancel each other out producing a case of destructive interference, and vice versa, when signal replicas arrive in-phase they tend to add up to each other; that is, a case of constructive interference. Due to the random nature of amplitude and phase of the multipath component, the subsequent composition produces random variations in a received signal envelope. These variations are called 'fading'.

The term mobile radio 'channel' refers to impact from propagation media on characteristics of the received signal; that is, it can be thought of as a filter acting on the transmitted signal. The radio channel then can be modelled with the transfer function of the filter that describes the channel in the time or, alternatively, in the frequency domain.

The common approach is to distinguish two types of radio channel; a *narrowband* or 'flat fading' channel versus a *wideband* or frequency selective channel. In the narrowband channel, multipath fading comes about as a result of relatively small path differences between signals coming from scatterers in the near vicinity of the mobile. The term 'small' relates to equivalent symbol length in the space domain. These differences and individual phase shifts contributed to by the scatterers, in the order of a few wavelengths, lead to significant phase differences in received components of the signal arriving via a different path.

By contrast, if strong scatterers exist well off of the direct or shortest path between the base and mobile, the time differences may be significant. If the relative delays are large compared to the basic unit of information transmitted on the channel (usually a symbol

Introduction to Mobile Network Engineering: GSM, 3G-WCDMA, LTE and the Road to 5G,
First Edition. Alexander Kukushkin.
© 2018 John Wiley & Sons Ltd. Published 2018 by John Wiley & Sons Ltd.

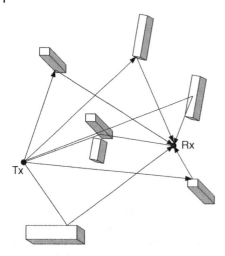

Figure 5.1 Multipath propagation.

or a bit), the signal will then experience significant distortion, which varies across the channel bandwidth. The channel is then a *wideband channel* and system design needs to take into account these effects.

Some of the basic characteristics of the channel can be seen in a simplified two-ray channel model. Consider the superposition of rays $A\, e^{if_1 t_1}$ and $A\, e^{if_2 t_2}$ arriving at different frequencies f_1, f_2 and different times t_1, t_2, respectively. Introduce new variables: delay $\tau = t_1 - t_2$, mean arrival time $t = {t_1+t_2}/{2}$, frequency offset $\Delta f = f_1 - f_2$ and centre carrier frequency $f_0 = {f_1+f_2}/{2}$. The sum of two components is then given by

$$2e^{i\omega_0 t + i\frac{\Delta\omega\tau}{4}}\left(e^{i\omega_0\frac{\tau}{2}+i\frac{\Delta\omega t}{2}} + e^{-i\omega_0\frac{\tau}{2}-i\frac{\Delta\omega t}{2}}\right) = 2e^{i\omega_0 t+i\frac{\Delta\omega\tau}{4}}\cos\left(\omega_0\frac{\tau}{2}+\frac{\Delta\omega t}{2}\right) \tag{5.1}$$

where $\omega_0 = 2\pi f_0$, $\Delta\omega = 2\pi\Delta f$.

Considering only frequency shift, we may assume $\tau = 0$. Then the two-ray model is truncated to

$$s(t) = 2e^{i\omega_0 t}\cos\left(\frac{\Delta\omega t}{2}\right) \tag{5.2}$$

Equation (5.2) represents the so-called 'beating' diagram for superposition of two oscillations with different frequencies, see Figure 5.2. The correlation interval τ_c can be estimated as a time interval between maximum and minimum of signal envelope, or half of the beating period $T = {1}/{\Delta f}$.

5.1 Channel Characterization

Statistical characteristics of the radio channel in general and the wideband channel in particular are commonly described at the level of second-order statistics; that is, correlation function or power spectrum density. The comprehensive analysis of the channel models and related statistics is not part of this overview and can be found in numerous books on that topic. Various statistics characteristics are used in system design depending on technology and applications. Nonetheless, there are several basic

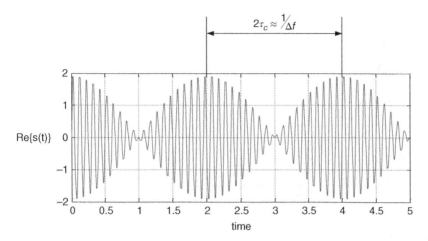

Figure 5.2 Beating diagram in a two-ray channel model.

statistical characteristics and parameters that are commonly used in design for all communication systems:

- Probability density function of fading → mean, variance of fading
- Doppler spectrum → maximum Doppler shift
- Power delay profile → delay spread
- Correlation function → correlation length in space and time domain
- Frequency response of the channel → coherence bandwidth

One of the main characteristics of the wideband channel is a power delay profile (PDP) of the channel defined as the variation of mean power in the channel with delay. The PDP is commonly presented by a discrete time model in the delay dimension to yield individual taps of power, as shown in Figure 5.3.

The most important parameter of the power delay profile is the delay spread or RMS delay spread. The delay spread $\Delta\tau$ is a square root of the difference between the mean of delay squared and the square of the mean delay.

$$\Delta_\tau = \sqrt{\overline{\tau^2} - \overline{\tau}^2} = \frac{1}{P_T}\sqrt{\sum_{m=1}^{n} P_m \tau_m^2 - \tau_0^2} \qquad (5.3)$$

where

$$\tau_0 = \frac{\sum_{i=1}^{n} P_i \tau_i}{\sum_{i=1}^{n} P_i} \qquad (5.4)$$

is a mean delay; that is, the delay corresponding to the 'centre of gravity' of the power delay profile.

The delay spread Δ_τ is independent of the mean delay and hence of the actual path length, being defined only by the relative path delays. The value of RMS delay is weighted with relative powers of the taps as well as their delays, making it a better indicator of system performance than the other parameters.

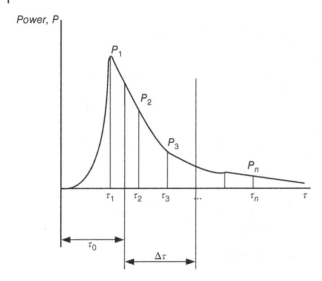

Power, P

Figure 5.3 Power delay profile.

Another important parameter for characterization of the random frequency selective mobile channel is a frequency response of the channel; that is, a Fourier Transform of the impulse response. The fundamental conclusion of channel characterization is that the delay spread $\Delta\tau$ and coherence bandwidth Δf_c are related via an 'uncertainty relationship':

$$\Delta f_c \cong 1/\Delta\tau \qquad (5.5)$$

The coherence bandwidth is a quantitative estimate of frequency shift when variation in signal strength at frequencies f and $f + \Delta f_c$ became uncorrelated. This leads to classification of the channels to frequency selective or frequency flat. Introducing the signal bandwidth Δf, we can define the frequency selective (in other words, time dispersive) channel by the inequality:

$$\Delta f/\Delta f_c \gg 1 \qquad (5.6)$$

In the opposite case where the coherence bandwidth is much larger than signal bandwidth, the frequency content of the signal is not distorted by the channel and the channel fading is said to be frequency flat.

In case of a flat fading channel, when $\Delta f/\Delta f_c \ll 1$, the signal symbol duration T_s is greater than delay spread:

$$T_s = 1/\Delta f \gg \Delta\tau \qquad (5.7)$$

As a consequence, different paths of the scattered/reflected signal cannot be resolved and this results in flat Rayleigh fading of the signal components. In the case of a frequency selective channel, the symbol duration is much smaller than delay spread,

$$T_s \ll \Delta\tau' \qquad (5.8)$$

This means that multiple replicas of the signal experience uncorrelated fading, which ensures that multiple paths could potentially be resolved.

Figure 5.4 Phasor diagram in the narrowband channel.

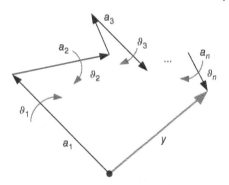

5.1.1 Narrowband Flat Channel

The signal received at the mobile, y, will be composed of a sum of waves from all scatterers, whose phase and amplitude depend on the reflection coefficient $a_m e^{j\theta_m}$ and scattering characteristics of the scatterer, and whose time delay is τ_m, $m = 1,2,\ldots$ Thus:

$$y = a_1 e^{j(\theta_1 + \omega\tau_1)} + a_2 e^{j(\theta_2 + \omega\tau_2)} + \ldots \tag{5.9}$$

Each of the components in this expression constitutes an 'echo' of the transmitted signal.

In the narrowband channel, differential delays of arriving waves are small compared to the symbol duration. The delays of the arriving waves could be considered approximately equal, so the amplitude does not depend upon the carrier frequency:

$$y = e^{j\omega\tau}(a_1 e^{j\theta_1} + a_2 e^{j\theta_2} + \ldots) \tag{5.10}$$

All frequencies in the received signal are therefore affected in the same way by the channel, and the channel can be represented by a single multiplicative component. The resulting signal vector, the term in parenthesis in equation (5.10), is composed by a vector addition of the multipath components, as shown in Figure 5.4. In every instant, the partial waves take different (complex) random values for each $a_m e^{j\theta_m}$, thereby composing the resulting random vector.

Such variations of received signal in narrowband channel are described by *Rayleigh* or Rician statistics and normally referred to as a 'flat fading'. The term 'flat' means here that a channel transfer function does not depend on frequency; that is, is flat. In that channel, the received signal became uncorrelated in receiving points separated at $\sim \lambda/2$ in distance, where λ is a signal wavelength.

The more precise determination of 'relatively small' differential delays of scattered components is provided in the next sections.

5.1.2 Wideband Frequency Selective Channel

In an opposite case of a wideband or frequency selective channel, the major impact from the channel is a signal dispersion in time. In other words, the impulse response of the channel is not a single delta pulse but rather a sequence of pulses (corresponding to different multipath components), each of which has a distinct arrival time in addition to having a different amplitude and phase. This is illustrated in Figure 5.5 with transmitted

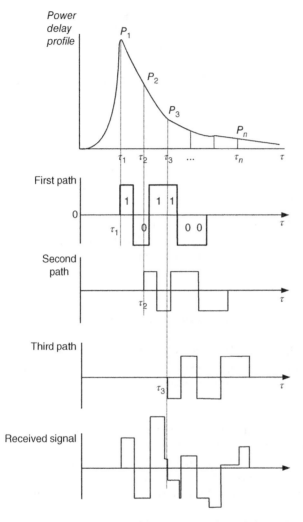

Figure 5.5 Inter-symbol interference in the wideband channel.

symbols presented in a simplified bipolar modulation format. As observed, the symbol duration in much less than delay spread; that is, $T_s \ll \Delta\tau$.

This signal dispersion leads to Inter-Symbol Interference (ISI) at the receiver, as illustrated in Figure 5.5.

One may obtain a simple estimate of the reduction in energy of the received signal caused by inter-symbol interference of multipath signals with delay spread $\sim \Delta\tau$. Consider a BPSK modulated bit stream of 'zeros' and 'ones' and assume that the received signal is composed by two rays with a relative delay τ, as shown in Figure 5.6. As can be seen from Figure 5.6, at the time intervals shown as dashed areas each of the signal components have the opposite sign being phase shifted BPSK symbols; that is, the phase of delayed component at interval τ is opposite to a phase of signal at the consecutive interval $T_s - \Delta\tau$. As a result, the energy of the received symbol will be reduced at a value of proportional to 2τ, see Figure 5.6. We may estimate the average value of delay in

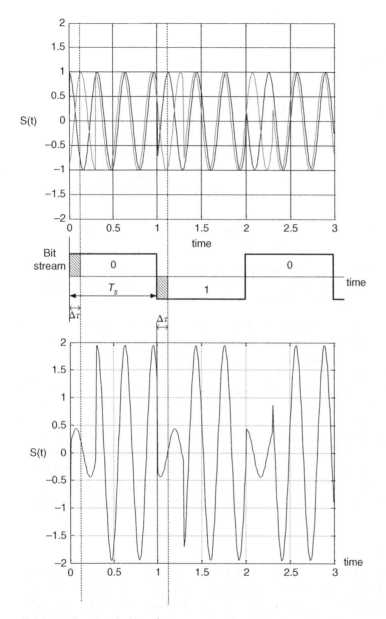

Figure 5.6 Inter-symbol interference.

respective multipath components as $\tau \approx {}^{\Delta\tau}\!/_2$. The energy per symbol in multipath component can then be estimated as $P_m \cdot (T_s - \Delta\tau)$, where P_m is a total power of multipath component. The energy per symbol to noise density ratio is given by

$$\frac{P_m \cdot (T_s - \Delta\tau)}{N_0} = \frac{P_m \cdot T_s \left(1 - {}^{\Delta\tau}\!/_{T_s}\right)}{N_0} = \frac{E_s}{N_0}\left(1 - {}^{\Delta\tau}\!/_{T_s}\right) \tag{5.11}$$

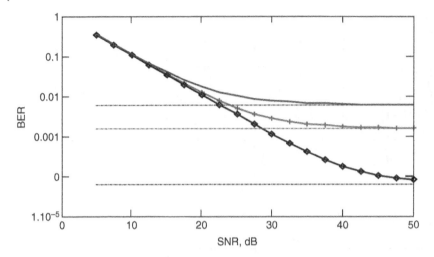

Figure 5.7 Average total BER versus SNR for different values of delay spread in Rayleigh channel: the solid line corresponds to a delay spread of 0.1-bit duration, +s to 0.05 and ⋄ s to 0.01, respectively; horizontal lines (dashes) correspond to irreducible BER values.

Therefore, the losses in E_s/N_0 due to the inter-symbol interference are proportional to a factor $1/(1 - \Delta\tau/T_s)$. When delay spread is of same order as symbol duration, correct demodulation of the signal could be possible only with additional signal processing allowing to mitigate the effect of ISI.

The error rate performance of a digital system in a wideband channel tends to level off at high signal-to-noise ratios in contrast to the *flat* or narrowband fading case where error rates decrease without limit (Figure 5.7). This arises because the ISI, rather than the random AWGN noise, is the dominant source of errors. As the signal level is increased, the ISI increases proportionately, so demodulation performance stays the same. The final value of the error rate is known as an error floor, or sometimes the irreducible error rate, although the 'irreducible' is a misnomer only meaning no dependence on signal-to-noise ratio; in fact, the error floor can be removed using equalization and other techniques.

5.1.3 Doppler Shift

The movement of the receiver or transmitter leads to a shift of the received frequency called the Doppler shift, the phenomena well-known from physics. Consider a simple case of a single sinusoidal wave reaching the receiver Rx. If Rx moves away from the transmitter Tx with a speed v, then the distance d between Tx and Rx increases with that speed. Thus:

$$E(t) = E_0 \cos(2\pi f_c t - k_0[d_0 + vt]) = E_0 \cos\left(2\pi t \left[f_c - \frac{v}{\lambda}\right] - k_0 d_0\right), \qquad (5.12)$$

where d_0 is the distance at time $t = 0$. The frequency of the received oscillation is thus decreased by v/λ; in other words, the Doppler shift is given by:

$$v = -\frac{v}{\lambda} = -f_c \frac{v}{c}, \qquad (5.13)$$

where c is the speed of light.

Figure 5.8 Projection of velocity vector |v| onto the direction of propagation *k*.

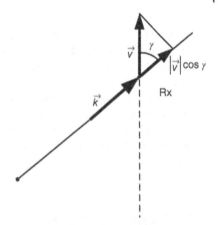

The Doppler shift is negative when Tx and Rx move away from each other. Since the speed of the movement is always small compared with the speed of light, Doppler shifts are relatively small. When the direction of Rx movement is not aligned with the direction of wave propagation, the Doppler shift is determined by the cosine rule or speed of movement *in the direction of wave propagation,* $v \cos(\gamma)$ (see Figure 5.8). The Doppler shift is then

$$v = -\frac{v}{\lambda} \cos \gamma = -v_{max} \cos \gamma \qquad (5.14)$$

In a mobile communications environment, the typical values of Doppler shift are in a range of less than 1 KHz.

If all multipath waves were Doppler shifted by the same amount – the effect on radio link performance would really be negligible – the local oscillator in the RX could easily compensate for such a shift. The important point is, however, that the different multipath components arrive with different Doppler shifts. Each pair of multipath components can be represented by a two-ray model described by equation (5.12). The superposition of several Doppler-shifted waves creates a sequence of fading dips. Doppler frequency is thus an important parameter of the channel, even though it is so small. Furthermore, the superposition of many slightly Doppler-shifted signals leads to phase shifts of the total received signal that can impair the reception of angle-modulated signals. These phase shifts lead to frequency dispersion, random Frequency Modulation (FM) of the received signal, which is especially important for signals with low bit rates where the channel delay spread is not the main source of distortion.

The sensitivity to the random frequency modulation depends on the modulation order. The QPSK modulation can tolerate up to $v_{max}T_s \leq 0.05$ whereas 64-QAM requires $v_{max}T_s \leq 0.01$, where T_s is a symbol duration and v_{max} is the estimate of maximum speed of relative movement Rx, Tx and/or scatterers.

It is important to note that the main propagation characteristics that should be taken into account when designing an OFDM system are expected delay spread, maximum Doppler frequency and, in the case of cellular systems, the targeted cell size.

Composition of the number of multipath components with different Doppler shifts results in a Doppler power spectrum of the received signal. The Doppler spectrum has two important interpretations:

1) It describes *frequency dispersion*. For narrowband systems, as well as Orthogonal Frequency Division Multiplexing (OFDM), such frequency dispersion can lead to transmission errors. It has, however, no direct impact on most other wideband systems (like single-carrier Time-Division Multiple-Access (TDMA) or Code Division Multiple-Access (CDMA) systems).

2) It is a measure for the *temporal variability* of the channel. As such, it is important for *all* systems. Assume that Doppler power spectrum has a bandwidth of $\Delta f \sim v_{max} = f_c{}^{v_{max}}/c$ and introduce a 'stationary' interval of the channel; that is, the time interval when the channel can be regarded as constant. The duration of stationary interval can be estimated as $\tau_c = \frac{1}{2\Delta f}$. This parameter is important for designing the size of equalizer window.

5.2 Worked Examples

5.2.1 Problem 1

Evaluate the maximum Doppler shifts in these cases:

1) a communication link with an airplane cruising at a speed of 850 km/h and carrier frequency 900
   MHz, and
2) communications at 2 GHz carrier with a user terminal in a high-speed train moving at 250 km/h.

Solution:
The maximum Doppler shift is given by $v_{max} = f_c{}^{v}/c$. Substituting the input values, we have
   1. $v_{max} = 700$ Hz and 2. $v_{max} = 480$ Hz.

5.2.2 Problem 2

What is the minimum symbol rate to avoid mitigation of the effects of Doppler spread in a mobile system operating at 450 MHz with a maximum speed of 100 km/h? Assume that the 'stationary' interval is given by the formula $\tau_c = \frac{1}{2v_{max}}$, where v_{max} is a maximum Doppler shift.

Solution:
The maximum Doppler shift $v_{max} = 42$ Hz. The stationary time interval for the channel is then $\tau_c \approx \frac{1}{2v_{max}} \approx 11$ ms. This is the maximum symbol duration, so the minimum symbol rate for undistorted symbols is the reciprocal of this, 84 symbols per second. The symbol rate is so small that most systems will rarely encounter significant effects in this form.

5.3 Fading

Random fluctuations in received signal can be estimated using statistical or random distributions. There are two major contributors to randomness of the received signal:

1) The level of a received wanted radio signal or an interfering signal can only in part be predicted deterministically. As discussed in Section 4.1, a global mean is commonly associated with the deterministic component of the received signal. The two additional components, namely 'shadowing' and 'fast fading', are both random but have different scales of variations. In this section, we consider the second and third component of radio propagation loss: slow fading or shadowing and fast fading components.
2) Input noise in the receiving system. The noise largely consists of radiated input noise in the antenna and internal noise in the receiver. It is totally stochastic or random.

5.3.1 Shadowing/Slow Fading

Referring to Figure 4.3, we may associate shadowing or slow fading with a second component of received signal strength. As discussed in Section 4.3, this component has a spatial scale of variations in order of several tens wavelengths. The shadowing component is especially important in radio coverage prediction where we need to ensure that a signal strength received at certain location, normally at the edge of the projected radio cell, will exceed certain threshold level. The statistical characteristics of shadowing can be described in terms of probability density function.

Consider the generic probability distribution plot in Figure 5.9. In this particular plot, the distribution is normalized at specific value of *global mean* (see Figure 4.3), say this value is the lowest median operating signal level that a radio receiver can operate with to achieve a given degree of performance.

With level of performance corresponding to a global mean, we do not take into account random variations in received signal. The signal would be acceptable for 50% of samples and not acceptable for the remainder. This is a level of performance that does not meet service requirements. We determine the limit of acceptable performance as being some value higher than the minimum operating level; that is, creating some margin of performance for acceptable signal level. This would have the effect of setting

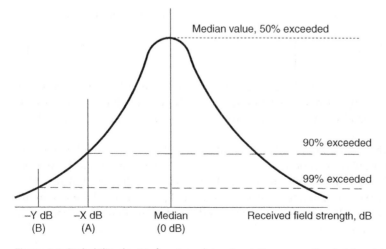

Figure 5.9 Probability density function of signal variations normalized at the global mean.

the minimum median operating level to some point on the left-hand side of the curve, such as point (A) or (B), see Figure 5.9.

As a simple illustration, if the value X is 10 dB, then shifting the reference value by 10 dB would change the percentage of signals above the reference value from 50 to 90%. If the median value was −90 dBm for example, and if we set a limit of −80 dB, then we can be sure that 90% of the signals experienced would be above −90 dBm, even taking account of the variability of the signal.

When shadowing is included, path loss L is composed of deterministic (L_{50}) and random (L_s) components

$$L = L_{50} + L_s \tag{5.15}$$

at 50% of locations at a given distance, as predicted by any standard path loss model (the median path loss or *global mean*).

The L_s is the shadowing component that follows a *lognormal* statistical distribution of signal power; that is, the *signal measured in decibels has a normal distribution* (i.e. is a zero-mean Gaussian random variable with standard deviation dB). The variation in L_s occurs over distances comparable to the widths of buildings and hills in the region of the mobile, usually tens or hundreds of metres.

Application of a lognormal distribution for shadowing models can be seen as follows. If contributions to the signal attenuation along the propagation path are considered to act independently, then the total attenuation A, as a power ratio due to N individual contributions $A1,..., AN$ will be simply the product of the contributions:

$$A = A_1 \times A_2 \times A_3 \cdots A_N \tag{5.16}$$

If this is expressed in decibels, the result is the sum of the individual losses in decibels:

$$L_s = L_1 + L_2 + L_3 + \cdots + L_N. \tag{5.17}$$

All of the contributions in (5.17) are taken as random variables, then, with $N \gg 1$, the central limit theorem holds and L behaves as a Gaussian random variable. Hence, A must be lognormal.

Probability density function of L_s is given by the standard Gaussian formula

$$p(L_s) = \frac{1}{2\sqrt{\pi}\sigma_L} exp \left[-\frac{L_s^2}{2\sigma_L^2} \right], \tag{5.18}$$

where σ_L is a standard deviation of received power in dB. Figure 5.10 represents typical path loss measured along the distance from the base station. The solid curve represents the global mean or signal strength incorporating slow fading/shadowing; that is, the L_s component.

In Figure 5.10, the cell range would be around 8 km if the effect of shadowing were neglected, then only 50% of locations at the edge of the cell would be properly covered. By adding the fade margin, the cell radius is reduced to around 5.3 km, but the reliability is greatly increased as a much smaller proportion of points exceed the maximum acceptable path loss. In other words, this margin is defined to maintain coverage with a certain *location probability*.

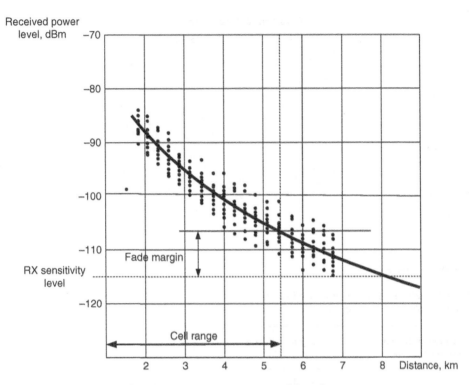

Figure 5.10 Effect of slow fading/shadowing on estimation of the cell range.

Another widely accepted name for *location probability* is *accessibility*. The common definition for location probability of X% is that the path loss is less than maximum acceptable value in X% of locations at the cell edge. Based on statistic properties of lognormal fading the location probability $Pr(L_s > z)$ with a given margin z can be calculated using the standard error function

$$Pr(L_s > z) = Q\left(\frac{z}{\sigma_L}\right) \qquad (5.19)$$

and

$$Q(t) = \frac{1}{\sqrt{2\pi}} \int_{x=t}^{\infty} exp\left(-\frac{x^2}{2}\right) dx = \frac{1}{2} erfc\left(\frac{t}{\sqrt{2}}\right) \qquad (5.20)$$

The plot of standard error function $Q(t)$ is provided in Figure 5.11. It can be used to evaluate the shadowing margin z needed for any location probability with substitute $z = t \cdot \sigma_L$. The lognormal (LNF) margin values versus cell edge probability are presented in Table 5.1.

Illustration of the lognormal fading model is given in Figure 5.12. The drop-outs in signal strength, also called outage A, are considered to occur when the local mean falls below a critical level, see Figure 5.12. The accessibility is then defined as $1 - A = Pr(L_s > z)$; that is, location probability.

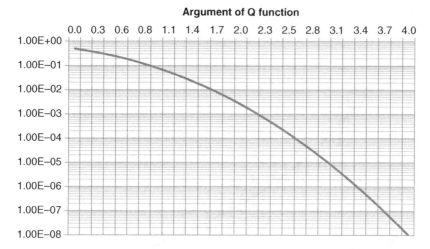

Figure 5.11 The Q function.

Table 5.1 Tabularized LNF margin in a function of location probability at cell border.

Location probability at the cell border, %	Lognormal fade margin, dB
50	$\sigma \times 0$
60	$\sigma \times 0.25$
70	$\sigma \times 0.52$
80	$\sigma \times 0.84$
90	$\sigma \times 1.28$
91	$\sigma \times 1.34$
92	$\sigma \times 1.4$
93	$\sigma \times 1.48$
94	$\sigma \times 1.55$
95	$\sigma \times 1.64$
96	$\sigma \times 1.75$
97	$\sigma \times 1.89$
98	$\sigma \times 2.05$
99	$\sigma \times 2.33$

5.3.2 Fast Fading/Rayleigh Fading

While mean path loss and shadowing could be predicted for RF coverage in particular locations, there are still significant variations in the received signal as the mobile moves over distances that are small compared with the shadowing correlation distance. This phenomenon is fast fading, presented by a third component in the received signal in Figure 4.3. The fast random variations in the signal are caused by multipath propagation in the neighbourhood of mobile unit. A typical situation corresponds to the obscured

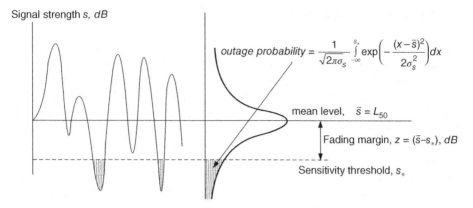

Signal strength s, dB

$$\text{outage probability} = \frac{1}{\sqrt{2\pi}\sigma_s} \int_{-\infty}^{s_*} \exp\left(-\frac{(x-\bar{s})^2}{2\sigma_s^2}\right) dx$$

mean level, $\bar{s} = L_{50}$

Fading margin, $z = (\bar{s}-s_*)$, dB

Sensitivity threshold, s_*

Figure 5.12 Illustration of fading margin and accessibility.

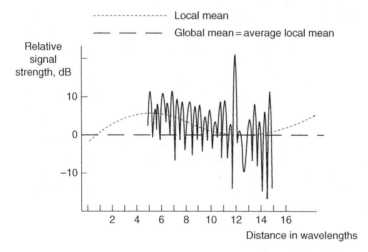

----------- Local mean

— — — Global mean = average local mean

Relative signal strength, dB

Distance in wavelengths

Figure 5.13 Rayleigh fading.

line of sight (LOS) between mobile unit and base station. A Raleigh fading is commonly assumed to describe statistics of the signal fluctuations. A typical recording of field-strength variations as a mobile unit is moving along a street is shown in Figure 5.13. The distance between adjacent fading dips is approximately one wavelength. During fading dips the instantaneous signal level drops 10–15 dB below the average level (local mean, i.e. the value after averaging over slow fading). Therefore, a fading margin between the local mean and the sensitivity threshold of the receiver is required to ensure that drop-outs caused by the input signal falling below the threshold will occur rarely.

The location probability in the case where the Rayleigh fading is taken into account results in additional margin. Whether or not this margin is to be taken into account in link budget calculation depends on the system design and specific situation that will be considered later.

5.4 Diversity to Mitigate Multipath Fading

When there is fading caused by multipath propagation, a considerable fading margin is needed to prevent signal drop-outs or outages during the deepest fading dips. Diversity can be used to bridge over the fading dips, reducing the necessary fading margin and thus producing a diversity gain. In a diversity arrangement, the signals from two or more propagation paths whose fading is only partially correlated (completely uncorrelated in ideal case) are combined. Several different types of diversity can be used in digital radio transmission, such as spatial diversity, frequency diversity and polarization diversity.

The underlying principle of diversity reception is based on statistical independence of variations in received electromagnetic field in antennas (diversity branches) that are separated either in distance or polarization. Combing the uncorrelated signals from diversity branches results in statistical (diversity) gain since fading dip in one branch not necessarily lead to fading deep in another branch and, instead, may coincide with a peak in signal in the other branch.

The frequency diversity implies transmission on two (or many) frequencies separated in the frequency domain at intervals exceeding coherence bandwidth of the propagation channel. Such a frequency offset ensures that signal fluctuations at two frequencies are uncorrelated in a common receiving point. A typical example is frequency hopping used in the GSM system that will be discussed in Section 7.3.5.4.

The time diversity principle in its primitive form implies that same sequence of bits is repeated over time intervals approximately equal to the correlation time interval of fading. The intended goal is to ensure uncorrelated variations in the received signal bit stream. In an advanced form, time diversity is employed by interleaving a coded data block into several time frames separated over the correlation time interval of fading. In the worst-case scenario, the only part of coded data might be lost in a fading dip allowing recovery of data by means of channel coding. In practice, this approach results in reduction of errors making them statistically independent and thus significantly improving performance of error-correction coding.

5.4.1 Space and Polarization Diversity

The original form of diversity is antenna or space diversity, which can be used in both analogue and digital transmission. Other types of antenna diversity are polarization diversity and antenna pattern diversity.

Consider the geometry of scattering shown in Figure 5.14 with two antennas located in point 1 and 2. Assume that the scattering angle, that is the size of angular sector of direction of waves arriving in antennas 1 and 2, equals to ϑ. Then the path difference Δ of the waves arriving at either point 1 and 2 is given by $\Delta = d \sin(\vartheta/2)$. When $\Delta = \lambda/2$, the signals in points 1 and 2 arrive in opposite phase. The value of $\Delta = \lambda/2$ can serve as an estimate of the space correlation scale. Therefore, the antenna spacing d for space diversity has to be equal to the space correlation scale, that is

$$d = \frac{\lambda}{2 \sin(\vartheta/2)} \tag{5.21}$$

Consider the downlink transmission from base station to mobile terminal. In the neighbourhood of the mobile, station scattering has an isotropic nature meaning that

Figure 5.14 Geometry of scattering.

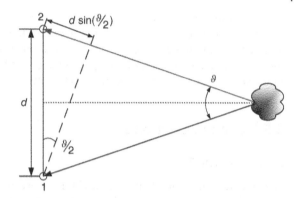

waves arrive within an angular sector of ± 180°; note that the angles from 180 to 360° create the same path difference with the opposite sign. Therefore, the spatial correlation scale equals half a wavelength $\lambda/2$ that corresponds to a Rayleigh fading case.

In the opposite case of waves arriving at the base-station antenna from an individual mobile station, arriving waves are concentrated in a narrow angular sector thus exemplifying a plane wave. The scattering volume is concentrated in the vicinity of the mobile station. It forms a remote scattering volume of limited size that can be regarded as being located in a far-zone relative to a base-station antenna.

This is a simplified view at the uplink wave from mobile unit reaching the base-station antenna. The uplink scattering angle is highly dependent on the mobile environment and is difficult to predict. From a practical point-of-view, antenna separation of several wavelengths is needed for fairly independent fading in a base-station site.

In practice, polarization diversity is used instead. The mechanism of polarization diversity is based on the observed fact that polarization of a radiated wave becomes random due to multiple scattering in the multipath channel. For example, ideally vertical polarized radiation at the source produces both vertical and horizontal components in the receiver antenna as the received electromagnetic field is formed by a superposition of randomly polarized waves carried by multipath components. Combing the uncorrelated signals from diversity branches results in statistical (diversity) gain since fading dip in one branch does not necessarily lead to fading deep in another branch and, instead, may coincide with a peak in signal in the other branch.

Diversity gain with Rayleigh fading depends on the diversity arrangement, how many diversity branches are combined and the level of correlation between them. Most of the potential diversity gain is achieved with two antennas. In a commercial mobile system, most common diversity implementation in a base-station antenna is based on two cross-polar antennas. There are several types of diversity combining that could be distinguished in pre-detection and post-detection diversity. In post-detection diversity, the signals from diversity branches are added coherently; that is, their mutual phases must be adjusted continuously. Such a mechanism enables the signal voltages to be added together thus producing a maximum diversity gain of 6 dB when two signals of equal strength are combined. In pre-detector diversity, there is a choice of maximum ratio and equal-gain combination. The optimum arrangement in theory is the maximum ratio, whereby the signal-to-noise ratio for the different diversity branches is measured and the gain of the branches is set to maximize the signal-to-noise ratio after addition. This means that greater weight is given to the diversity branch with the best signal-to-noise ratio. In an equal-gain combination, there is no setting of branch amplification. Equal-gain combination is the most widely used pre-detector arrangement with a maximum achievable gain of about 3 dB. Some vendor specific proprietary improvements may increase the gain up to 5 dB.

5.5 Worked Examples

5.5.1 Problem 1

Consider transmission of information bits with a duration of $T_b = 1$ ms over a multipath fading channel with an average fading rate of $R_f = 1/(8 \cdot T_b)$. Assume the channel encoder can only use a repetition of symbol frame as a means of mitigation against fading. Introduce a structure of mapping coded words into a channel symbol stream. Estimate the channel rate sufficient to deploy a repetition scheme.

Solution:
   The average fade duration or interval between fading dips is $T_f = \frac{1}{R_f} = 8 \cdot T_{\text{inf bit}}$. The correlation scale of fading in time domain can be estimated as $\tau_c = T_f/2 = 4 \cdot T_{\text{inf bit}}$, see Figure 5.15. The channel encoder should repeat the code word in a time interval of the order τ_c. This way the fades over time τ_c are uncorrelated and at least one of two replicas of the code word could likely be received with sufficient signal strength. Given that channel encoder should be adjusted to the incoming information bit stream, the repeated code word size should be $8 \cdot T_{\text{inf bit}}$ and channel rate should be twice of the rate of the information bit rate, as shown in Figure 5.15.

5.5.2 Problem 2

The measured delay spread of multipath channel is $\Delta \tau = 5\,\mu s$. Assume that frequency hopping system transmit next coded frame at the frequency is $f = f_c + m \cdot \Delta f_h$, shifted at

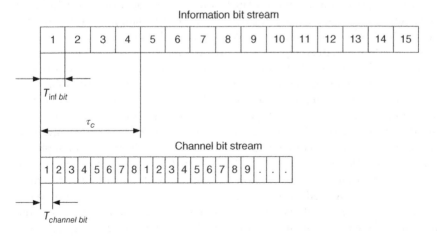

Figure 5.15 Channel coding with repetition.

$m\cdot\Delta f_h$ according to a frequency hopping pattern. What is the minimum values of Δf_h given that delay spread of the communications channel is $\Delta\tau = 5\,\mu s$?

Solution:

The objective of frequency hopping system is to deploy a frequency diversity scheme; that is, make fading at consecutive coded frames uncorrelated. The minimum frequency hop value should then be no less than the coherence frequency bandwidth of the channel $\Delta f_c = 1/\Delta\tau$. Therefore, $\Delta f_h = \Delta f_c = 200$ kHz.

5.5.3   Problem 3

Consider a mobile communication system operating at 1800 MHz in an urban environment. Consider the following:

1) assuming a vehicular speed of 60 km/h, estimate a value of time correlation of the fades;
2) estimate value of the channel stationary interval impacted by vehicular movement.

Solution:

1) the space correlation interval is given by $l = \lambda/2\sin(\vartheta_s/2)$, where ϑ_s is an angular size of scattering sector. Assuming a typical mobile neighbourhood with absence of line of sight, we have the case for isotropic scattering, $\vartheta_s = 180°$. The space correlation interval is $l = \lambda/2$, time correlation interval is $t_c = 1/_{v}=\lambda/2v = 5$ ms.
2) the maximum Doppler shift is $\Delta f = v/\lambda$. The stationary interval is $\tau_c = 1/2\Delta f = \lambda/2v = 5$ ms, that is, equal to correlation time interval of Rayleigh fading.

5.6   Receiver Noise Factor (Noise Figure)

The impact of noise on the system needs to be calculated to determine the system performance. The major noise contributions in mobile wireless systems usually come from the receiver itself. In any case, the total noise associated with the system can be calculated by assuming that the system consists of a two-port network, with a single input and a single output as shown in Figure 5.16. The network is characterized by a gain G, being the ratio of the signal power at the output to the signal power at its input, and by a noise factor F. The noise factor is the ratio between the output noise power of the element divided by gain G (i.e. referred to the input) and the input noise.

The noise in the input of the receiver may consist of several components, such as man-made industrial noise, spurious emission of electrical and electronic devices and thermal noise. Thermal noise is non-avoidable, it is always present. Thermal noise imposes the fundamental limit on radio communications; that is, thermal noise floor.

Figure 5.16 Noisy two-port network.

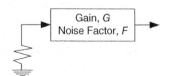

The power spectral density of thermal noise is given by

$$N_0 = k_B\, T \tag{5.22}$$

where kB is Boltzmann's constant, $k_B = 1.38 \cdot 10^{-23}\,\text{J/K} = 1.38 \cdot 10^{-23}\,W \cdot Hz^{-1} \cdot K^{-1}$, T is an environmental temperature of the noise source in Kelvin, and the noise power (noise floor) is

$$N = N_0 B \tag{5.23}$$

where B is a receiver bandwidth (in units of Hz). It is common to write equation (5.23) using logarithmic units (power P expressed in units of dBm is $10\log_{10}(P/1mW)$). Assuming room temperature, the noise spectral power density has the value:

$$N_0 = -174\,\text{dBm} \tag{5.24}$$

This means that the noise power contained in a 1-Hz bandwidth is -174 dBm. The noise power contained in bandwidth B is

$$N = -174 + 10\log_{10}B\,,\ \text{dB} \tag{5.25}$$

Noise power density N_0 can also be normalized to MHz bandwidth and then approximated to 114 dBm/MHz as an easier figure to work with. Any physically realizable system introduces additional noise on the top of the thermal noise floor. That additional noise component defines the noise figure F of network element of receiver system. The noise figure is defined as ratio of SNR at the input to SNR at the output of the receiver

$$F = \frac{S_{in}/N_{in}}{S_{out}/N_{out}} = \frac{S_{in}/k_B T \cdot B}{G \cdot S_{in}/N_{out}} = \frac{N_{out}/G}{k_B T \cdot B} \tag{5.26}$$

We assume that the network in Figure 5.16 is impedance matched to resistance. The noise figure is then

$$F = \frac{N_{out}}{N_{in}} = \frac{N_{out}}{k_B T \cdot B} \tag{5.27}$$

where N_{out} is the output noise power of the network element referred to input; that is, the actual noise output power divided by G. Noise figure F depends on the design and physical construction of the network. Its value in decibels is the noise figure of the network,

$$F_{dB} = 10\log_{10}F \tag{5.28}$$

The numerical value of the noise power [dBW] can be expressed approximately as

$$N_{out} = F_{dB} - 204 + 10\log_{10}B \tag{5.29}$$

and, respectively, in dBm:

$$N_{out} = F_{dB} - 174 + 10\log_{10}B. \tag{5.30}$$

The value of noise power given by equations (5.29) or (5.30) is known as a *receiver noise floor*, please note that it differs from thermal noise floor at noise figure F_{dB}. For successful recovery of the received data the signal level must exceed the noise floor with sufficient margin, at least, for non-CDMA systems. This margin is given by the signal-to-noise ratio required for decoding/demodulation of the received data stream with performance

Figure 5.17 Cascade network.

levels specified in terms of Bit-Error Rate (BER) or other parameters such as SINAD, MOS and so on.

The receiver sensitivity level is determined by addition of required SNR on the top of receiver noise floor:

$$Rx_{sens} = receiver\ noise\ floor + SNR = N_{out} + SNR \tag{5.31}$$

Receiver sensitivity level varies according to modulation scheme and desired level of performance, often quoted in terms of bit-error-rate (BER). If this value is, say, 9 dB above noise (or noise and distortion), then the minimum receiver sensitivity for a 270-kHz system with a 8 dB noise figure would be:

$$-114 - 5.68 + 8 + 9 = -102.68\ dBm$$

In general, a complete system can be characterized by a cascade of two-port elements, where the *i*th element has gain *Gk* and noise factor *Fk* (Figure 5.17). Each element could consist of an individual module within a receiver, such as an amplifier or filter, or one of the elements within the channel such as the antenna, feeder or some source of external noise.

The gain *G* of the complete network is then simply given by

$$G = G_1 \cdot G_2 \cdots G_N, \tag{5.32}$$

while the overall noise factor is given by the Friis formula:

$$F = F_1 + \frac{F_2 - 1}{G_1} + \frac{F_3 - 1}{G_1 \cdot G_2} + \cdots + \frac{F_N - 1}{G_1 \cdot G_2 \cdots G_N} \tag{5.33}$$

where noise figures and gains in (5.33) are in absolute units, not in dB. As observed from (5.33), the noise from the first element adds directly to the noise of the complete network, while subsequent contributions are divided by the gains of the earlier elements. It is therefore important that the first element in the series has a low noise factor and a high gain, since this will dominate the noise in the whole system. As a result, receiver systems often have a separate *low noise amplifier* (LNA) placed close to the antenna, often at the top of the mast and sometimes directly attached to the feed of a dish antenna in order to overcome the impact of feeder loss.

6

Radio Network Planning

In terms of the geographic coverage of the network, often the limit of coverage will be determined by one of the following factors:

- Coverage or noise limited, in which the environmental noise and the minimum signal level needed to provide the desired level of availability will be the limiting factors.
- Interference limited, in which the presence of other radio systems will increase the level of signal that will be required to achieve the necessary level of performance.

In some cases, the network configuration can be driven by capacity limitations that impact the cell size and cluster size in systems with frequency reuse. The example of an interference limited system is a 3G-WCDMA system where the interference is a major factor affecting both capacity and cell size. The GSM system exemplifies a mixed case when both noise and co-channel interference can limit a cell range depending on service (voice or data) and respective performance criteria.

One difference between interference and noise lies in the fact that interference suffers from fading, while the noise power is typically constant (averaged over a short time interval). The statistical nature of interfering signal as well as of the wanted signal is taken into account by adding effective fade margin and, possibly, additional correction factors into link budget. The effective fading margin can be estimated by calculating location probability at the cell edge in the presence of multiple interfering servers. This approach uses joint multivariate statistics of lognormal fading of the signals from multiple servers (co-channel interferers).

When planning needs to be more detailed, a Monte-Carlo simulation of interference in multi-server environment could be used as an advanced planning tool.

6.1 Generic Link Budget

A link budget is to be compiled before start of the dimensioning of the radio network. In the link budget, different design criteria for coverage (e.g. outdoor, indoor, in-car) are determined. In the link budget, factors such as receiver sensitivity and different margins are considered. It should be noted that all the values for margins discussed are estimates only.

The specific margins and gains in link budget calculation may or may not be included depending on system technology under consideration and details of implementation. The goal of link budget compilation is to estimate a cell range for both uplink and

Introduction to Mobile Network Engineering: GSM, 3G-WCDMA, LTE and the Road to 5G,
First Edition. Alexander Kukushkin.
© 2018 John Wiley & Sons Ltd. Published 2018 by John Wiley & Sons Ltd.

downlink. As an example, we next consider the link budget for downlink in GSM system; that is, from base station to mobile.

6.1.1 Receiver Sensitivity Level

Sensitivity level of the mobile station's receiver MS_{sens} varies with its class and manufacturer and is given in specs provided by the manufacturer. The receiver sensitivity depends on the required signal carrier to noise ratio (SNR). This level is defined for a specified BER in a Gaussian noise background, in a so-called static channel. When frequencies are reused, the received carrier power must be large enough to combat both noise and interference, that means C/(N+I) must exceed the receiver threshold. Recalling that required CIR (Carrier-to-Interference Ratio) is defined at the edge in the frequency reuse system, the additional CIR margin can just add up on the top of the receiver noise floor plus SNR or can replace SNR in the case of CIR dominance.

To account for the dynamic propagation channel, some margins have to be added to compensate for Rayleigh fading, interference and body loss. The obtained signal strength P_{req} is a least but not last estimate for required power in the receiver to process a phone call. P_{req} is rather independent of the environment since Rayleigh fading is supposed to be present in any real mobile environment.

$$P_{req} = MS_{sens} + M_R + M_I + M_b, \tag{6.1}$$

where

MS_{sens} = Mobile Station sensitivity level, dBm;

M_R = Rayleigh fading margin, dB;

M_I = Interference margin, dB;

M_b = Body loss, dB that normally taken into account RF coverage in pedestrian environments only.

In some cases, the mobile station sensitivity level MS_{sens} is defined in a dynamic channel taking Rayleigh fading into account, opposite to the MS_{sens} level in a static channel. In this case, the M_R margin should be omitted in calculations of required signal level.

6.1.2 Design Level

Extra margins have to be added to P_{req} to handle lognormal fading as well as different types of penetration losses. These margins depend on the environment and on the desired area of coverage. The resultant signal strength is used when planning the system and will be referred to as the design level, P_{design}. That signal strength is the value that should be obtained on the cell border. The design level can be calculated from

$$P_{design} = P_{req} + M_{LNF} + M_{BPL} + M_{car} \tag{6.2}$$

where

M_{LNF} – lognormal fading margin,

M_{car} – car penetration loss,

M_{BPL} – building penetration loss.

6.1.2.1 Rayleigh Fading Margin

The required sensitivity performance of the system in terms of Frame-Error Rate or BER is normally specified for each type of system channel. The sensitivity is defined as the level where the required quality performance is achieved. Sometimes receiver sensitivity level MS_{sens} provided in specifications has Rayleigh fading already taken into account. The practical value of Rayleigh fading margin is $M_R = 3$ dB.

6.1.2.2 Lognormal Fading Margin

The values of the lognormal fading margin were given in Table 5.1. Typical values of the standard deviation depending on clutter type for lognormal fading are given in Table 6.1.

6.1.2.3 Body Loss

The human body has several effects on MS performance compared to a free-standing mobile phone; for instance, the head absorbs energy and antenna efficiency of some MSs can be reduced. Other effects may be a change of the lobe direction and the polarization. These effects can be neglected in the link budget since:

1) no mobile antenna gain is used and
2) X-polarized antennas are standard equipment today.

In this case the polarization loss is included in the downlink link budget and in the uplink, both polarizations can be received. The body loss M_b recommended by ETSI is 3 dB for 1800 MHz, 5 dB for 900 MHz.

6.1.2.4 Car Penetration Loss

When the MS is situated in a car without an external antenna, an extra margin has to be added in order to cope with the penetration loss of the car. This extra margin is approximately $M_{car} = 6$ dB.

6.1.2.5 Design Level

Given the values of the margins listed previously, design level for outdoor coverage in the GSM-1800 system can be calculated as follows:

$$P_{design} = MS_{sens} + M_{LNF} + M_{car} = -105 \text{ dBm} + 10 \text{ dB} + 5 \text{ dB} = -90 \text{ dBm}$$

Table 6.1 Standard deviation of lognormal distribution in different environments.

Environment	Standard deviation, σ, dB
Dense urban	10
Urban	8
Suburban	6
Rural	6

6.1.2.6 Building Penetration Loss

It is known from measurements that building penetration loss to floors higher up in a building generally decreases. This effect is known as height gain. This is actually an effect of the building penetration loss definition and not of the building structure.

Building penetration loss is defined as the difference between the average signal strength immediately outside the building and the average signal strength over the ground floor of the building. The building penetration loss for different buildings is lognormally distributed with a standard deviation: σ_{BPL}.

Variations of the loss over the ground floor could be described by a stochastic variable, which is lognormally distributed with a zero-mean value and a standard deviation of σ_{floor}. In this calculation, σ_{BPL} and σ_{floor} are lumped together by adding the two as if they were standard deviations in two independent lognormally distributed processes.

The resulting standard deviation, σ_{indoor}, could be calculated as the square root of the sum of the squares:

$$\sigma_{indoor} = \sqrt{\sigma_{BPL}^2 + \sigma_{floor}^2} \tag{6.3}$$

6.1.2.7 Outdoor-to-Indoor Design Level

The indoor design level when the base station is situated outdoors is calculated according to

$$P_{design} = P_{req} + M_{LNF} + M_{BPL} \tag{6.4}$$

where $M_{BPL} = BPL_{mean} + \sigma_{indoor}$ can be seen as the indoor margin. BPL_{mean} is the mean value of the building penetration loss. Some values of BPL_{mean} and σ_{indoor} are given in Table 6.2.

In the case of indoor base station location, the coverage is calculated based on signal-to-noise criteria rather than CIR criteria since interference from the other cell is efficiently blocked by the walls of the building. The major factor is a lack of coverage.

6.1.3 Power Link Budget

The next step is to calculate all gains and losses associated with an antenna system on the downlink and uplink. The definitions are as follows:

- Downlink (DL) is the direction from the BS to the MS. The downlink budget estimates the received signal strength in the MS.
- Uplink (UL) is the direction from the MS to the BS. The uplink budget estimates the power level received in the base station.

Table 6.2 Typical values of building penetration loss.

	BPL_{mean}	σ_{indoor}
Dense urban	18	9
Urban	18	9
Suburban	12	8

The path balance implies that the coverage of the downlink is equal to the coverage of the uplink. The power budget calculated for both uplink and downlink shows whether the uplink or the downlink is the weak link. When the downlink is stronger, the EIRP/ERP used in the prediction should be based on the balanced BTS output power. When the uplink is stronger, the maximum BTS output power is used instead.

In the calculations that follow, the antenna gain in the MS and the MS feeder loss are both zero and therefore omitted. It is also assumed that the antenna gain and the feeder loss are the same for the transmitter and receiver side of the BTS.

The following abbreviations are used:

$P_{Rx,MS}$ = Received power in MS, dBm

MS_{sens} = Receiver sensitivity level in MS, dBm

BS_{sens} = Base-station receiver sensitivity level, dBm

$P_{Tx,MS}$ = MS transmit power, dBm

$P_{Tx,BS}$ = BS transmit power, dBm;

L_{comb} = BS combiner loss, dB;

L_{dip} = external diplexer loss, dB that should only be taken into account if an external duplexer is used.

L_{feed} = feeder loss, dB;

$G_{ant,BS}$ = antenna gain in BS, dBi;

$G_{ant,MS}$ = antenna gain in MS, dBi;

G_{div} = diversity gain, dB.

The losses in the antenna-feeder system depend on the base-station site configuration. Figure 6.1 shows base-station antenna system configuration for a dual band GSM900/1800 system when installation of the 1800 MHz has been done with diplexers utilizing the existing feeder system. As shown, the base station has a dual band dual polarization two-port antenna. Figure 6.2 shows a configuration with a four-port dual band antenna and separate feeder sets for 900 and 1800 MHz base stations. The 1800 MHz configuration commonly uses the Masthead amplifier with a low noise figure to compensate for uplink feeder loss. In the case of a four-port dual band antenna or two separate single band antennas (not considered here) and single set of shared feeders, one needs another set of diplexers at the top after feeder before the antenna itself.

6.1.4   Power Balance

Consider the configuration shown in Figure 6.1 and assume $G_{ant,MS} = 0$ dBi. Applying relevant gains and losses either to the downlink or uplink we obtain:

$$P_{in,MS} = P_{Tx,BS} - L_{dip} - L_{feed} + G_{ant,BS} - L - M_{LNF}, \text{ downlink} \tag{6.5}$$

$$P_{Rx,BS} = P_{Tx,MS} + G_{ant,BS} + G_{div} - L_{dip} - L_{feed} - L - M_{LNF}, \text{ uplink} \tag{6.6}$$

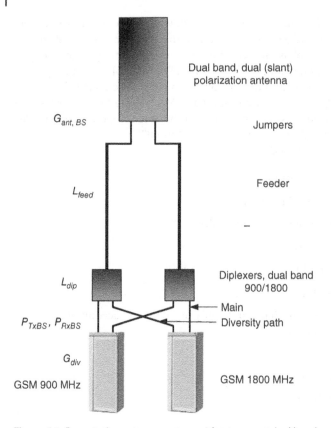

Dual band, dual (slant) polarization antenna

$G_{ant, BS}$

Jumpers

Feeder

L_{feed}

–

Diplexers, dual band 900/1800

L_{dip}

Main
Diversity path

P_{TxBS}, P_{RxBS}

G_{div}

GSM 900 MHz

GSM 1800 MHz

Figure 6.1 Base-station antenna system with a two-port dual band cross-polarization antenna and diplexer for combining 900 and 1800 MHz RF paths.

We have assumed a pedestrian environment and lognormal fading is taken into account. In order to ensure a good reception of the uplink signal, received signal level should exceed sensitivity threshold, $P_{Rx,BS} \geq BS_{sens}$. It is important to note that the path loss is reciprocal; that is, path loss for downlink is the same as for uplink. The maximum allowed path loss (MAPL) can be calculated from the power budget. For the downlink, we obtain the following equation:

$$MAPL_{DL} = EIRP_{BS} - P_{design, MS}, \text{ for downlink} \qquad (6.7)$$

From (6.5) we have

$$MAPL_{DL} = L(downlink) = P_{Tx,BS} + G_{ant,BS} - L_{dip} - L_{feed} - M_{LNF} - MS_{sens} \qquad (6.8)$$

To achieve power balance, it is necessary to make sure that the required minimum signal level by the MS (downlink) is reached at the same point as the required minimum by the Base Transceiver Station (BTS) (uplink).

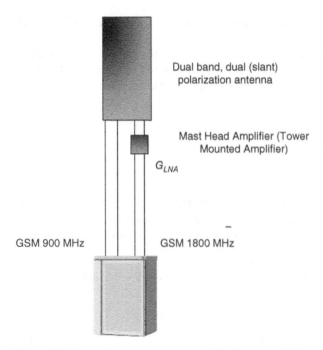

Dual band, dual (slant)
polarization antenna

Mast Head Amplifier (Tower
Mounted Amplifier)

G_{LNA}

GSM 900 MHz   GSM 1800 MHz

Figure 6.2 Base-station antenna system with a four-port dual band cross-polarization antenna and MHA for 1800 MHz.

That means that $MAPL_{DL} = MAPL_{UL} = MAPL$, where $MAPL_{UL}$ can be obtained from (6.6):

$$MAPL_{UL} = L(uplink) = P_{Tx,MS} + G_{ant,BS} + G_{div} - L_{dip} - L_{feed} - M_{LNF} - BS_{sens}.$$
$$(6.9)$$

In some configurations, the uplink becomes weaker than the downlink and power budget is unbalanced. The unbalanced budget will lead to situation when signal at weaker path cannot be decoded. When uplink is weaker, the mobile station can receive base-station signal on downlink, but the base cannot 'hear' the mobile. A common approach is then to use a Tower Mounted Amplifier (TMA), another term to describe the same is MHA (Mast Head Amplifier), which is a low noise amplifier (LNA) that is mounted close to the receiving antenna, see Figure 6.2. The TMA improves the uplink sensitivity and extended coverage can be achieved.

In the following calculations, we consider power balance without TMA, otherwise the TMA gain (G_{LNA}) should be added in the uplink power budget. The balanced budget can be used to adjust the maximum transmit power of the base station, $P_{Tx,BS}$. Assuming $MAPL_{DL} = MAPL_{UL}$, we have

$$P_{Tx,BS} = P_{Tx,MS} + G_{div} - BS_{sens} + MS_{sens}$$
$$(6.10)$$

Apparently, in case of a weaker uplink path we should take $MAPL = MAPL_{UL}$ as allowable path loss.

Given allowable path loss we can estimate a cell range using a suitable propagation model, as described in Section 6.4 or in the literature where numerous models have been developed for a path loss prediction, such as Okumura–Hata, COST-231, Ikegami–Walsh model and so on.

6.2 Worked Examples

6.2.1 Problem 1

A 900 MHz mobile system is designed with a location probability of 95% at the cell edge with a shadowing variance of 6 dB. The propagation environment is assumed to be modelled by the Okumura–Hata propagation model for medium size cities. The mean path losses are given by the equation

$$L_{50} = 43.6 + 33.9 \cdot \log_{10}f - 13.82\log_{10}h_{BS} - a(h_{MS)} + (44.9 - 6.55\log_{10}h_{BS})\log_{10}d \tag{6.11}$$

where

$$a(h_{MS)} = h_{MS}(1.1 \cdot \log_{10}f - 0.7) - 1.56 \cdot \log_{10}f + 0.8 \tag{6.12}$$

The units are as follows:

Frequency f in MHz, heights h_{MS}, h_{BS}– in m, distance d in km, $h_{MS} = 1.5$ m and $h_{BS} = 30$ m.

The maximum allowable path loss, MAPL, is estimated as 146 dB. Determine the range of the cell with given accessibility. How will an increase in shadowing variance of 2 dB affect the estimated cell range?

Solution:

The maximum allowable path loss is given by the sum of the global mean and shadowing component; that is

$$L = L_{50} + L_s \Rightarrow MAPL = L_{50} + M_{LNF} \tag{6.13}$$

The global mean L_{50} is described by the Okumura–Hata model, while the second term in equation (6.13) is a lognormal (shadowing) margin M_{LNF} defined by the required level of accessibility. To find margin M_{LNF} for L_s, we may consider error function $Q(t)$ or Table 5.1. With a given location probability 95% and variance 6 dB we find $M_{LNF} = \sigma \cdot 1.64 = 9.84$ dB from Table 5.1.

Therefore, the global mean component L_{50} in $MAPL$ is

$$L_{50} = MAPL - M_{LNF} = 146 - 9.84 = 136.16 \text{ dB} \tag{6.14}$$

The cell range r_{cell} can be found applying the Okumura–Hata model to the global mean component of the $MAPL$:

$$43.6 + 33.9 \cdot \log_{10}f - 13.82\log_{10}h_{BS} - a(h_{MS)} + (44.9 - 6.55\log_{10}h_{BS})\log_{10}r_{cell}$$
$$= 136.16$$

Given the values for the heights and frequency, we obtain the cell range $r_{cell} = 2.45$ km.

With a change of fading variance to 8 dB, the fading margin required for 95% location probability rises to 13 dB and cell range to 2 km.

6.2.2 Problem 2

Analyse hypothetical communications with an aircraft using a cellular system on the ground. Establish a link budget assuming that the plane is flying at 10 km altitude over an area where the mobile communication service has been provided by a cellular system designed with a cell radius of 20 km. Assume the following:

- cell radius has been calculated for propagation exponent $n = 4$,
- 10 dB penetration loss through the aircraft hull and
- −10 dB antenna gain of the BS antenna in the direction of the aircraft.

Solution:
Recall the formula for the path loss $L = L_0 - \alpha \log R$, where α is a propagation exponent and $L_0 = 33.44 + 20 \log f (MHz) + 20 \log R(km)$. The component L_0 describes the path loss in a free space at distances $R < R_0$, where R_0 is a break point or assured distance at which the line of sight between the mobile and base station is observed. For simplicity, assume $R_0 = 0.1$ km, and fix the value of L_0 at the break point R_0, that is

$$L_0 = 33.4 + 20\log_{10}f + 20\log_{10}R_0$$

From a definition of cell range, maximum allowable path loss,

$$\text{MAPL} = L_0 + \alpha\log_{10}(R_{cell}) = EIRP - P_{design}$$

The effective radiated power in direction of airplane is reduced according to antenna pattern

$$EIRP_{air} = EIRP - 10 \text{ dB}$$

L_0 has the same value, design level P_{design} should be increased to cope with penetration loss through aircraft hull, that is

$$P_{design\_air} = P_{design} + 5 \text{ dB}$$

The maximum allowable path loss in ground-to-air communication is then

$$MAPL_{air} = EIRP_{air} - P_{design\_air} = EIRP - 10 - P_{design} - 10 = MAPL - 20 \text{ dB}$$

On the other hand, the maximum path loss in direction of aircraft in zenith is

$$L_{air} = L_0 + 20\log_{10}(R_{air})$$

The difference between $MAPL_{air}$ and L_{air} should ensure a positive margin that determines whether the air-ground link in the zenith direction is sustainable.

$$\Delta L = MAPL_{air} - L_{air} = 40\log_{10}(R_{g\_cell}) - 20\log_{10}(R_{air}) - 20$$
$$= 32.04 - 20 = 12 \text{ dB} > 0$$

The horizon for the line of site to aircraft flying at the height of 10 km is approximately 41.2 km. Substituting the horizon distance into this equation we obtain

$$\Delta L = MAPL_{air} - L_{air} = 19.7 - 20 = 0.3 \text{ dB} < 0$$

that is, the negative margin for a link availability towards aircraft at the horizon. However, this estimate is rather coarse since it does not take into account variation in antenna pattern across elevation angle and increase in path loss under sliding angles to the surface.

6.2.3   Problem 3

Consider a trunked radio type communication system at 400 MHz. The base transmits 10 W, channel bandwidth is $BW = 12.5$ KHz, base-station antenna gain is 10 dBi, mobile receiver antenna has a gain of 3 dBi and noise figure $NF = 6$ dB. Assume the propagation model is given by equation (3.12) with a breaking point for L_0 at 1 km and propagation exponent $\alpha = 3.5$. Find the coverage range r_c given with a required SNR $= 15$ dB and 95% of location probability with lognormal fading variance of $\sigma = 10$ dB.

Solution:
Find the receiver noise floor using equation (5.30)

$$\text{Noise floor} = -174 + 10 \log_{10}BW + NF = -127 \text{ dBm}$$

Adding the SNR, we obtain the receiver sensitivity level

$$Rx_{sens} = \text{Noise floor} + \text{SNR} = -112 \text{ dBm}$$

From Table 5.1, the fading margin M_{LNF} for 95% accessibility

$$M_{LNF} = 1.645 * \sigma = 16.45 \text{ dB}$$

The receiver design level is given by

$$P_{MS,design} = Rx_{sens} + M_{LNF} = -95.5 \text{ dBm}$$

Applying the propagation model, the received signal strength is calculated from

$$P_{Rx,MS} = P_{Tx,BS} + G_{ant,BS} + G_{ant,MS} - L,$$

where $L = 33.42 + 20\log_{10}f + 20\log_{10}d, \ d \le d_0$

$$L = L_0 + 35 \cdot \log_{10}(d), \ d > d_0,$$
$$L_0 = L(d = d_0)$$

Assuming that coverage range sets beyond the breakpoint distance d_0, we may consider path loss at the distances $d > d_0$, that is

$$L = L_0 + 35 * \log_{10}(d) = 85.5 + 35\log_{10}(d) \text{ dB}$$

The received signal strength $P_{Rx,MS}$ should be above the design level $P_{MS,design}$ that defines a coverage threshold:

$$P_{MS,design} = P_{Rx,MS} = P_{Tx,BS} + G_{ant,BS} + G_{ant,MS} - L$$

Then the MAPL is given by

$$MAPL = 85.5 + 35 \cdot \log_{10}(r_c) = P_{Tx,BS} + G_{ant,BS} + G_{ant,MS} - P_{MS,design} \quad (6.15)$$

After substitution, all given system parameters into equation (6.15) we obtain

$$P_{MS,design} = -95.5 \text{ dBm},$$

$35 \cdot \log_{10}(r_c) = 63$ dB, and coverage range $r_c \approx 63$ km with accessibility 95%.

7

Global System Mobile, GSM, 2G

7.1 General Concept for GSM System Development

Development of the fundamental technology for GSM started in the Nordic countries in the 1980s, led by Ericsson and then transferred to a working group of standardization body 'Groupe Special Mobile' (GSM) within the Conference Europeenne des Postes et Telecommunications (CEPT) standard committees. Since the standardization of the basic GSM900, the GSM system has experienced extensive modifications in order to address increasing end-user requirements presented by mobile operators participating in standardization bodies. The main part of further standardization of the GSM system was conducted in the European Telecommunications Standards Institute (ETSI) Special Mobile Group (SMG) until 2000 and since then by the 3rd Generation Partnership Project (3GPP).

One of the major objectives of the GSM standard was to create a digital system suitable for low-cost mass production. The criteria stipulated that the GSM system should provide at least the same or better speech quality and better spectrum efficiency compared to existing (to date) analogue mobile systems.

That system, named GSM-Global System Mobile, should provide ISDN services on the fixed side in addition to a number of specific services, named GSM specific services, including:

- Global roaming (initially Pan-European)
- Authentication (fraud control)
- Ciphering (speech, data, signalling)
- User confidentiality (ciphered subscriber number on-air-interface).

GSM achieved global acceptance by end of the 1990s. While the share of GSM in the modern mobile network is decreasing giving way to new generations, GSM still signifies a considerable source of revenue for mobile operators.

7.2 GSM System Architecture

The components of the GSM network are shown in Figure 7.1. The GSM network can be split into three subnetworks: the radio access network (RAN), the core network and the management network. These subnetworks are called subsystems in the GSM standard.

Introduction to Mobile Network Engineering: GSM, 3G-WCDMA, LTE and the Road to 5G,
First Edition. Alexander Kukushkin.
© 2018 John Wiley & Sons Ltd. Published 2018 by John Wiley & Sons Ltd.

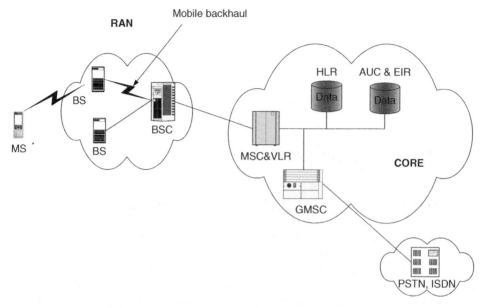

Figure 7.1 GSM System architecture.

These respective three subsystems can also be called the Base-Station Subsystem (BSS), the Network Switching Subsystem (NSS) and the Operation Support Subsystem (OSS).

The abbreviations in Figure 7.1 are as follows:

MS – Mobile Station (mobile phone)   HLR – Home Location Register

BS – Base Station (site)   VLR – Visited Location Register

BSC – Base-Station Controller   AuC – Authentication Centre

MSC – Mobile Switching Centre   EIR – Equipment Identity Register

GMSC – Gateway MSC   PSTN – Public Switching Telephone Network

User equipment (Mobile Station: MS) is normally a handheld terminal that communicates over the air with a base station, called the Base Transceiver Station (BS) in GSM. The BS transceiver is installed on some outdoor or indoor site together with additional site infrastructure that includes antennas, power supply and transmission equipment for connection to a Base-Station Controller (BSC).

- The logical object related to BS is a radio cell, which is a set control and traffic channels. The cell or logical Base Station is defined by presence of Broadcast Control Channel (BCCH). A number of BSs are controlled by one BSC.
- The BSC manages radio resources on in base stations, it is responsible for RF channel allocation and takes part in call setup, manages handovers.
- The base stations and BSC are connected by fixed lines or point-to-point radio links, this part of system infrastructure is named Mobile Backhaul.
- The BSs, BSCs and mobile backhaul together form the radio access network, RAN.

The BS and BSC perform different tasks in support of communications over the air interface, the task distribution between nodes is given in Table 7.1

Table 7.1 Functionalities of the base station and controller.

Main Function	BS	BSC
Management of radio channels		x
Mapping of upper layers to radio channels		x
Channel coding and rate adaptation	x	
Authentication		x
Encryption	x	x
Frequency hopping	x	
Uplink signal measurement	x	
Traffic measurement		x
Paging	x	x
Handover management		x
Location update		x

The RAN is connected to the Core network, which is comprised of a Mobile Switching Centre (MSC), Home Location Register (HLR) and number of logical network nodes including Gateway MSC (GMSC), Equipment Identity Register (EIR), Authentication Centre (AuC). Additional network elements may include components of the Value Added Services (VAS) platform.

The MSC performs all of the switching functions including path search, data forwarding and service feature processing. The main difference between an ISDN switch and the MSC is that the MSC also has to consider mobility of users. The MSC has to provide additional functions for location registration of users as well as manage the handover of a connection when a user moves from cell to cell. A cellular network may have several MSCs with each being responsible for some part of the network called the Location Area (LA).

Calls originating from or terminating in the fixed network are handled by a dedicated Gateway MSC (GMSC). The interworking of a cellular network and a fixed network (e.g. PSTN, ISDN) is performed by the Interworking Function (IWF). It is needed to map the protocols of the cellular network onto those of the respective fixed network. Either GMSC or IWF can be implemented as a standalone node or as a SW functionality with some HW interfaces in the MSC.

The Home Location Register (HLR) and the Visited Location Register (VLR) store the current location of a mobile user. Normally, the VLR is a logical node implemented in MSC. HLR and VLR databases store the profiles of users, which are required for charging and billing and other administrative issues. The HLR database is a root database where the provisioning of new subscribers is made. Given the importance of HLR database for operator revenue it often has a redundant standby node, sometimes geographically distributed.

Two other databases perform security functions: the Authentication Centre (AuC) stores security-related data such as keys used for authentication and encryption; the Equipment Identity Register (EIR) registers equipment data.

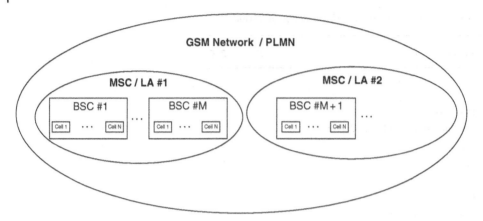

Figure 7.2 GSM system hierarchy.

The network management is centralized in Operation and Maintenance Centre (OMC). OMC functions include the administration of subscribers, terminals, charging data, network configuration, operation, performance monitoring and network maintenance.

Figure 7.2 summarizes the hierarchical relationship between the network components MSC, BSC and BS. Each MSC relates to a Location Area (LA), which comprises several BSCs and respective radio cells/base stations.

Each cell group is assigned to a BSC connected to via mobile backhaul. For each LA there exists at least one BSC, but cells of one BSC may belong to different LAs. The exact partitioning of the network area, with respect to LAs, BSCs and MSCs, is decided by the network operator. Each location area has a unique identifier (a Location Area Identity or LAI) that is broadcast regularly by the base station via a control channel. The mobile station monitors the broadcast and stores the current LAI. When the MS changes its location to another LA, the broadcasted LAI changes. The MS notices the change and requests a location update (in VLR/HLR) from the MSC.

The Public Land Mobile Networks (PLMN) run by different operators constitute islands in the Public Switched Telephone Network (PSTN). When the PSTN initiates a call to a mobile terminal belonging to a PLMN, the call request is fed to the interface between the PSTN and the PLMN. The interface consists of the operator's Gateway Mobile Switching Centre (GMSC). Details of all the subscribers belonging to the PLMN are contained in the Home Location Register (HLR) database.

7.2.1 Location Area Identity (LAI)

Each LA of a cellular network has its own identifier. The LAI is also structured hierarchically and internationally unique, with the LAI again consisting of an internationally standardized part and an operator-dependent part:

- CC, three digits;
- MNC, two digits;
- Location Area Code (LAC), a maximum of five digits or a maximum of 2×8 bits, coded in hexadecimal.

This LAI is broadcast regularly by the base station on the Broadcast Control Channel (BCCH). Thus, each cell is identified uniquely on the radio channel as belonging to an LA, and each MS can determine its current location through the LAI.

If the LAI that is 'heard' by the MS changes, the MS notices this LA change and requests an update to its location information in the VLR and HLR (location update). The significance for GSM networks is that the MS itself rather than the network is responsible for monitoring the local conditions of signal reception, to select the base station that can be received best and to register with the VLR of the LA that the current base station belongs to. The LAI is requested from the VLR if the connection for an incoming call has been routed to the current MSC using the MSRN. The LAI determines the precise location of the MS where the mobile can be subsequently paged. When the MS answers, the exact cell and therefore also the base station become known; this information can then be used to switch the call through.

7.2.2 The SIM Concept

GSM was the first mobile network technology that introduced a personal chip card, the Subscriber Identity Module (SIM). The SIM card turns a handset into a mobile station (MS) with a set of network services allowed for use by subscription. The SIM concept allows to distinguish between equipment mobility and subscriber mobility. In general, a subscriber can register to the locally available network with their SIM card using different handsets. This enables international roaming independent of mobile equipment and network technology, provided that the air-interface standard in visited network is supported by mobile terminal.

In addition to subscriber-specific data like an optional Personal Identification Number (PIN) and address book with names and telephone numbers, the SIM can also store network-specific data; for example, lists of carrier frequencies used by the network to broadcast system information periodically. The SIM also takes over security functions: all of the cryptographic algorithms are realized on the SIM, which implements important functions for authentication and user data encryption based on the subscriber identity and secret keys.

7.2.3 User Addressing in the GSM Network

All mobile users in the GSM network must be assigned a certain addresses or identities in order to identify, authenticate and localize them. The obvious address is a 'real' telephone number of the user provided with subscription. This number is called the Mobile Subscriber ISDN Number (MSISDN). In addition to the telephone number, several other identifiers have been defined; they are needed for management of user mobility and for addressing network elements. GSM distinguishes explicitly between a user and mobile equipment. The user identities are stored on the SIM; the equipment identities on the mobile equipment.

7.2.4 International Mobile Station Equipment Identity (IMEI)

The International Mobile Station Equipment Identity (IMEI) is a kind of serial number that uniquely identifies a mobile station, manufacturer and the date of manufacturing.

The IMEI is registered by the network operator, who stores it in the EIR. The IMEI can be used to identify stolen or non-functional equipment.

7.2.5   International Mobile Subscriber Identity (IMSI)

When registering for service with a mobile network operator, each subscriber receives a unique identifier, the International Mobile Subscriber Identity (IMSI). This IMSI is *stored in the SIM*. The IMSI uses a maximum of 15 decimal digits and consists of three parts:

1) Mobile Country Code (MCC): three digits, internationally standardized;
2) Mobile Network Code (MNC): two digits, for unique identification of mobile networks within a country;
3) Mobile Subscriber Identification Number (MSIN): a maximum of 10 digits, identification number of the subscriber in their mobile home network.

A three-digit MCC has been assigned to each of the GSM countries and two-digit MNCs have been assigned within countries (e.g. 505 as the MCC for Australia and MNC 01, 02 and 03 for the networks of Telstra, Optus and Vodafone, respectively).

7.2.6   Different Roles of MSISDN and IMSI

The distinction between call number (MSISDN) and subscriber identity (IMSI) primarily serves to protect the confidentiality of the IMSI and subsequently the user. Opposite to the MSISDN, the IMSI is a private identifier. The association of IMSI and MSISDN is stored in the HLR; that is, in the internal database of the operator network.

The MSISDN composition follows the international ISDN numbering plan with the following structure:

- Country Code (CC), up to three digits;
- National Destination Code (NDC), typically two or three digits;
- Subscriber Number (SN), a maximum of 10 digits.

The CCs are internationally standardized, complying with the ITU-T recommendation E.164.

There are country codes with one, two or three digits; for example, the country code for the USA is 1, for the UK 44 and for Australia 61. The national operator or regulatory administration assigns the NDC as well as the SN, which may have a variable length. The NDCs of the mobile networks in Australia have one digit (i.e. 4), while the SN has eight digits. The MSISDN is stored centrally in the HLR.

7.2.7   Mobile Station Routing Number

The Mobile Station Routing Number (MSRN) is a temporary location-dependent ISDN number. It is assigned by the locally responsible VLR to each MS in its area. Calls are routed to the MS by using the MSRN. On request, the MSRN is passed from the HLR to the GMSC.

The MSRN has the same structure as the MSISDN:

- CC of the visited network;

- NDC of the visited network;
- SN in the current mobile network.

The components CC and NDC are determined by the network visited and depend on the current location. The SN is assigned by the current VLR and is unique within the mobile network. An MSRN is assigned in such a way that the currently responsible switching node MSC in the visited network can be determined from the subscriber number, which allows routing decisions to be made.

7.2.8 Calls to Mobile Terminals

The call follows the standard calling procedure on the PSTN (or ISDN) number just entering the MSISDN number of a B-party. By means of the country code and NDC, the fixed network (e.g. the PSTN or ISDN) establishes a connection with the operator's gateway (GMSC). The GMSC invokes the HLR in order to get B-party location to route the call to correct MSC. The location of the mobile subscriber is known to HLR from latest location update or registration procedure performed by the MS (B-party). When a mobile subscriber registers in the network, it obtains the MSRN number that identifies a user in a specific location area of MSC/VLR. The MSRN number is then stored in the HLR against the user IMSI and MSISDN. When the MS moves to another LA, the MSRN number changes accordingly and the HLR is updated. In response to the GMSC request, the HLR provides a current MSRN number for the associated MSISDN and GMSC then routes the call correct MSC, see Figure 7.3.

The next step is that the MSC initiates paging to the MS within a known location area. The immediate location of the MS is defined by a LAI. The LAI is retrieved from the VLR when incoming call is routed to the current MSC using the MSRN. The paging will use the TMSI (Temporary Mobile Subscriber Identity) number to identify the MS. In BSS, the call processing follows the procedure described in Section 7.9. The role of the TMSI is explained in Section 7.2.9.

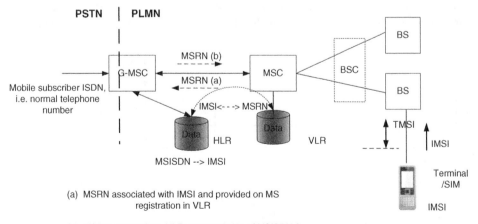

(a) MSRN associated with IMSI and provided on MS
 registration in VLR

(b) MSRN retrived from HLR on association of MSISDN &
 IMSI & MSRN and used for routing call to MSC.

Figure 7.3 Connection to the public telephone network.

The specific feature of GSM and next generation mobile networks is that the MS takes responsibility for selecting the best server (Base Station) by monitoring received signal quality and then registers with VLR/MSC of the Location Area that the selected BS belongs to.

7.2.9   Temporary Mobile Subscriber Identity (TMSI)

After subscriber registration in the network, the network starts to use a temporary identity, TMSI instead of an IMSI, for communication via the air interface. The TMSI is sent to the MS over the encrypted channel. The TMSI is assigned by the VLR and only has local significance in the Location Area handled by the VLR, the TMSI has not passed to the HLR. With a location update, the network provides a new TMSI after re-authentication of the mobile. All ongoing communication over the air interface is performed using the two-tuple (TMSI, LAI) instead of IMSI.

7.2.10   Security-Related Network Functions: Authentication and Encryption

The GSM security is based on information stored in the SIM. The SIM contains a 128-bit permanent secret key (Ki) associated with the subscriber identity (IMSI). The GSM subscriber authentication is done by a cryptographic challenge-response protocol based on the permanent key (Ki). The challenge response and integrated key generation protocol A3 are both implemented. A smart card inside the SIM contains a key generation protocol A3 processing the challenge response. In addition to the user's SIM card, the secret permanent key Ki for each MS is stored in a second location, namely the Authentication Centre (AuC). Authentication procedure checks the access to the correct secret key Ki by the mobile user. During call setup, the AuC initiates an authentication request to the mobile by sending a random 128-bit string RAND to the terminal, see Figure 7.4. It also generates a signed response, SRES, using the A3 algorithm and Ki security key.

The MS responds by using RAND and Ki, computing with A3 and returning the 32-bit output SRES to the network. Authentication is successful when two SRES are identical. After successful authentication, the network sends the TMSI to the MS in an authentication response. In addition to RAND and SRES, the Authentication Centre also generates

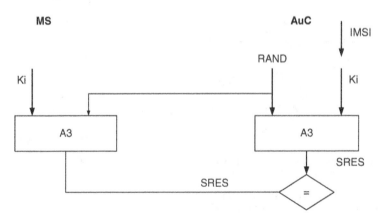

Figure 7.4 Principle of subscriber authentication.

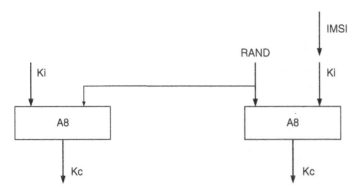

Figure 7.5 Generation of the cipher key Kc.

a temporary session key Kc using algorithm A8 with the same input parameters, namely Ki and RAND; see Figure 7.5. The session key Kc is used for encryption of the air inter-face. It is important to note that neither the Base-Station Subsystem nor the MSC/VLR have access to secret key, Ki. The authentication triplet (RAND, SRES and Ki) is gen-erated by the AuC and is sent to the serving nodes MSC/VLR (or SGSN in the case of GPRS) from the Authentication Centre.

7.2.11 Call Security

Ciphering is one of the security procedures designed to protect the subscriber's identity and data. It is an optional procedure in GSM. When the ciphering is active, all informa-tion exchanged between the mobile and the network on the dedicated radio channels is encrypted. A key previously set between the network and the mobile station (MS) is used to encipher and to decipher the encrypted information burst by burst. The A5 type encryption algorithm [1] applied to a GSM network is a 3GPP standardized ciphering method. It has several modifications, A5/1, A5/3 with an increased level of protection against eavesdropping. A5/3 ciphering is based on the Kazumi F8 algorithm [1] and used in both GSM and WCDMA.

Encryption is performed at the transmitting side after channel coding and interleav-ing and immediately preceding modulation. On the receiving side, decryption directly follows the demodulation of the data stream. The encryption of both signalling and user data is performed at the MS as well as at the base station (see Figure 7.6). This is a case of symmetric encryption; that is, ciphering and deciphering are performed with the same key Kc and the A5 algorithm.

Based on the secret key Ki stored in the network, the cipher key Kc for a con-nection or signalling transaction can be generated at both sides, and the BS and MS can decipher each other's data. Signalling and user data are encrypted together (TCH/SACCH/FACCH); for dedicated signalling channels (SDCCH) the same method as for traffic channels is used.

This process is also called a *stream cipher*; that is, ciphering uses a bit stream that is added bitwise to the data to be enciphered. Deciphering consists of performing an additional EXCLUSIVE OR operation of the enciphered data stream with the ciphering stream. The FN of the current TDMA frame within a hyperframe) is another input for

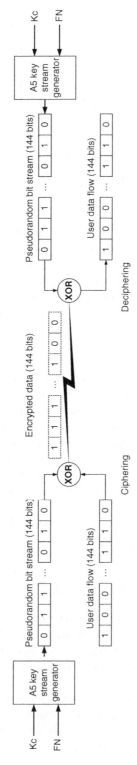

Figure 7.6 Combining payload data stream and ciphering stream.

the A5 algorithm besides the key Kc, which is generated anew for each connection or transaction. The current frame number is broadcast on the SCH and is thus available any time to all mobile stations currently in the cell. Synchronization between ciphering and deciphering processes is thus performed through FN. Only the actual data sequence is encrypted – not the training sequence. Encryption and decryption is implemented through by Modulo-2 addition of the identical encryption sequences to the transmitted and received signals.

7.2.12   Operation and Maintenance Security

To ensure security for the Base Station, an administrative authority system is implemented in the man-machine interface (MMI) system for operation and maintenance. The MMI system can be operated only by a person with a defined user ID and password. Internal security is ensured by restricting the rights of the users; that is, some users may use only certain man-machine language (MML) commands.

7.3   Radio Specifications

GSM is a combination of FDMA and TDMA. The 'primary band' of GSM includes two subbands of 25 MHz each:

- 890–915 MHz uplink
- 935–960 MHz downlink

The primary GSM900 frequency band comprises 2×25 MHz in a duplex arrangement with 124 duplex channels with 200 kHz channel spacing. This channel spacing allows a system data rate of about 270 kbps with GMSK modulation and a modest adjacent channel selectivity requirement (9 dB). The rate of 270 kbps corresponds to a bit (symbol) length of 4.6928 µs. Each RF carrier frequency pair is assigned an Absolute Radio Frequency Channel Number (ARFCN). The modulation spectrum for GMSK is wider than 200 kHz, resulting in some level of interference on adjacent channels.

The intra-cell interference is normally avoided by means of frequency planning. Therefore, adjacent channel interference to services other than GSM could be important mainly near the band borders. The border frequencies are therefore usually avoided. If there is no special agreement with the users of the adjacent bands, the normal practice is not to use channels 0 and 124. As a consequence, the number of carriers could be limited to 122. The specific frequency plan may be different in different areas and controlled by relevant spectrum management agencies. The GSM spectrum allocation is given in Table 7.2 and the summary of GSM RF characteristics is given in Table 7.3.

7.3.1   Spectrum Efficiency

One of the outcomes of the studies conducted in the 1980s was that digital radio transmission could eventually provide better speech quality and higher spectrum efficiency in a commercially deployed mobile communication system compared with existing up-to-date analogue systems. Overall spectrum efficiency is determined by

Table 7.2 GSM frequency bands.

Type	Uplink (MHz)	Downlink (MHz)
GSM450 (Europe)	450.4–457.6	460–467.6
	478.8–486	488.8–496
GSM850 (Americas)	824–849	869–894
GSM900	880–915	925–960
Classical	890–915	935–960
Extended	880–915	925–960
GSM1800	1710–1785	1805–1880
GSM1900 (Americas)	1850–1910	1930–1990

Table 7.3 GSM characteristics.

Parameters	Characteristics
Operating frequency, MHz	935–960 uplink, 890–915 downlink
Channel bandwidth, kHz	200
Channel rate, kbps	22.8
Duplex separation, MHz	45
Speech rate, kbps	13[a]
Speech coding	RPE – LPT[a] (Regular Pulse Excitation – Long Term Prediction)
Control channels	standalone and associated (embedded)
In-call control channel rate, kbps	4.8 (+4.6 CRC)
Total channel rate (traffic channel), kbps	270
Duplexing technique	FDMA
TDMA	8 time slots per frame
Carrier usage	124 duplex carriers
Bits per TDMA slot	156
Time slot duration, μs	577
Frame duration (8 time slots), ms	4.615
Modulation	GMSK
Modulation index	0.3
Frequency hopping	slow hopping, 217 hops/s = 1200 b/hop

a) In the original design, full-rate speech coders with a data rate of 13 kbps are used. However, at a later stage GSM used also half-rate speech coders by accommodation of both full-rate and half-rate traffic channels. As well as speech, data transmission at different speeds can also take place over full-rate or half-rate traffic. In the latest releases GSM uses AMR (Adaptive Multi-Rate) and Wideband AMR (WB-AMR) speech coding. It is important to note that a deployment of AMR in GSM avoids transcoding of the voice call during intersystem handovers between GSM and WCDMA, since the AMR speech coder is a standard speech coder in the 3G system.

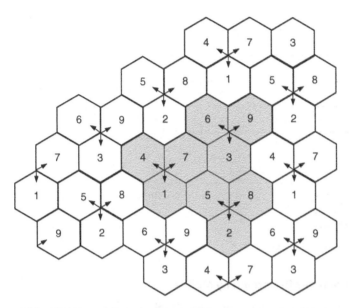

Figure 7.7 Cell structure with cluster size 3 × 3.

required reuse distance between co-channel cells (cluster size), which, in turn, depends on the local mean of the protection ratio that is a minimum required C/I over fast fading. Digital modulation with FEC channel coding used in GSM achieves spectrum efficiency three times higher than the analogue cellular systems of earliest generation. A minimum cluster size used in GSM system planning is 3 × 3, which is shown in Figure 7.7. With 90% of availability the local mean of required C/I is about 9 dB.

7.3.2 Access Technology

The GSM air-interface standard is based on a multicarrier TDMA and FDMA. Each radio frequency, GSM carrier, is repeatedly transmitted over TDMA frames of 4.615 ms. Each TDMA frame is subdivided into eight time slots, see Figure 7.8. Each of these slots can be assigned to a full-rate (FR) traffic channel (TCH-FR), two half-rate traffic channels (TCH-HRs) or one of the control channels.

The FDMA implementation with a duplex separation of 45 MHz implies the following:

- the downlink TCH from the base station to mobile is transmitted at the frequency carrier f_n,
- the uplink TCH from mobile to base station is transmitted at the frequency carrier $f_n - 45$ MHz.

One or more timeslots can be granted to a GSM user during communication session, the number of allocated time slots (TS) depends on service requested: voice service requires one TS, while several TSs may be allocated for packet data services. Up to eight simultaneous communication sessions can be supported by a single frequency carrier over eight time slots.

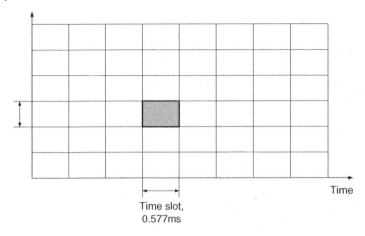

Figure 7.8 The multiple-access scheme in GSM.

7.3.3 MAHO and Measurements Performed by Mobile

One user is never allowed to occupy all eight time slots. On the contrary, the user experiences an idle period during the time frame. The possibility of listening or transmitting in another frequency within idle periods brings important TDMA system benefits. Such periods can be used for system signalling, preparing for handover and repetition of data blocks in packet-service sessions (see Section 8.5.4).

The effective utilization of the idle period leads to a distinctive (compared to the legacy systems existed at the time of GSM development) feature of GSM, a Mobile Assisted HandOver (MAHO), that essentially means that a handover is performed under command from base-station controller, but the controller's decision is based on the information received from the mobile. The mobile must measure C/I and C/N for signals from neighbour cells and transmit this information to the base.

The following measurements are performed by a mobile within idle periods during a time frame:

- Signal strength and Frame-Error Rate (FER) on the used channel
- Signal strength from neighbour cells and validation of neighbour list
- Averaging received signal field-strength measurements over many frames to get a local average over fast fading
- Check that the signal comes from neighbour cell

This information is then transmitted to the base station.

TDMA also allows a terminal to transmit and receive in different time slots (time duplex). This eliminates relatively expensive and bulky duplex filters at the terminals and removes the necessity for the MS to transmit and receive simultaneously. The latter was especially important in earlier releases of GSM.

The TDMA/FDMA operation means that:

1) MS will receive a downlink burst from the BS,
2) retune to the uplink frequency and
3) transmit an uplink burst three timeslots later.

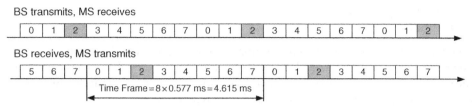

BS transmits, MS receives

| 0 | 1 | 2 | 3 | 4 | 5 | 6 | 7 | 0 | 1 | 2 | 3 | 4 | 5 | 6 | 7 | 0 | 1 | 2 |

BS receives, MS transmits

| 5 | 6 | 7 | 0 | 1 | 2 | 3 | 4 | 5 | 6 | 7 | 0 | 1 | 2 | 3 | 4 | 5 | 6 | 7 |

Time Frame = 8 × 0.577 ms = 4.615 ms

Figure 7.9 Burst schedule at the BS. Time slot 2 is assigned to mobile user.

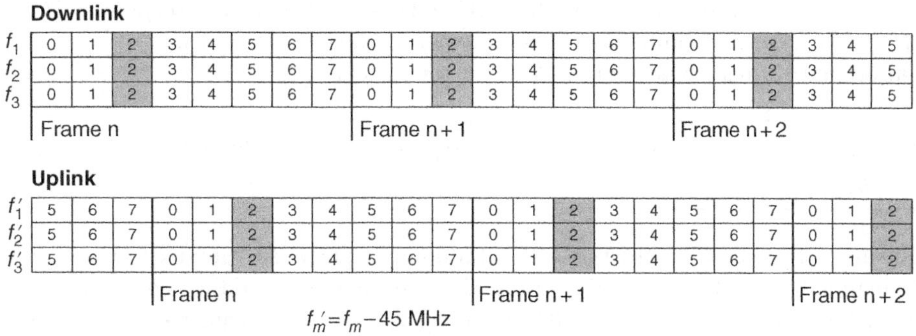

Downlink

f_1	0	1	2	3	4	5	6	7	0	1	2	3	4	5	6	7	0	1	2	3	4	5
f_2	0	1	2	3	4	5	6	7	0	1	2	3	4	5	6	7	0	1	2	3	4	5
f_3	0	1	2	3	4	5	6	7	0	1	2	3	4	5	6	7	0	1	2	3	4	5

Frame n   Frame n + 1   Frame n + 2

Uplink

f'_1	5	6	7	0	1	2	3	4	5	6	7	0	1	2	3	4	5	6	7	0	1	2
f'_2	5	6	7	0	1	2	3	4	5	6	7	0	1	2	3	4	5	6	7	0	1	2
f'_3	5	6	7	0	1	2	3	4	5	6	7	0	1	2	3	4	5	6	7	0	1	2

Frame n   Frame n + 1   Frame n + 2

$$f'_m = f_m - 45 \text{ MHz}$$

Figure 7.10 Duplex arrangement with three frequency channels allocated to a sector/cell. Total number of available channels is $3 \times 8 = 24$ channels.

The timing schedule at the BS is shown in Figure 7.9.

Each time slot within a TDMA frame is numbered from zero to seven and these numbers repeat for each consecutive frame. The time slot and frame durations are derived from the fact that traffic channels are arranged in 26 TDMA frames and so 26 TDMA frames are transmitted in 120 ms. Then frame duration is defined as ratio $\frac{120}{26} = 4.615$ ms and time slot duration is $\frac{120}{26 \cdot 8} = 0.577$ ms, respectively. Figure 7.10 shows the duplex arrangements at BS side where three frequency carriers have been deployed. The two time slots corresponding to a two-way traffic channel are mutually displaced in time to a three-time-slot interval. The figure corresponds to the case in which a base station not using frequency hopping has been allocated four carriers, each of which carries eight physical channels in a TDMA frame. The mutual displacement of the time slots for the downlink and uplink directions corresponds to a quasi-time-duplex arrangement (even if FDD is used by the system, some TDD advantages are obtained).

No duplex filter is required in the terminals; instead there is a fast Transmit/Receive switch that alternately connects the transmitter and receiver to the antenna.

7.3.4 Time Slot and Burst

The contents of a time slot are called a *burst*. This section considers how the control information required for successful decoding is embedded into different bursts. There are five different types of bursts in GSM:

- Normal burst
- Synchronization burst
- Access burst

- Frequency correction burst
- Dummy burst

Different bursts contain different information that can be packed in different format depending on the type of channel it belongs to. The channel classification is provided in the next section, however, association between burst type and the channel is emphasized here. The duration of burst is the same as duration of timeslot.

7.3.4.1 Normal Burst
This is used to carry information on the TCH and Control Channels (except for RACH, SCH and FCCH, see definitions of these channels in the later sections).

Duration 0.577 ms = bits + Guard Period = 156.25 bits.

TB	Encrypted bits	Training sequence	Encrypted bits	TB	GP = 8.25 bits
3	58	26	58	3	

The normal burst (NB) consists of a 26-bit training sequence surrounded by two 58-bit information blocks. Three tail bits are added at the beginning and the end of the burst. The total duration of the burst is 148 bits leaving a guard period equivalent in duration to 8.25 bits.

Encrypted bits are 57 bits of data/speech + 1 'stealing flag'; this leads to 114 bits of user data per normal burst. The stealing flag indicates the type of information carried by $2 \times 57 = 114$ bits choosing between user data or signalling data associated with traffic channel.

The training sequence is a known bit pattern used by an equalizer to estimate an instant channel response to be then used in conjunction with a Viterbi decoder. The tail bits (TB) in the normal burst are always set to zero (000) to ensure that the Viterbi decoder begins and ends in a known state. The training sequence is inserted in the middle of the burst in order to maximize the distance from the power ramp and minimize the distance from information bits. It also called 'midamble'.

The eight training sequences have been specified. *Distinct training sequences* are allocated to *channels* using the *same frequencies* in the cells that are close enough to interfere with one another. This process is managed by the network. The training sequence is then used for both additional time acquisition and to obtain a channel response used then by equalizer.

7.3.4.2 Frequency Correction Burst (FB)
The frequency correction burst is used for frequency synchronization of the MS. It is broadcasted in downlink on Broadcast Common Control Channel (BCCH). Repetitions of these bursts compose FCCH (Frequency Correction Channel).

Duration 0.577 ms = bits + Guard Period = 156.25 bits

TB	Fixed bit pattern	TB	GP = 8.25 bits
3	142	3	

Fixed bits are all zeros. The frequency correction burst is also used by MSs as a frequency reference for their internal time bases. Every bit in the frequency correction burst (including the tail bits) is set to zero and, after GMSK modulation, this results in a pure sine wave at a frequency around 68 kHz (1625/24 kHz) higher than the RF carrier centre frequency.

7.3.4.3 Synchronization Burst

The Synchronization Burst is used for time synchronization of the MS. It carries the TDMA Frame Number and Base-Station Identity Code (BSIC). Synchronization burst is used in the *downlink* direction only. *This is the first burst that MS needs to demodulate.*
Duration 0.577 ms = bits + Guard Period = 156.25 bits

TB	Information bits	Sync sequence	Information bits	TB	GP = 8.25 bits
3	39	64	39	3	

The synchronization burst (SB) carries 78 bits of coded data formed into two blocks of 39 bits on either side of a 64-bit training sequence. As its name suggests, this burst carries details of the GSM frame structure and allows an MS to fully synchronize with the BS. The synchronization burst is the first burst that the MS has to demodulate and, for this reason, the training sequence is extended to 64 bits. This extended sequence provides a larger autocorrelation peak than the 26-bit sequence of the normal burst. It also allows larger multipath delay spreads to be resolved. All synchronization bursts use the same training sequence.

7.3.4.4 Access Burst

This is used for random access on the RACH or after handover on the new TCH.
Duration 0.577 ms = bits + Guard Period = 156.25 bits

TB	Sync sequence	Encrypted bits	TB	GP = 68.25 bits
8	41	36	3	

The access burst is only used in the *uplink* direction. The access burst is used by mobile at the first access attempt on RACH. This is the first burst from MS that the Base Station needs to demodulate. The access burst is much shorter than the other bursts, therefore, a larger guard period of 68.25 bit is included. That long guard period compensates for the propagation delay between the MS and BS. The guard period of 68.25 bit periods extends over 252 μs, allowing the MS to be up to 38 km from the BS before its uplink bursts will spill into the next time slot.

7.3.4.5 Dummy Burst

The dummy burst is used as a filter burst. The format is the same as a normal burst but the dummy burst carries no information. The dummy burst contains a training sequence and some predefined bit pattern instead of encrypted information bits on either side of the training sequence. The dummy burst is used to fill inactive time slots, which must be transmitted continuously and at a constant power.

7.3.5 GSM Adaptation to a Wideband Propagation Channel

Due to the scattering and reflections in propagation media the signal arrives at the receiver along multiple paths. The length of those paths is different, the phase of each signal replica is different and arrival time is different, as well. This phenomenon called time dispersion. As discussed in Section 7.5, time dispersion may cause intersymbol interference (ISI). The single information bit (or symbol) is composed by its few replicas that may cause the problem with correct demodulation of the signal.

The GSM system has to operate in different environments, from rural to dense urban. The system specifications recommend a minimum set of qualitative criteria in terms of frame (FER) and bit-error-rate (BER) by specifying the performance of the MS and BS receivers over a wide range of different operational environments. To this end, the GSM specifications define several different channel models and these are used to specify the performances of the MS and BS receivers.

Each channel model consists of a number of independently fading impulses, or paths, at different time delays. In practice, the power delay profile of mobile radio channel cannot be separated into its different paths; however, the channel models have been defined with a discrete delay profile in a way that they can be easily implemented for equipment testing. The example of GSM channel models is shown in Figure 7.11 for urban (TU) and rural (RA) channels for GSM that have excess delay spreads of 5 μs and 0.5 μs, respectively. Using relationships (5.5) and (5.7) and value of GSM symbol duration $T_s \approx 3.69$ μs, one may conclude that urban channel (TU) is a frequency selective channel while the rural channel is rather frequency flat.

Depending on the speed of the terminal impulse, the response of the radio channel can change drastically during a frame of 4.6 ms. This means that, for each time slot, the receiver must carry out bit synchronization and set the channel equalizer. A certain trade-off between burst duration, bit duration and complexity of equalizer has to be considered. With the training sequence placed in the middle of the burst, and considering the maximum terminal speed and the radio frequency, it has been estimated that the maximum burst length over which the channel was nearly stationary would be about 0.5 ms. Long bursts carry less overheads relative to the user part.

7.3.5.1 Training Sequence and Equalization

The training sequence is used to 'sound' the radio channel and produce an estimate of its impulse response at the receiver. This estimate is used in the demodulation process

Environment	Channel shape	Channel parameters	
Typical urban	One exponential cluster consisting of 20 Rayleigh-fading paths	1 cluster P1 = 1 τ1 = 0 μs Delay spread = 0.5 μs	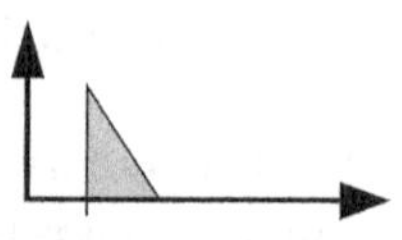
Rural area	One exponential cluster consisting of 19 Rayleigh-fading paths and 1 non-fading path.	1 cluster P1 = 1 τ1 = 0 μs Delay spread = 0.14 μs	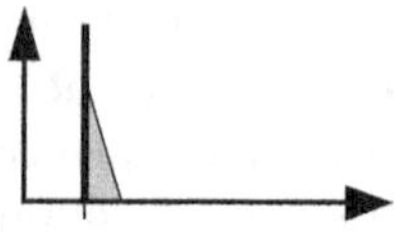

Figure 7.11 The GSM channel models [2].

Figure 7.12 Structure of training sequence.

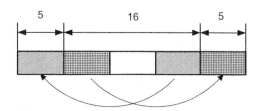

to 'equalize' the effects of multipath propagation. The impulse response can sometimes change even during a burst. It may happen if the terminal velocity is very high and also in a higher frequency band such as 1800 MHz. Therefore, if the setting of the channel equalizer was optimized with respect to the impulse response at the beginning of the burst, the equalization may be suboptimal for the last part of the burst. This results in an increased error rate. The degradation becomes larger for a large width of the Doppler spectrum (depending on the terminal speed and the radio frequency).

To avoid the complication of having to adapt the channel equalizer to variations in the impulse response of the propagation channel during a time slot, short slots are used and, in addition, the training sequence is placed in the middle of the burst. Consequently, the first section of the burst must be stored before demodulation can proceed. The training sequence consists of a 16-bit sequence extended in both directions by copying the first five bits at the end of the sequence and the last five bits at the beginning, see Figure 7.12.

The specifications [3] define eight different training sequences ('colour codes') for use in the normal burst, each with low cross-correlation properties following GMSK modulation. This reduces the risk of synchronizing to a distant strong co-channel carrier.

7.3.5.2   The Channel Equalization

The normal burst, which contains both the data and the training sequence, is passed through a baseband modulator at the transmitter and then through the mobile radio channel. The received waveform will contain an ISI caused by the multipath propagation in the radio transmission channel.

At the receiver, the burst is de-multiplexed into two bit streams to separate the training sequence and the data bits. The received training sequence is used to estimate the impulse response of the radio channel in the channel estimator. The channel estimator is used to produce waveforms for possible combinations of the sequence of data bits. Since the channel estimator contains some ambiguity function due to a time windowing of estimation processing, further processing is done according to the Viterbi algorithm of soft decoding.

In practice, the time window for channel estimation is limited in size. On one hand, it has to contain the most significant multipath components; that is, it should accommodate an excess delay spread of the channel. On the other hand, the size of the window corresponds to a number of states in a soft decoder, which directly affects the complexity of the Viterbi channel equalizer. A consecutive increase in delay (size of the processing window) corresponding to a bit duration doubles the number of states in the equalizer. The maximum reasonable complexity at the time of GSM development was considered to be a limit equalizer window of up to four radio symbols.

An almost worst-case time dispersion in the mobile environment was estimated at about 16 μs that approximately triples the delay spread of the urban channel, see

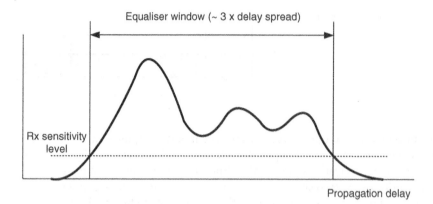

Figure 7.13 Equalizer window versus delay spread.

Figure 7.13. The equalizer window with a size of 16 μs should accommodate four radio symbols, thus imposing a symbol length of about 4 μs.

Recall that the training sequence consists of a 16-bit sounding sequence with five bits appended to either end. Those bits allow derivation of channel estimates for delay spreads (width of impulse response of the channel) of five bit periods. This is adequate for the specified urban and rural channels for GSM that have excess delay spreads of 5 and 0.5 μs, respectively. As a consequence, perfect autocorrelation of the training sequence is ensured over a shift of up to five bits. Delays in excess of 5 bits of preamble may cause errors. These errors could be removed by the de-interleaver and FEC decoder. It could be noted that GMSK signal may experience severe inter-symbol interference when the delay spread in the channel is greater than just 10% of a symbol duration. For speech transmission, a BER of order 0.3% is tolerable, so an equalizer with a resolution of about half a bit duration is appropriate for these purposes. However, requirements for data transmission are much greater and one of the possible strategies is to deploy Automatic Repeat Request protocol (ARQ).

7.3.5.3   Diversity Against Fast Fading

When either the receiver or scattering objects move with time, the received signal experiences time fluctuations caused by fading. At the input of the decoder/demodulator the signal is composed by slow fading + fast (Rayleigh) fading + noise components. The time scale of the noise fluctuations t_N is determined by the receiver bandwidth B, $t_N = 1/B$, while the time scale t_f of Rayleigh fading is given by $t_f = \lambda/2v$, where v is a speed of mobile unit or typical speed of scattering objects.

Roughly, the fading time scale is the time that elapses between two fading dips and gives an estimate of minimal duration between fades; that is, free of fade interval. Actual duration between fades will depend on the fade's amplitude. For the GSM-band (900 MHz), the spatial distance between two dips is about 15 cm so, if a mobile is travelling with a speed of $v = 50$ km/h, the time between two dips is

$$t_f = \frac{0.3\ m}{2 \cdot 14\ m/s} = 10.7\ ms$$

that lasts approximately two GSM frames (4.615 ms).

The fade duration is an important parameter for the design of *channel coding* and *interleaving* schemes, as well as ARQ protocol. Spatial correlation properties of the fades are needed for design of the space diversity systems.

Instead of antenna diversity in handheld terminals, a combination of channel coding, interleaving and coordinated frequency hopping is used to obtain a diversity gain in respect to the multipath fading. A necessary condition is that the propagation channel has fairly large time dispersion. Together, these features of GSM give such high diversity and coding gains that the required protection ratio (the local mean over the fast fading) will typically be 9–10 dB.

Interleaving a *full-rate traffic channel* means that the 456 bits in a 20-ms speech frame are split up into 57-bit sequences that are spread out over eight TDMA frames; that is, over 40 ms (see Section 7.9.2). If the duration of a fading dip is not more than few milliseconds, typically only one user time slot (one TDMA frame) is affected. The de-interleaver will then change the error burst to a relatively random error sequence spread over eight code words. Thus, one-eighth of the bits in each code word will be subject to a BER of about 50%. In a most situations, Forwards Error-Correction coding is sufficient to mitigate these errors.

7.3.5.4 Frequency Hopping

Fading dips that are longer than channel coding with interleaving can cope with, occurring over quasi-stationary propagation paths, something that affects portable terminals in a pedestrian environment. In this case, a fading dip could affect many consecutive TDMA frames reducing the effect of interleaving. In other words, time diversity does not work while frequency diversity may help for slow-moving terminal.

GSM's method of frequency diversity is frequency hopping, whereby each physical channel is switched between different radio channels. For each TDMA frame the carrier frequency is changed, see Figure 7.14. The size of a typical frequency hop should ideally be large enough to give almost uncorrelated fast fading in the different frequency slots. In practice, coherence bandwidth of GSM channels can vary from several hundred kHz to a few MHz. Given that an operator typically owns only a few MHz of spectrum, and only a subset of frequencies can be used instantly in each cell, fading in channels used for hopping can still be correlated.

Nonetheless, frequency hopping provides certain a advantage even in this case: co-channel interference from other cells, in particular, is whitened and leads to less interference on average and tighter frequency reuse becomes possible.

While frequency hopping was specified only as an option, it is a norm in commercial mobile networks. Without frequency hopping, a far higher dB protection ratio would be needed for portable terminals. Frequency hopping cannot be used for the main signalling radio channel, the Broadcast Carrier. The Broadcast Carrier must be on a fixed frequency known by the terminals. In some system implementations, a separate and larger cluster size is used for the Broadcast Carrier compared with the rest of the frequency hopping traffic channels. A traffic channel or dedicated control channel may hop into the broadcast carrier frequency as part of its hopping sequence, but their power is then adjusted to the level of Broadcast Carrier.

Frequency hopping could be implemented in two ways: as a baseband or RF synthesizer hopping. In baseband hopping, each transmitter operates on a fixed frequency.

Downlink (mobile receives)

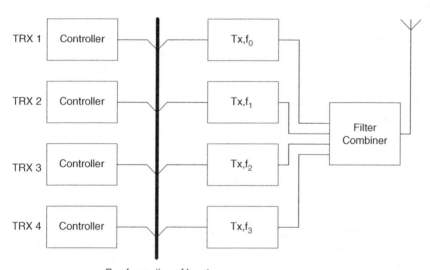

Figure 7.14 Frequency hopping concept.

Figure 7.15 Baseband frequency hopping.

At transmission, all bursts, irrespective of which connection they belong to, are routed to the transmitter of the proper frequency, see Figure 7.15.

The advantage with this mode is that narrowband filter combiners can be used with very small insertion loss. This makes it possible to use many transceivers without having to connect several combiners in cascade. The disadvantage is that it is not possible to use a larger number of frequencies in the hopping sequence than the number of available transmitters in the cell. *RF Synthesizer* hopping means that one transmitter handles all bursts that belong to a specific connection. The bursts are sent 'straight on forwards'

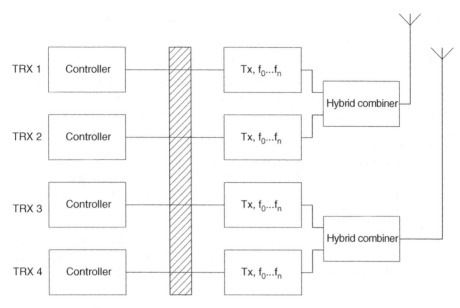

Figure 7.16 RF synthesizer frequency hopping.

and not routed by the bus in contrast to baseband hopping. The transmitter tunes to the correct frequency at transmission of each burst, see Figure 7.16.

The advantage is that the number of frequencies that can be used for hopping is not dependent on the number of transmitters. It is possible to hop over a lot of frequencies even if only a few transceivers are installed. The gain from frequency hopping can thereby be increased. This concept is often called fractional loading. A disadvantage with synthesizer hopping is that wideband hybrid combiners have to be used. This type of combiner has approximately 3 dB loss making more than two combiners in a cascade impractical.

7.4 Background for the Choice of Radio Parameters

The symbol rate over the radio channel and the primary TDMA structure are chosen as a result of a compromise between acceptable transmission performance over the worst specified propagation channel and implementation/cost limitations. The following factors have impacted the TDMA structure:

- Time for measurements on nearby cells (Mobile Assisted HandOver: MAHO)
- Time for frequency switching in the mobile terminal
- Low TDMA overhead (long bursts, narrow guard slots)
- Low transmission delay due to TDMA formatting (small frame length: high system data rate)
- Large range for a given transmitter peak power (portable terminal, moderate slot duration per frame)
- Moderate equalizer complexity (equalizing window covering a small number of symbols)

- Short burst length: smaller than correlation time of channel; that is, delay spread of the channel.

These factors result in the following design constraints and trade-offs:

- The equalizer window size of 16 μs limits equalizing range to four radio symbols and a minimum symbol length of 4 μs, thus limiting the information rate to 250 kbaud.
- The burst length is limited by the period during which the impulse response of the channel is stable. In a worst-case scenario on a high-speed train, the impulse response is stationary over an interval of ~0.25 ms (for high-speed trains). With a training sequence in the middle of the burst, the maximum burst length is ~0.5 ms.
- The design constraints for the length of TDMA frame are shown in Figure 7.17. As observed, the major impact to frame length comes from the switching time of frequency synthesizer. The mobile has to be able to transmit, receive and listen at variable frequencies in line with the duplex arrangement, frequency hopping pattern and neighbour cell frequency list. At the time of GSM development, the switching time of frequency synthesizer was about 1 ms, together with requirements for MAHO, that is, listening and measuring the signal level of neighbour cells; this limits the minimum length of TDMA frame to >4 ms.

Apparently, a limitation coming from the switching time of synthesizer was an issue of a commercial nature rather than a technical one: two synthesizers could be deployed in the mobile terminal, one active at the current frequency, another one settled for the next channel to transmit or receive.

- Given the burst duration constraints, ~ 0.5 ms and ~4 ms for frame length, we have another limitation for the maximum number of time slots per TDMA frame; that is, eight time slots per TDMA frame.
- Another design parameter is a total transmission delay due to the overall data processing and interleaving procedure in particular. The depth of interleaving should be greater than the fade duration but not too large in order to not extend a burst

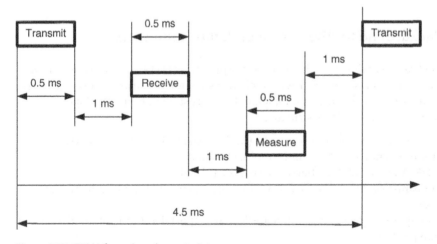

Figure 7.17 TDMA frame length constraints.

transmission delay. The result is a reasonable compromise between total transmission delay due to TDMA formatting and interleaving of each burst over eight time frames ~40 ms and requirements for listening and measuring the neighbour cells during idle slot periods for MAHO preparation.

7.4.1   Guard Period, Timing Advance

To prevent data bursts from different terminals overlapping in the input to the base-station receivers, a guard period (GP) with a time of 8.25 bits ~31 μs has been introduced. The GP copes with variations in the two-way propagation time to terminals at different distances from the base. The guard period corresponds to a two-way propagation path of about 4.5 km. This is considerably less than the maximum specified range (cell radius) of 35 km. Therefore, to prevent data bursts from different terminals overlapping in the input to the base-station receivers, the base-station instructs the terminals to insert a suitable delay between received and transmitted data bursts. The delay is adjusted such that a transmitted burst from the terminal reaches the base-station receiver at the right position relative to the time slot structure. Regardless of how far the terminal is away from the base, the bursts arriving to the base receiver must arrive close to the middle of the allocated time slot. The closer a terminal to the base station, the greater the delay inserted. This is illustrated in Figure 7.18.

The timing advance parameter is transmitted as a six-bit number providing 64 timing advance steps each corresponding to the period of one bit, namely 4.69 μs. This means that the system may apply a maximum timing advance of $63 \times 4.69 \, \mu s = 232 \, \mu s$, which corresponds to a round trip propagation distance of approximately 70 km and a maximum distance from the base station to the cell boundary of 35 km.

The timing advance values might be used as one of the parameters determining when handover between cells should take place. It could also be used for used to calculate the distance between the terminals and the base, which will be an important added value service.

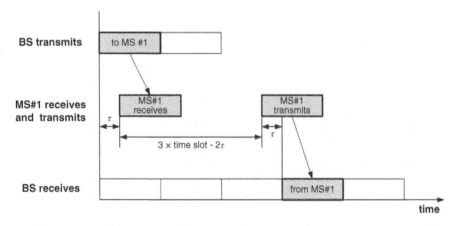

Figure 7.18 Timing alignment.

7.5 Communication Channels in GSM

There are physical and logical channels in GSM radio system. A physical channel stands for one timeslot at one frequency carrier, while a logical channel refers to the specific type of information carried by the physical channel. Different kinds of information are carried by different logical channels, which are then mapped or multiplexed on physical channels, see Figure 7.19.

Logical channels can be classified either as Common Channels or Dedicated Channels:

- Common channels are of type point-to-multipoint. All mobiles can overhear them; that is, they are not specifically addressed to a certain mobile.
- Dedicated channels are of the point-to-point type. They are used for a dedicated and bi-directional communication between the base station and the mobile.

Logical channels can also be divided in two groups: Traffic Channels (TCH) and Control (Signalling) Channels.

7.5.1 Traffic Channels (TCHs)

Payload data are transmitted via the traffic channels. The payload might consist of encoded voice or raw data. A logical traffic channel for speech and circuit-switched data in a GSM is abbreviated to TCH, which can be either full rate (TCH/FR) or half rate (TCH/HR). A logical traffic channel for packet-switched (PS) data is referred to as a packet data traffic channel (PDCH). Two half-rate channels are mapped to the same timeslot but in alternating frames. In this way, the full TCH is split into two half-rate channels that can be allocated to different subscribers, thus increasing traffic capacity of the GSM carrier.

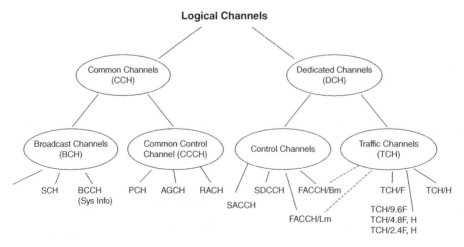

Figure 7.19 Logical channels. The abbreviations TCH/H and TCH/F correspond to full- and half-rate traffic channels, respectively. The numbers 9.6, 4.8 and 2.4 refer to specific data service rates over circuit switch traffic channels.

For TCH/FRs:

- Full-rate voice channels: the output data rate of the voice encoder is 13 kbps. Channel coding increases the effective transmission rate to 22.8 kbps.
- Full-rate data channels: the payload data with data rates of 9.6, 4.8 or 2.4 kbps are encoded with Forwards Error Correction (FEC) codes and transmitted with an effective data rate of 22.8 kbps.
- With full rate TCH, the analogue voice is digitized into a 20 ms frame that is coded and mapped into each consecutive timeslot.

For TCH/HRs:

- Half-rate voice channels: voice encoding with a data rate as low as 6.5 kbps is feasible. Channel coding increases the transmitted data rate to 14.4 kbps.
- A digitized analogue voice signal in every 20 ms block is coded and mapped into each second timeslot. This effectively makes it possible to 'fit' two simultaneous user TCHs into one physical channel.

It should be noted that half-rate voice coding does not deliver a good quality voice over mobile propagation channel. It was widely used in initial deployment of the GSM system offering price efficiency of service. At present, this service no longer exists and modern mobile phones do not support half-rate voice coding.

7.5.2 Control Channels

7.5.2.1 Common Control Channels

Broadcast Channel (BCH). The BCH belongs to a group of Common Channels and, transmitted at downlink, these are point-to-multipoint channels. They contain general information about the network and the broadcasting cell. There are three types of broadcast channels:

1) *Frequency Correction Channel* (FCCH) contains an unmodulated sinusoidal pulse with a frequency offset related to a nominal carrier. By acquiring the FCCH, the mobile tunes its frequency to a BCCH carrier.
2) *Synchronization Channel* (SCH) contains the Base-Station Identity Code (BSIC) and TDMA frame number.
3) *Broadcast Control Channel* (BCCH) transmits a system information message containing detailed network and cell-specific information such as: frequencies used in the particular cell and neighbour cells, frequency hopping sequence, channel combination, which informs the mobile station about the mapping method used in the particular cell, paging groups and information on neighbour cells. *The BCCH should be cleared of interference.* The BCCH is always transmitted on a constant power level. Frequency Hopping or Power Control *never* happen on BCCH.

Paging Channel (PCH) downlink common control channel used for paging to find a specific MS.

Random Access Channel (RAC) is the uplink control channel accessed by the MS using a slotted Aloha protocol in request for dedicated resources.

Access Grant Channel (AGCH) is used in the downlink to assign a dedicated signalling channel (SDCCH) to a granted MS.

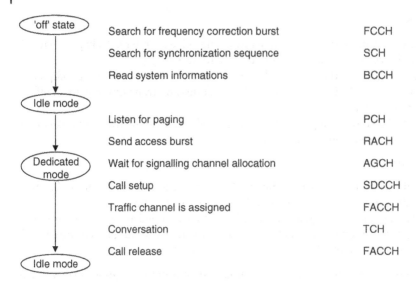

Search for frequency correction burst	FCCH
Search for synchronization sequence	SCH
Read system informations	BCCH
Listen for paging	PCH
Send access burst	RACH
Wait for signalling channel allocation	AGCH
Call setup	SDCCH
Traffic channel is assigned	FACCH
Conversation	TCH
Call release	FACCH

Figure 7.20 Utilization of logical channels during call setup.

7.5.2.2 Dedicated Control Channels

There are three dedicated control channels used for signalling between mobile and base station, namely, the Slow Associated Control Channel (SACCH), Fast Associated Control Channel (FACCH) and Standalone Dedicated Control Channel (SDCCH). The *SDCCH* is used during the call setup for a traffic channel allocation. Both the *SACCH* and *FACCH* are used for signalling during the call, signalling is always associated with the allocated traffic channel.

The following signalling messages are sent on the downlink SACCH:

- power command,
- time advancement,
- frequency hopping sequence,
- frequencies used by adjacent channel.

The uplink SACCH contains values of:

- Frame-Error Rate (FER) of the downlink traffic channel,
- Received signal level from neighbour cells.

Different logical channels are used in different stages of the call setup procedure, see Figure 7.20. Mobiles in idle mode listen to certain common channels. Call setup signalling is always initiated on common signalling channels. At a certain stage in call setup, the communication is switched from common channels to dedicated channels.

7.6 Mapping the Logical Channels onto Physical Channels

A physical channel is defined by timeslot number and frequency (ARFCN). The mapping of logical channels onto physical channels is done in the frequency and time domain. The mapping of a logical channel onto a physical channel in the frequency domain is based on the ARFCFN and the rules for optional frequency hopping. In the time domain, logical channels are transported in the corresponding time slots of the physical channel. Logical channels are mapped onto physical channels in certain time-multiplexed combinations where they can occupy a complete physical channel or just part of a physical channel.

7.6.1 Frame Format

As discussed in previous sections, the eight timeslots with a duration of 577 µs each are combined in a frame. The duration of the frame, 4.61 ms, is the basic period of a GSM system. A total of 26 of these frames are combined into a *multiframe*, which has a duration of 120 ms. The period of 120 ms is an exact one, it is chosen to be a multiple of 20 ms to obtain some synchronization with a fixed network, ISDN in particular. This definition leads to a value of the burst period that is exactly $120/26/8 = 15/26$ ms $= 0.577$ ms. The next period is a *superframe* that has a duration of 6.12 s. It contains 51 of 26 multiframes. Finally, 2048 of these superframes are combined into a *hyper frame*, which lasts 3 hours and 28 min, see Figure 7.21. The hyperframe is implemented for cryptographic reasons; the GSM encryption algorithm has a cyclic nature and the encryption period is exactly the length of one hyperframe.

In fact, two kinds of multiframes are defined in GSM:

- a multiframe consisting of 26 TDMA frames (payload – speech and data – frames) and
- a multiframe of 51 TDMA frames (signalling frames).

As observed from Figure 7.21, the superframe can be either composed of 51×26 or 26×51 basic frames. Each 26 subsequent TDMA frames form a multiframe that multiplexes two logical channels, a TCH and the SACCH, onto the physical channel, see Figure 7.22. If the traffic consists of full-rate channels (8 traffic channels per carrier), 24 frames of the multiframe are used for user traffic and one frame for associated signalling. The eight time slots in the thirteenth frame (see Figure 7.22) carry SACCH signalling information associated with eight traffic channels, one slot per traffic channel.

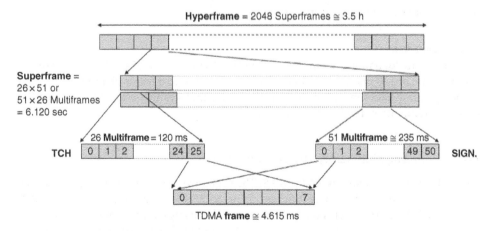

Figure 7.21 GSM frame formats.

Introduction to Mobile Network Engineering

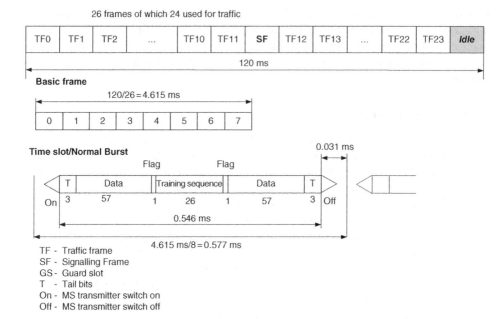

Figure 7.22 TDMA structure for traffic channels, a 26-frame multiframe.

One frame is not used for neither signalling nor traffic. This frame is used by the terminals to read the base identity ('BSIC') of carriers from other cells

7.6.2 Transmission of User Information: Fast Associated Control Channel

Each data burst comprises two user sequences of 57 bits each. A *flag bit* is associated with each 57-bit sequence and this denotes whether the sequence contains normal speech information or if the sequence is instead being used for system signalling (FACCH). A short break in speech transmission is hardly noticeable, since the speech coder fills out the slot with information.

7.6.2.1 Data Rates
Gross data rate for a full-rate physical channel is 114 bits/0.004615s = 24.7 kbps. During a multiframe of 120 ms, 24 or 12 bursts will be allocated to a traffic channel.

The signalling data sent in only one time slot over multiframe period, therefore SACH signalling data rate is 114 bits/0.120s ~ 1 kbps.

7.6.3 Signalling Multiframe, 51-Frame Multiframe

The cycle of control channels is 51 × 8 TS. The sequence repeats itself after IDLE frame. This cycle is chosen deliberately different from the cycle of the traffic channels, which is 26 × 8 TS. These two cycles were chosen not to have any common divider in order to allow MS in dedicated mode to listen to the SCH and FCCH. By listening to those control channels, the MS stays pre-synchronized with neighbour BSs and performs signal

BCCH + CCCH (downlink)

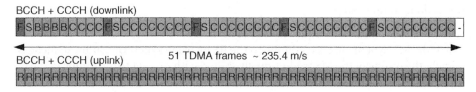

BCCH + CCCH (uplink)   51 TDMA frames ~ 235.4 m/s

Figure 7.23 Example of signalling channel allocation in a 51-multiframe.

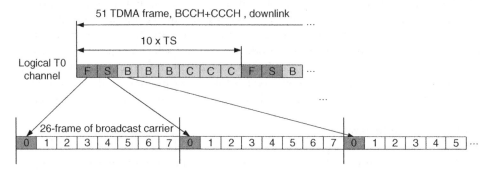

F - Frequency Correction Channel (FCCH)
S - Synchronization Control Channel (SCH)
B - Broadcast Control Channel (BCCH)
C - Common Control Channel (CCCH)

Figure 7.24 Mapping the control channels onto broadcast carrier with signalling over Time Slot 0.

strength measurements. The MS listens to the SCH and FCCH of the neighbour cells during the IDLE frame in a traffic channel cycle of a 26-frame multiframe.

Some of control channels are unidirectional, which results in different structures for uplink and downlink. Common Channels are always arranged in the groups and only certain combinations can be transmitted over a 51-frame multiframe according to GSM specification 05.02.

As seen in Figure 7.23, frames 1, 11, 21,... are FCCH frames, and the subsequent frames 2, 12, 22,... form SCH. Frames 3, 4, 5 and 6 of the 51-frame BCCH multiframe transport the appropriate BCCH information, whereas the remaining frames may contain different combinations of logical channels. Some blocks of CCCH may include PCH and AGCH and SDCCH.

According to GSM specifications, there are various ways to map the control channel leading to different channel arrangements for specific cell. The most common approach is to reserve a TS0 on BCCH carrier for common control channels leaving TS1 to TS7 to traffic channels. In some situations, additional TS may be allocated for SDCCH mapping. The example of mapping the control channels from a 51-frame onto a physical channel in time slot TS 0 is shown in Figure 7.24.

7.6.4 Synchronization

Two kinds of synchronization are distinguished: frequency synchronization and time synchronization of the bits and frames.

This is done in three steps:

1) The MS tunes its carrier frequency to that of the BS.
2) Next, the MS synchronizes its timing to the BS by using synchronization sequences.
3) Finally, the timing of the MS is additionally shifted with respect to the timing of the BS to compensate for the runtime of the signal between the BS and MS (timing advance).

7.6.4.1 Frequency Synchronization

The BS is time synchronized via a fixed network, as a backup it also uses very precise rubidium clocks as frequency references. Due to space and cost limitations, the oscillators at the MS have much lower precision that at the BS. The BS can transmit its high-precision frequency reference periodically and the MS can adjust its local oscillator based on this received reference. Transmission of the reference frequency is done via the FCCH. The reference frequency equals the carrier frequency offset by the MSK modulation frequency and is represented by a sequence of zeros in modulated signal.

7.6.4.2 Time Synchronization

Time synchronization information is transmitted from the BS to the MS via the SCH. SCH bursts contain information regarding the current index of the hyperframe, superframe and multiframe. The MS uses the transmitted reference numbers regarding the multiframes and so on to set its internal counter. The MS then transmits the RACH burst relative to this internal reference. Based on the reception of the RACH, the BS can estimate the roundtrip time between the BS and the MS and use this information for timing advance (described in the next section). In addition to synchronization data with respect to the TDMA structure, the SCH also transmits the Base-Station Identity Code (BSIC), which consists of the network code and the base station's colour code. The terminal then receives information on the BCCH. The BCCH continuously transmits information on the identity of the network (operator) and cell, and on the channel allocations for the cell and the broadcast carriers for the six adjacent cells.

7.6.5 Signalling Procedures over the Air Interface

7.6.5.1 Synchronization to the Base Station

When a SIM card is inserted after switching the power on, the first step for the MS is to adjust to the local radio environment.

1) The MS starts the search for BCCH carriers. Normally, the station has a stored list of up to 32 carriers of the operator network. Signal level (RXLEV) measurements are performed on each of these frequencies. Alternatively, in case of roaming, if no list is available all GSM frequencies have to be measured to find best potential BCCH carrier. The MS uses path loss criteria and stored threshold data to create a preference list of potential carriers.
2) Locking on the preferred carriers the MS searches for a FCCH signal in order to perform frequency adjustment and to start time synchronization between the MS and the network.
3) The MS reads the SCH using an SCH training sequence of 64 bits for fine tuning of the frequency correction and time synchronization. After that the MS is able to read and decode synchronization data from the SCH, the BSIC and the Frame Number and

other information messages from BCCH and CCCH. Validating correct BSIC and path loss criteria, the MS can select the candidate cell.

4) The terminal then receives information on the BCCH. The BCCH continuously transmits information on the identity of the network (operator) and cell, and on the channel configuration for the cell as well as neighbour list with neighbour carriers.

5) The MS performs steps 1 to 4 in order to synchronize with the six neighbour cells with the strongest signal level (RXLEV) and read out their BCCH/SCH information.

6) Steps 1 to 5 are repeated at the time interval set by the network. When measured parameters significantly change (fall out of the range of defined thresholds), the MS starts the cell reselection process.

7.6.5.2   Registering With the Base Station

After locking on to the Broadcast Carrier in selected cell, the MS is able to register. Through the registration, the network stores information about the location area in which the MS is camped. Therefore, the MS is known to the network so that it can be reached by calls initiated by the network. Registration takes place either when the terminal is powered up inside radio coverage area and when it enters a new location area. As part of the registration procedure, the GSM may check the status of terminal equipment by sending enquiry to the Equipment Identity Register (EIR).

The process of registration is as follows:

1) The MS pages the BS via RACCH.
2) The BS assign the registration channel SDCCH to the MS via message over AGCH.
3) The signalling over SDCCH includes authentication of the mobile user, location registration. A successful registration results in the Visiting Location Register sending a message to the subscriber's Home Location Register with details of the subscriber's ID and where they should be paged for incoming calls.
4) The MS starts monitor paging channel, PCH.

If the terminal moves into another cell, synchronization to a new broadcast carrier takes place. If the cell belongs to a different location area, reregistration takes place so that the network can transfer paging signalling on the PCH to the new location area. (The size of a location area is a trade-off between heavy registration signalling in small location areas and heavy paging signalling in large location areas containing many cells.)

7.6.5.3   Call Setup

Setting up calls to or from a terminal requires extensive signalling for transfer of address information and allocation of a radio channel/time slot. Initially, signalling is via the CCCH (Common Control CH) and, subsequently, the SDCCH. The outcome is that the call is assigned to a traffic channel. The procedure makes it possible for the network to determine in which cell within the location area the terminal is situated.

The setting up of a call *to a terminal* is initiated by the network paging the terminal over the PCH (Paging CH) in all cells belonging to the location area. The terminal acknowledges the call on the RACH (slotted Aloha). The procedure makes it possible to determine in which cell of the location areas the terminal is situated. Thereafter, the network sends a message to the terminal via the Access Grant Channel (AGCH), instructing it to switch over to a given SDCCH, which has an associated SACCH. The SDCCH is

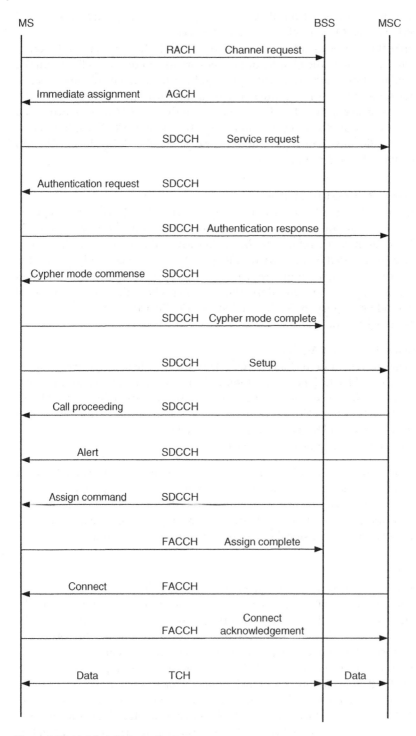

Figure 7.25 Mobile-initiated call setup.

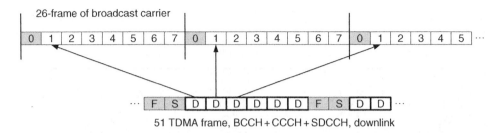

Figure 7.26 Signalling on dedicated downlink control channel during the call setup.

used for transmission of the calling and called-party numbers, for authentication, for sending encryption keys and so on. Finally, a traffic channel is allocated to the terminal.

For setting up a call *from a terminal*, the terminal sends a call request via the RACH. The network sends back details of the allocated SDCCH on the AGCH. Further signalling takes place as described previously, the procedure is shown in Figure 7.25 and mapping of signalling frame to a broadcast carrier is shown in Figure 7.26.

7.7   Signalling During a Call

The in-call signalling performed using two control channels, SACCH and FACH. The respective split of control commands is as follows:

- on SACCH
  from base:
    power command
    time advancement
    frequency hop structure
    frequencies used by adjacent channel
  from terminal
    BER on traffic channel
    signal levels from adjacent cells
    comfort noise
- on FACCH
  from base:
  command to switch channel (frequency and time slot)
- on SACCH
  Different training sequences ('colour codes')
  Mobile Assisted Hand Over (MAHO)
  Discontinuous transmission

7.7.1   Measuring the Signal Levels from Adjacent Cells

The terminal receives information over BCCH from the current base station about the frequencies of the broadcast carriers from the adjacent cells. There is a time period

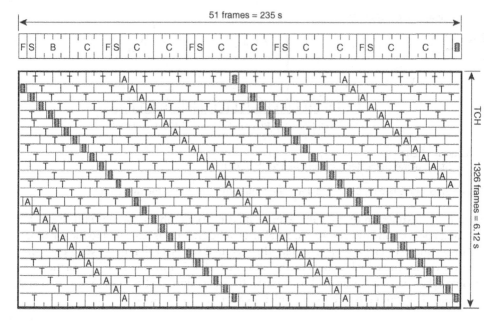

Figure 7.27 Sliding frames [4].

during each TDMA frame available for the MS to measure the level of a carrier from an adjacent cell. To ensure reliable readings averaging must be carried out over the fast fading. Several measurements are made of each carrier before the mean values are sent over the SACCH to the base. The terminal also needs to associate the BSIC for the measured neighbour cell. It means that terminal has to read the BSIC (Base-Station Identity Codec) transmitted on the neighbour SCH. The terminal measures and reads a neighbour carrier during the last, idle TDMA frame in the multiframe.

A complication here is that the base stations may not be mutually time synchronized. On the other hand, the MS cannot read the neighbour BSIC if not synchronized to neighbour SCH. This means that the terminal has to listen to the BCCH on the broadcast carrier from an adjacent cell for an entire TDMA frame to capture a 0 time slot. This alone is not enough since the SCH uses only 1 in 10 of the 0 time slots. For this reason, two different multiframes are used in GSM. The inward and outward traffic channels use multiframe A (26 basic TDMA frames), whereas the broadcast channels use multiframe B (51 frames). This means that the idle TDMA frame that is used for listening will slide over the TDMA frames in multiframe B, thus ensuring that, after a number of A multiframes, the terminal will have reached the correct 0 time slot in the broadcast channel of an adjacent cell, see Figure 7.27 [4].

7.7.2 Handover

Handover is a key feature of a cellular network supporting mobility of mobile user when it moves from the coverage area of one cell to that of another. The network switches an air-interface communication link from one cell to another by allocating radio and other traffic resources in a new cell and possibly BSC and MSC depending on the type of handover. The reasons for handover decisions could be based either on poor signal quality or traffic congestion in serving cell (or selected cell during call setup).

7.7.2.1 Intra-Cell and Inter-Cell Handover
In case of an intra-cell handover, the MS does not leave the cell at all but is allocated to another carrier in the same cell. With an inter-cell handover, the MS is moved from one cell to another.

7.7.2.2 Intra- and Inter-BSC Handover
An inter-cell handover can be an intra or an inter-BSC handover. In the first case, the target cell and the source cell are controlled by the same BSC. In the latter, source and target cells are located in different BSC areas. The inter-BSC handover has to be handled by the MSC, but nevertheless the decision about it is made by the BSC controlling the source cell.

7.7.2.3 Intra- and Inter-MSC Handover
In case of an inter-BSC handover, the target cell might be located in a different MSC area to the source cell. For such an inter-MSC handover, the current MSC/VLR must contact the target MSC/VLR and then transfer the call to it. In case of an intra-MSC handover, source and target cells belong to the same MSC area.

7.7.2.4 Intra- and Inter-PLMN Handover
Finally, the MS might be moved even from one PLMN to another one. This is called inter-PLMN handover or roaming.

7.7.2.5 Handover Triggering
The decision for a handover is made by the BSC on the basis of the uplink and downlink measurements taken by the MS and BS, respectively. Since the MS partly provides the necessary data or handover decision, the GSM handover was named the Mobile Assisted Handover (MAHO). When the decision about the MAHO is made, the BSC looks for a suitable target cell and moves the ongoing call to a new cell or transceiver (intra-cell handover).

The measured data on local radio conditions at the MS are transmitted to the BS via the SACCH. The following information is used by the MSC when deciding the best cell for handover:

- Carrier level and connection quality measured in Frame-Error Rate probability (both BS and MS) averaged over 12 s.
- Signal levels of neighbour carriers in the MS receiver (MS).
- Distance from BS calculated from the timing advance (BS).
- Interference level in the base receiver in idle time slots (BS).

During the call, the BSC continuously monitors these measurements taken by the MS and BS. Handover becomes mandatory when the signal level in the serving cell drops below a certain threshold. The next step in the handover process is to select a suitable target neighbour cell with a good downlink signal level. Given the fact that the radio signal

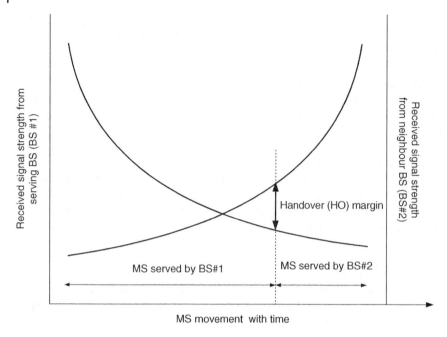

Figure 7.28 Handover decision.

fluctuates due to shadowing that is an important factor at the borders of the cell, some Handover Margin should be introduced in decision factors, as shown in Figure 7.28. This is the case of power budget handover. The handover can also be triggered via timing advance data when a mobile moves too far from the base and the base is not able further adjust timing advance.

7.7.3 Power Control

Power control in the base and terminal transmitters reduces the average interference level to other cells and optimizes power consumption of mobile terminal. Power control decisions to increase or reduce transmit power are based on the carrier level and transmission quality at both the terminal and base. In addition to power control, GSM uses Discontinuous Transmission and Reception (DRX). During the natural gaps in incoming speech there is no need to use entire signal processing chain and transmit at the same power level as in presence of speech. The system use voice activity detector in both BS and the MS and generate a background noise at lowest but comfortable level during the pauses in speech. The 'comfortable' noise is transmitted via SACCH instead of traffic channel.

The DRX is also used to replace speech frames with too-low transmission quality. The process is controlled by a channel decoder. When an isolated speech frame is discarded, the previous frame is repeated. If conditions persist, several repetitions of the speech frame will cause significant degradation of quality. In that case, after each repetition the output level is progressively reduced down to 0 and comfort noise is inserted instead.

7.8 Signal Processing Chain

The following operations take place on the transmitting side:

- *Source coding*. Converts the analogue speech signal into a digital equivalent.
- *Channel coding*. Adds extra bits to the data flow. In this way, redundancy is introduced into the data flow increasing its rate by adding information calculated from the source data in order to allow detection or even correction of bit errors that might be introduced during transmission.
- *Interleaving*. Consists of mixing up the bits of the coded data blocks. The goal is to have adjacent bits in the modulated signal spread out over several data blocks. The error probability of successive bits in the modulated stream is typically highly correlated and the channel coding performance is better when errors are decorrelated. Therefore, interleaving improves the coding performance by decorrelating errors and their position in the coded blocks.
- *Ciphering*. Modifies the contents of these blocks through a secret code known only by the mobile station and the base station.
- *Burst formatting*. Adds synchronization and equalization information to the ciphered data. Part of this is the addition of a training sequence.
- *Modulation*. Transforms the binary signal into an analogue signal at the right frequency. Thereby the signal can be transmitted as radio waves.

The receiver side performs the reverse operations as follows:

- *Demodulation*. Transforms the radio signal received at the antenna into a binary signal. Today, most demodulators also deliver an estimated probability of correctness for each bit. This extra information is referred to as soft decision or soft information.
- *Deciphering*. Modifies the bits by reversing the ciphering code.
- *Deinterleaving*. Puts the bits of the different bursts back in order to rebuild the original code words.
- *Channel decoding*. Tries to reconstruct the source information from the output of the demodulator using the added coding bits to detect or correct possible errors caused between the coding and the decoding.
- *Source decoding*. Converts the digitally decoded source information into an analogue signal to produce speech.

The processing chain is depicted in Figure 7.29.

7.8.1 Speech and Channel Coding

The speech coder in first-generation GSM outputs a binary signal in the form of sequences of 260-bits at a rate of 50 blocks per second (20 ms per block).

From a full-rate speech encoder: 260 bits every 20 ms = 13 kbps

The channel coding introduces redundancy into the data flow by increasing the bit rate. For the TCH/FS mode, there are three bit-coding classes. The class 1a bits have a three-bit cyclic redundancy check (CRC) and all class 1 bits are encoded by a convolution code. The class 2 bits remain unprotected. The bits are classified according to their sensitivity to transmission errors.

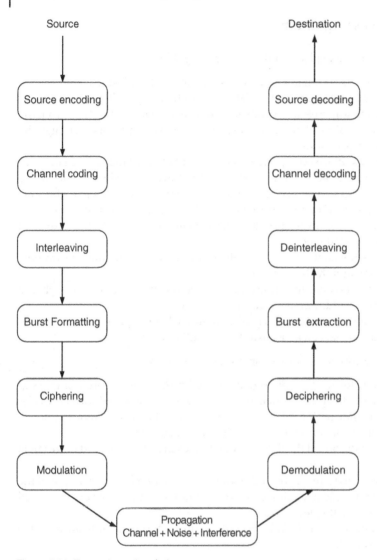

Figure 7.29 Transmit receive chain.

Class 1a includes bits for which transmission errors result in a strongly disrupted output signal from the speech decoder. If transmission errors occur in the class 1a group (despite FEC channel coding), the 20 ms frame will be replaced by the preceding frame from the speech coder ('frame erasure'). To enable error detection, three parity bits are inserted into the class Ia group (Cyclic Redundancy Check, CRC).

Assigned to class 1b are bits for which transmission errors result in fairly large degradation of the speech quality. Therefore, FEC is used. Class Ia (including the parity bits) and class 1b bits are then combined with four tail bits (since a convolution code with

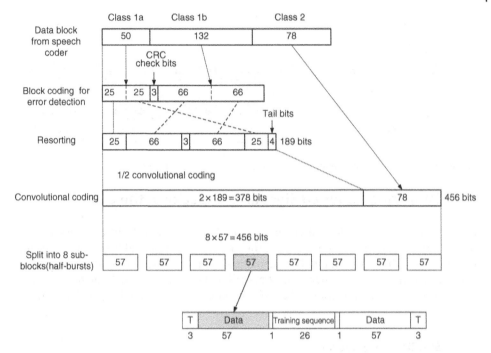

Figure 7.30 Channel coding.

constraint length of 5 is used). This gives a block comprising a total of 189 bits, which is coded in a half-rate convolution coder. The coder outputs 378 bits (2×189).

The remaining 78 bits from the speech coder are assigned to class 2. Because these are relatively non-critical with regards to the impact of transmission errors on speech quality, they are not given any protection against bit errors through FEC. Thus, the channel coder arrangement outputs a total of 456 bits every 20 ms, which matches to the capacity of a full-rate channel; see Figure 7.30.

7.8.2 Reordering and Interleaving of the TCH

While convolutional channel coding is a powerful mitigation tool against statistically independent errors in a memoryless channel, it alone cannot fix burst errors; that is, a long sequence of correlated errors occurred during fading dips. As discussed in Section 7.5, deep fading can extend over the whole frame. In mobile communication channels, the channel coding is used together with the interleaving technique. In this way, the channel encoded data block is distributed over several frames and burst errors occurred uniformly across transmitted codewords. That, in turn, significantly increases efficiency of decoding and demodulation in the receiver.

The reordering and interleaving process mixes the encoded data block of 456 bits and groups the bits into eight sub-blocks (half bursts). The eight sub-blocks are transmitted on eight successive bursts (interleaving depth equals 8) in eight frames, see Figure 7.31.

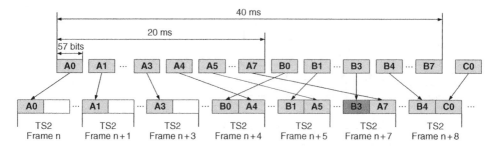

Figure 7.31 Reordering and interleaving for TCH allocated to a user in TS2.

7.9 Estimating Required Signalling Capacity in the Cell

The required capacity of logical channels should be calculated to fit the cell's specific needs. In the case of mismatch, it may be the case that a cell's signalling channels are too few to support the traffic channels. If signalling channel capacity is too high, the traffic channels may not be able to maintain all the call requests coming in. A balance of capacity for both signalling and traffic channels provides best call handling performance of a cell.

7.9.1 SDCCH Configuration

Given that BCCH together with other control channels always occupy the first time slot (TS0) in the first TRX in the cell, the main subject for capacity calculation is an SDCCH channel carrying the Short Message Service (SMS). A logical SDCCH channel can physically be located in a number of locations (TSLs) as defined in the GSM recommendations.

There are two options for defining SDCCH capacity:

1) Combined control channels
2) Non-combined control channels.

When the combined control channel configuration is used, both common control channels as well as the dedicated control channels are transmitted on TSL0 of the BCCH TRX of the cell. Additional SDCCH capacity may be assigned to any time slot of any transceiver (reducing capacity of TCHs, see Figure 7.32b).

When a non-combined control channel configuration is used, TSL0 of the BCCH TRX always carries the common control channel and some (one or more) other TSL has to be defined for dedicated control channels. The non-combined configuration provides extra capacity for PCH, AGCH, SDCCH and SACCH. In this case, TSL 0 is allocated for common control channels only, separate TSL(s) could to be configured for SDCCH

Figure 7.32 SDCCH configuration: (a) combined configuration and (b) separate SDCCH.

Table 7.4 Erlang capacity of the cell versus number of TRX.

TRX per cell	1	2	3	4	6
BCCH + CCCH	0.5	1	1	1	1
SDCCH	0.5	1	1	2	3
TCH	7	14	22	29	44
Erl (2% GOS)	2.94	8.2	14.9	21.04	34.68

and SACCH usage; see Figure 7.32(b). It is important to note that signalling channel GOS and signalling traffic load values could be different from respective figures for voice traffic.

An example of cell capacity is given in Table 7.4 where the Erlang capacity of the cell with 2% blocking probability varies with configuration of signalling and traffic channels on the cells with different number of transceivers.

7.9.2 Worked Example

7.9.2.1 Problem 1
Consider a GSM 900 deployment scenario with the following assumptions:

1) Area of deployment 900 MHz GSM/EDGE network is 200 km² with 25 250 sub-scribers.
2) Assume 25 mErl/sub for CS services with 2% GoS.
3) SDCCH load is 1.35 mErl/sub with 2% GoS.
4) All sites to be deployed are three-sector sites.

The BTS/MS parameters for the power budget are as follows:

- BTS transmit power = 40 dBm,
- antenna gain = 18 dBi,
- antenna-feeder system loss = 3.3 dB,
- fading margin = 13 dB, body loss = 1.6 dB,
- MS sensitivity = −110 dBm,
- MS antenna gain = 0 dBi,
- output MS transmit power = 23 dBm, diversity gain = 3 dB,
- BTS RX sensitivity = −110 dBm.

Assume that propagation path loss L is modelled as

$$L = 40 * (1 - 0.004 * \Delta h_B) * \log(R) - 18 * \log(\Delta h_B) + 21 * \log(f) + 80 \ dB$$

where is the distance between base station and mobile station in kilometres, f is the carrier frequency in megahertz and Δh_B is the base station antenna height, in metres, measured from the average rooftop level.

Assume BTS antenna height above the rooftops is 5 m and tree-sector cell area A is of circular shape, that is $A = \pi r^2$, where r is a cell radius (range).

The design task is as follows:

Calculate number of tree-sector sites required to serve the area. Provide the BTS configuration in terms of the number of TRX and control channels defined for a cell.

Solution:

Calculating link budget, we obtain allowable path loss on uplink and downlink estimates 140.1 dB and 154.1 dB, respectively. The MAPL is limited by the uplink and therefore MAPL = 140.1 dB.

Using the given propagation model, we obtain the following estimates:

- cell radius −1.87 km
- number of three-sector sites – 19, number of cells – 57
- number of subscribers/cell – 443
- SDCCH Erlang/cell required – 0.6 Erl that results in three time slots required for SDCCH per cell with 2% GoS

traffic Erlang /cell required – 25 250 sub · 0.025 Erl/57 cell = 11.075 Erl/cell

- number of TCHs with 2% blocking probability supporting 11.075 Erl – 18 TCHs
- therefore, signalling + traffic need is BCCH + SDCCH + TCH = 1 + 3 + 18 = 22 time slots/cell, which leads to the result that three TRX/cells are to be configured.

References

Forsberg, D. Horn, G., Moeller, W.-D. and Niemi, V., *LTE Security*, John Wiley & Sons, Ltd, Chichester, 2010.

Technical Specification Group Radio Access Network; Deployment aspects (Release 14), 3GPP TR 25.943 V14.0.0 (2017–03).

Technical Specification Group GSM/EDGE, Radio Access Network; Radio transmission and reception, (Release 1999), 3GPP TS 05.05 V8.20.0 (2005–11).

Eberspächer, J., Vögel, H.-J., Bettstetter C. and Hartmann, C., *GSM – Architecture, Protocols and Services*, John Wiley & Sons, Ltd, Chichester, 2009.

8

EGPRS: GPRS/EDGE

The GSM Packet Radio Service, abbreviated to GPRS, introduces the concept of sharing the pool of available channels in the cell between different users. The concept of sharing is as follows:

- Several timeslots in the PTCH on one carrier frequency may be allocated to one user – this is known as 'bundling' of timeslots. Timeslots can be bundled on the uplink (UL) and the downlink (DL). The allocation of timeslots may also be asymmetric between UL and DL.
- Opposite to circuit switch traffic, such as voice, a timeslot is not reserved exclusively for one user; that is, a timeslot may be shared by several users according to their priority or on round-robin basis.

In order to support the new concept of packet service, the GPRS system was deployed as an overlay of GSM with two new network nodes, SGSN and GGSN, new interfaces and new functionalities in Base-station controller, BSC, Figure 8.1.

EDGE stands for Enhanced Data rate for GSM Evolution. It is a further evolution of GPRS, giving the option for an increased system data rate using extended modulation schemes at the air interface with no impact to other parts and nodes of the system. The combination of GRS and EDGE is usually referred to as *EGPRS*.

8.1 GPRS Support Nodes

1) GGSN stands for Gateway GPRS Support Node. The GGSN routes incoming packets to the current location of the mobile. Therefore, it has to have interface to the HLR in order to obtain the required location information for mobile terminating packet transfers. The second node is known as the Serving GPRS Support Node (SGSN). The SGSN sets a mobility management context for an attached MS. The SGSN also does ciphering for packet-service traffic. This is different from circuit switch traffic that is encrypted between the MS and BSC.
   Both SGSN and GGSN are also known as GSN nodes and comprise the Packet Core Network. Both of the GPRS support nodes (GGSN, SGSN) also collect charging data. Details about the volume of data transferred by the user are collected by the SGSN.

Introduction to Mobile Network Engineering: GSM, 3G-WCDMA, LTE and the Road to 5G,
First Edition. Alexander Kukushkin.
© 2018 John Wiley & Sons Ltd. Published 2018 by John Wiley & Sons Ltd.

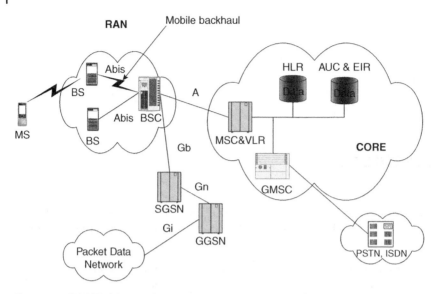

Figure 8.1 GPRS/EDGE network.

8.2 GPRS Interfaces

Over the air-interface, packet data share the same Abis interface with circuit switch (voice) data. On the other hand, in BSC, packet data are processed in a separate Packet Control Units (PCU), which are the new components in GSM BSC, and then transferred to the packet core network over the Gb interface.

The Gn interface connects SGSN and GGSN, and Gi is an interface between GGSN and Packet Data Network via an Access Point Node that normally contains a firewall and other elements of ISP infrastructure, such as DNS and DHSP; see Figure 8.2.

All G-interfaces are IP based. Physically, all Packet Core nodes are connected via a carrier grade IP backbone network. The GPRS tunnelling protocol (GTP) is used to tunnel both data and signalling between the GSNs. It could be noted that signalling and user planes in GPRS are not separated. They are both encapsulated in the GTP tunnel onto a single interface (Gn, Gp) instead.

8.3 GPRS Procedures in Packet Call Setups

The GPRS introduces new procedures and states in mobility management. Two procedures, *GPRS Attach* and *GPRS Detach*, are mobility management functions to establish and to terminate a connection with the GPRS network.

1) In order to access packet services the MS must be 'attached' in the GPRS network using a logical procedure between the MS and the SGSN as shown in Figure 8.3. With GPRS Attach, the mobile moves to the READY state and the mobility management context is established, the MS is authenticated, the ciphering key is generated, a ciphered link setup and the MS is allocated a Temporary Logical Link Identity (TLLI). The SGSN gets the subscriber information from the HLR.

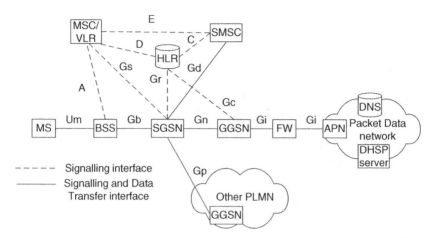

Figure 8.2 GPRS interfaces.

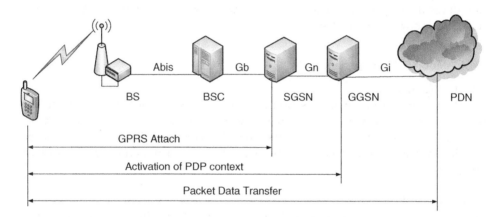

Figure 8.3 GPRS procedures.

2) After a GPRS Attach, the SGSN tracks the location of the MS. The MS can send and receive SMS, but no other data. To transfer other data, it has to first activate a *PDP context*.

The procedure GPRS Detach moves the MS to IDLE state and the mobility management context is removed. The MS can be detached from the GPRS when the mobile timer expires. The GPRS Detach is normally generated by the MS, but can also be generated by the network.

8.4 GPRS Mobility Management

The GPRS is an additional service provided by the mobile network. The mobile user can be registered in the network with user's location known to the VLR and HLR but still unknown to the packet core. In order to provide packet services, the GPRS network

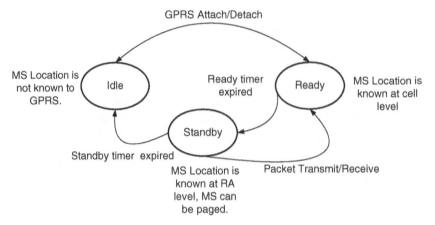

Figure 8.4 GPRS mobility management states.

introduces three specific mobility management states, namely, IDLE, STANDBY and READY, as shown in Figure 8.4.

8.4.1 Mobility Management States

8.4.1.1 IDLE State

The IDLE state is used when the subscriber (MS) is passive (not GPRS attached). The subscriber is not reachable by the GPRS network. The packet core network elements hold no valid context for the subscriber and the subscriber is not attached to the mobility management. In order to change state, the MS has to perform a GPRS attach procedure.

8.4.1.2 READY State

The subscriber is attached to the mobility management and the location of an MS is known on a cell level. The MS is capable of receiving PTM (Point To Multi-point) and PTP (Point-To-Point) data. The SGSN can send data to the MS without paging at any time and the MS can send data to the SGSN at any time. The network holds a valid *mobility management context* for the subscriber. If the READY timer expires, the MS moves to STANDBY state. If the MS performs a GPRS Detach procedure, the MS moves to the IDLE state and the mobility management context is removed. An MS in the READY state does not necessarily have radio resources reserved.

8.4.1.3 STANDBY State

The subscriber is *attached* to the mobility management and the location of an MS is known on a routing area level. The MS can be paged from the network. The network holds a valid mobility management context for the subscriber. If the MS sends data, the MS moves to the READY state. The MS or the network can initiate the GPRS Detach procedure to move to the IDLE state. After expiry of the MS reachable timer, the network can detach the MS.

8.4.2   PDP Context Activation

In the process of PDP context activation, the network provides an IP address for the MS. Two different options are possible:

- static allocation: the MS has a permanent IP address and
- dynamic allocation: the corresponding GGSN assigns a temporary IP address to the MS.

A PDP context is normally activated by the MS. In the case of mobile terminating packet transfer, the GGSN must initiate a paging to the MS that requests the PDP context activation in order to receive the packets: that is, the GGSN can only request a PDP context activation but not initiate one.

It is important to note here that a PDP context activation request from the GGSN is only possible with a static IP address has been allocated to the MS. This address must then be stored in the HLR's profile record for the respective MS.For each data transfer session, a PDP context must be created. It contains the PDP type (e.g. IPv4), the PDP address assigned to the MS (i.e. an IP address), the requested QoS class and the address of a GGSN that serves as the access point to the external network. This context is stored in all participating network entities: the MS, the SGSN and the GGSN. The call flow in Figure 8.5 shows the PDP context activation procedure initialized by the MS. As observed from Figure 8.5, while MS initiates the PDP context activation, the function of allocation and release of the PDP context belongs to GGSN. The GGSN creates an entry in PDP context table, which enables the GGSN to route data packets between the SGSN and the external PDN. It confirms that to the SGSN with a message CREATE PDP CONTEXT RESPONSE, which also contains the dynamic PDP address (if needed). Finally, the SGSN updates its PDP context table and confirms the activation of the new PDP context to the MS (ACTIVATE PDP CONTEXT ACCEPT).

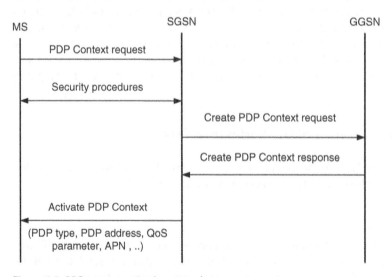

Figure 8.5 PDP context activation procedure.

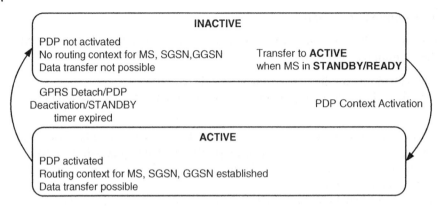

Figure 8.6 PDP states/phases.

One PDP context includes one address (usually a dynamic IP address) and one set of QoS attributes. Several PDP contexts may be activated simultaneously. When the subscriber has finished the use of the activated addresses, they have to be deactivated. As long as PDPs are activated, the MS has to be in the STANDBY or READY state. A return to GMM 'IDLE' automatically deactivates all active PDPs. Figure 8.6 shows the relation between phases of the packet services in relation to the states of the mobile station and respective procedures.

8.4.3   Location Management

Just as in circuit-switched GSM, the main task of location management is to keep track of the user's current location, so that incoming packets can be routed to their MS. For this purpose, the MS has to send location update messages to its SGSN. The location update frequency depends on the state in which the MS currently is. In IDLE state, no location update is performed; that is, the current location of the MS is unknown. If a MS is in READY state, it will inform its SGSN of every movement to a new cell. The GSM Location Area is usually divided into several Routing Areas (RAs) for packet services. When the MS is in a STANDBY state it must inform the SGSN when moving into new RA.

8.5   Layered Overview of the Radio Interface

The GPRS radio interface can be modelled as a hierarchy of logical layers with specific functions. An example of such layering is shown in Figure 8.7.

8.5.1   SNDP

The role of the Subnetwork Dependent Convergence Protocol is to convert the IP protocol used in GGSN and external Packet Data Network to a protocol stack used in the air interface of the GPRS network. This conversion is performed in SGSN.

LLC: Logical Link Control (LLC) provides a logical link between the MS and the SGSN. This link is encrypted between SGSN and MS.

Figure 8.7 GPRS MS: network reference model [1].

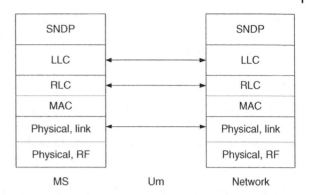

Data-Link Layer (RLC/MAC): The Medium Access Control (MAC) and Radio Link Control (RLC) layer operates above the Physical Link layer in the reference architecture. The RLC/MAC layer provides services for information transfer over the physical layer of the GPRS radio interface. These functions include backward error-correction procedures enabled by the selective retransmission of erroneous blocks. The MAC function arbitrates access to the shared medium between a multitude of MSs and the Network. The RLC/MAC layer uses the services of the Physical Link layer. The layer above RLC/MAC (i.e. LLC) uses the services of the RLC/MAC layer on the Um interface.

Physical: The physical layer has been separated into two distinct sublayers defined by their functions:

- *Physical RF* layer performs the modulation of the physical waveforms based on the sequence of bits received from the Physical Link layer. The Physical RF layer also demodulates received waveforms into a sequence of bits that are transferred to the Physical Link layer for interpretation.
- *Physical Link* layer provides services for information transfer over a physical channel between the MS and the network. These functions include data unit framing, data coding and the detection and correction of physical medium transmission errors. The Physical Link layer uses the services of the Physical RF layer.

8.5.2 Layer Services

The MAC function defines the procedures that enable multiple MSs to share a common transmission medium, which may consist of several physical channels; that is, time slots. The MAC function provides arbitration between multiple MSs attempting to transmit simultaneously and provides collision avoidance, detection and recovery procedures. The operations of the MAC function may allow a single MS to use several physical channels in parallel.

The RLC function defines the procedures for a bitmap selective retransmission of unsuccessfully delivered RLC data blocks.

The RLC/MAC function provides three modes of operation:

- Unacknowledged operation;
- Acknowledged operation; and
- Non-persistent operation. The transfer of RLC data blocks in non-persistent RLC/MAC mode is controlled by the numbering of the RLC data blocks within one

Temporary Block Flow (see Section 8.5.6) and may include retransmissions if block cannot be decoded. The retransmission is based on the Automatic Repeat request mechanism (ARQ).

8.5.3 Radio Link Layer

The Radio Link Control RLC (between the MS and the PCU) segments the LLC packets into smaller packets called 'radio blocks' for transmission over the radio interface and re-assembles the received radio blocks from the radio interface (and from the Abis interface) into LLC packets. The relations between the data sequences in the three layers are shown in Figure 8.8.

Figure 8.8 also shows that each layer adds its own header and check sequence in support of the layer functions.

8.5.3.1 RLC Block Structure

The Radio block includes a Block Header (BH) and parity bits for selective ARQ; that is, Block Check Sequence (BCS). The BH is a bit different for data block and signalling block. The latter has two parts: a MAC header and RLC header, while the data block only has a MAC header, see Figure 8.9. The MAC header comprises the Uplink State Flag (USF), T and PC fields:

- The USF is used in connection with the reservation of radio blocks on the inward traffic channel.
- The T field is a flag, which tells if a block is used for data transfer or for signalling.
- The PC field is used in connection with power control.

The RLC blocks are numbered with a *TFI* (Temporary Flow Identifier) and the receive side can request retransmission of erroneous blocks. The same TFI is included in every

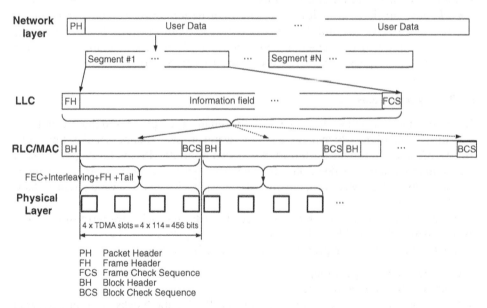

Figure 8.8 GPRS transformation data flow.

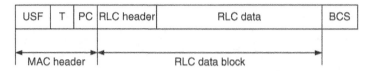

Figure 8.9 Radio block structure for GPRS data transfer.

RLC header belonging to a particular TBF as well as in the control messages associated to the LLC frame transfer (e.g. acknowledgements) in order to address the peer RLC entities.

Besides the ARQ based on the BCS, convolutional FEC coding can be applied for a Radio Data Block depending on the coding scheme used. The radio blocks are fed to the Physical Layer, which is based on slots in the normal eight-slot TDMA frame. Each of the eight slot positions constitutes one Packet Data Channel, which is multiplexed between traffic channels and different signalling channels.

One radio block is allocated four slots in successive frames, that is, one block comprises $4 \times 114 = 456$ bits, see Figure 4.8. Most of the slots in a Packet Traffic Channel (PTCH) are used for data transfer (PDTCH: Packet Data Traffic Channel) but a small fraction of the slots is used for signalling via a PACCH (Packet Associated Control Channel); that is, a signalling mechanism during the packet data call is similar to a circuit switch call in the original GSM.

8.5.4 GPRS Logical Channels

GPRS introduces several new logical channels, not all of them are mandatory. In particular, the system information message related to GPRS services can be broadcasted on BCCH instead of using the separate Packet Broadcast Control channel. Two most important GPRS logical channels are the Packet Data Traffic Channel (PDTCH) and Packet Timing advance Control Channel (PTCCH):

- The Packet Data Traffic Channel (PDTCH) is a channel resource allocated to a single MS on one physical channel for user data transmission. In multislot operation, one MS may use multiple time slots in parallel for individual packet transfer. PDCH is a unidirectional channel.
- The Packet Timing advance Control Channel (PTCCH) is used in the uplink direction for the transmission of access bursts to estimate the timing advance for one mobile. In the downlink direction, one PTCCH is used to transmit timing advance information to up to 16 MSs.

8.5.5 Mapping to Physical GPRS Channels

A physical channel allocated to carry packet logical channels, PDTCH, is called a Packet Data CHannel (PDCH). A PDCH carries only GPRS logical channels. The mapping in time of the logical channels is defined by a multiframe structure. The multiframe structure for PDCH in basic configuration consists of 52 TDMA frames, divided into 12 blocks (of four frames), two idle frames and two frames used for the PTCCH. As shown in Figure 8.10, PTCCH information is transmitted in positions 12 and 38 of the 52 multiframe structure.

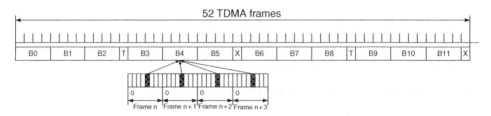

Figure 8.10 Multiframe structure for PDCH [1].

The timing of a 52 multiframe is $4.165 \times 52 = 24\,0$ms. The 'effective' time transmission interval (TTI) of a radio block consisting of 4 TDMA frames is then 240 ms/12 = 20 ms. This is different from the 'physical' duration of a radio block $4 \times 4.165 = 18.6$ ms, and arises as a consequence of adding four additional TDMA frames (signalling and idle frames) into 52 multiframes. The T frame is used for PTCCH (timing advance channel) and the 'X' frame is an idle frame that can be used by the MS for signal measurements and BSIC identification.

Mapping of logical channels onto the radio blocks is defined by means of the ordered list of blocks (B0, B6, B3, B9, B1, B7, B4, B10, B2, B8, B5, B11). Two frames are used for PTCCH and the two idle frames as well as PTCCH frames can be used by the MS for signal measurements and BSIC identification. When packet only service mode is used by a mobile terminal the synchronization is performed via a specific packet-service Compact Synchronization Channel (CSCH) that is similar to SCH but mapped on to a GPRS 52 multiframe. In this case, the mapping of PDCH to a multiframe is slightly different from the one given in Figure 8.10. Details can be found in [1].

8.5.6 Channel Sharing

In GPRS there are two distinct features different from GSM:

- Several timeslots in PTCH on one carrier frequency may be allocated to one user – this is known as 'bundling' of timeslots. Timeslots can be bundled on the UL and the DL. The allocation of timeslots may also be asymmetric. For instance, a subscriber who wants to download some data from the internet will receive more timeslots for the DL than for the UL.
- One timeslot is not reserved exclusively for one subscriber; that is, a timeslot may be shared by several subscribers. As a GPRS connection is packet switched, the packet transmissions from several users can be multiplexed through the same timeslots on the Um. In the case that several subscribers share a UL timeslot and all of them want to transmit, if none of them has a higher priority than the others the PCU allows each MS in turn to transmit some data. So the mobiles transmit one after the other until each transmission is complete.

There are two parameters that are important for the allocation of GPRS radio resources:

- the Temporary Flow Identifier (TFI)
- the Uplink State Flag (USF)

These parameters are sent to the mobile as part of the MAC protocol header and are used to give information about channel usage to the mobiles that are sharing a timeslot.

Any data transfer for a single user is associated with the establishment of Temporary Block Flow (TBF), as described in the next section.

8.5.6.1 Downlink Radio Channel

Several subscribers may share a radio channel on the DL. Therefore, in each DL radio block (see Section 8.5.7) a TFI is necessary for determining the owner of each packet. A TFI has 5 bits, so 32 different values are possible; that is, up to 32 subscribers may theoretically share a DL radio channel. Several mobiles may also share a radio channel on the UL. These mobiles must be informed when it is their turn to send. Therefore, an additional parameter is sent in each DL radio block: the USF. It indicates which subscriber can send next on the relevant

8.5.6.2 Uplink Radio Channel

An USF has 3 bits, so eight different values are possible; that is, up to eight subscribers may theoretically share a UL timeslot. If one UL timeslot has been configured (reserved) as a PRACH (Packet Random Access Channel – used by an MS to request a connection) then USF = 111 is reserved to identify the PRACH and the other seven values (000 to 110) remain to identify up to seven subscribers on this UL timeslot.

8.5.7 TBF

A TBF is a physical connection used by the two Radio Resource entities to support the unidirectional transfer of LLC PDUs on packet data physical channels. The TBF is assigned a radio resource on one or more PDCHs and comprises a number of RLC/MAC blocks carrying one or more LLC PDUs. A TBF is a temporary resource and is maintained only for the duration of the data transfer. The major function of TBF is an orderly transfer of a number of radio blocks (RLC/MAC) constituting a customer data packet across the air interface.

Each TBF is assigned a *Temporary Flow Identity* (TFI) by the network. The TFI is assigned in a resource assignment message and is part of the first octet of the RLC/MAC block. The TFI allows several MSs to share one RTSL.

INACTIVE PDP phase (Packet Idle state)
In packet idle mode, no temporary block flow exists.
ACTIVE PDP Phase (Packet Transfer Mode)
In *packet transfer mode*, the mobile station is allocated a radio resource providing a temporary block flow on one or more physical channels. Continuous transfer of one or more LLC PDUs is possible. Concurrent TBFs may (but do not have to) be established in opposite directions (as mentioned, TBFs are unidirectional and are not connected to the TBF in the opposite direction).

8.5.7.1 TBF Establishment

The schematic procedure for the data transfer session is shown in Figure 8.11.

8.5.7.2 DL TBF Establishment

The SGSN has to know the cell of the MS. It will send the LLC PDUs to the correct PCU. The PCU allocates one or more PDTCHs for the TBF and indicates that and the TFI to the MS in an assignment message.

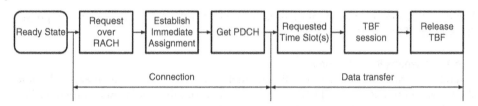

Figure 8.11 MS initiated DL TBF establishment.

A *DL TBF* may be established:

- on *PACCH* when a concurrent TBF exists (or timer is running in the MS).
  The PCU sends a PACKET_DOWNLINK_ASSIGNMENT or PACKET_TIMESLOT
  _RECONFIGURE message. The TBF mode (GPRS/EGPRS) is always the same as the
  existing UL TBF.
- on *PCCCH* when it is supported and no DL TBF exists.
  The PCU allocates one or more PDTCH for the TBF and send a PACKET_
  DOWNLINK_ASSIGNMENT message to the MS.
- on *CCCH* without PCCCH and no DL TBF exists.

First, the PCU allocates one PDTCH and sends a IMMEDIATE_ASSIGNMENT message. The possible multislot allocation is done later with a reallocation message.

The schematic signalling flow is shown in Figure 8.12. The reservation of time slots
(groups of four) for radio blocks is only given as long as there is information stored in
the buffer memory on the transmit side. As soon as all the buffered data bits have been
successfully transferred (including possible ARQ retransmissions) the channel allocation is released and can be used by other connections. When a new burst of data arrives
to the buffer, a new reservation of time slots must be made.

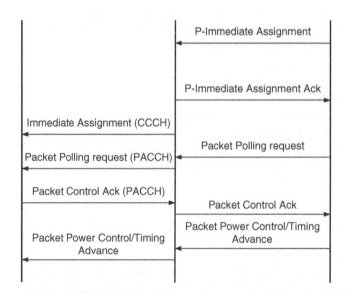

Figure 8.12 Downlink TBF assignment, MS monitors CCCH.

8.5.8 EGPRS Channel Coding and Modulation

The modulation is either GMSK or linear 8PSK. With 8PSK, the system data rate reaches 59.2 kbps per time slot, which is three times higher than the maximum data rate possible with GMSK modulation. The drawback of 8PSK is that it requires linearization in an amplifier since 8PSK is QAM type modulation. The pulse shaping is the same as for GMSK, resulting in the same spectrum envelope.

EGPRS deploys the adaptation of the code rate by allowing a choice between GMSK and 8PSK (a dynamic selection of suitable combinations of channel coding and modulation is used depending on the quality C/I and E/N0 of the radio channel). The net transmission data rate (including the effect of channel coding) is reduce or increased depending on the quality of the radio channel.

The set of bursts that results from a single-user data package is marked by a TFI, which is used at the receiving side to reassemble the user data package. When there are block errors, the radio blocks, which contain errors, are identified by their TFI numbers. The TFIs are sent back to the transmitter side so that corresponding blocks can be retransmitted. Positive acknowledgement is sent back for successfully received blocks. As the block errors are gradually cleared, the error control window can be moved forwards. Radio blocks without errors are fed to the LLC layer, so that LLC frames can be generated on the receive side.

The optimum code arrangement; that is, the code that gives the highest throughput, depends on the BER (Bit-Error-Rate) from the data demodulator at the receive side. This quality information must be transmitted back to the transmitter side producing a certain delay. Adaptation is therefore not possible to the variations due to fast fading. The measured BER values are therefore averaged over the fast fading. Practical evaluations indicated that this link adaptation did not perform fully satisfactorily. A large part of the link adaptation is obtained by an advanced ARQ procedure, called a type II hybrid-ARQ. The link adaptation used at the original GPRS, which was based on different modulation-and-coding schemes (MCS) in combination with simple ARQ, is called a type 1 hybrid-ARQ.

EGPRS obtains the best throughput only when the radio channel is of very good quality. This has two consequences:

1) An effective dynamic link adaptation is necessary that dynamically selects modulation and a code rate to the local average of C/I and C/N.
2) EGPRS was typically introduced in existing GSM cell structures. As higher protection ratios would be needed for the highest throughputs, the highest user data rates could only be used in part of each cell.

EGPRS has nine basic coding schemes, MCS-1…9, see Table 8.1. In general, a higher coding scheme has higher coding rate, and consequently higher peak throughput, but it also tolerates less noise or interference.

8.6 GPRS/GSM Territory in a Base-Station Transceiver

A timeslot can be used for circuit-switched (CS) traffic (then it is served by the BSC) or for packet-switched (PS) traffic (then it is served by the Packet Control Unit, PCU, in the BSC). All full or dual rate, traffic channels are capable of carrying PS traffic.

Table 8.1 EGPRS modulation coding schemes [1].

Scheme	Coderate	Modulation	Data rate per PDCH, kbps
MCS-9	1	8PSK	59.2
MCS-8	0.92	8PSK	54.4
MCS-7	0.76	8PSK	44.8
MCS-6	0.49	8PSK	29.6
MCS-5	0.37	8PSK	22.4
MCS-4	1	GMSK	17.6
MCS-3	0.85	GMSK	14.8
MCS-2	0.66	GMSK	11.2
MCS-1	0.53	GMSK	8.8

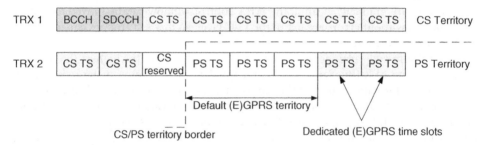

Figure 8.13 CS-PS borders in the BS transceiver.

All timeslots in one BS controlled by the PCU belong to the PS or EGPRS territory, all other to the CS territory. The border between CS and PS territory moves dynamically, depending on parameters set and on the required traffic, Figure 8.13.

Timeslots in the PS territory are classified into dedicated, default and additional time slots. Dedicated time slots, when present, are reserved for EGPRS traffic exclusively.

8.6.1 PS Capacity in the Base Station/Cell

Voice traffic always has priority over data traffic in EGPRS. For that reason, in the early stages of deployment it was possible to implement GPRS in an existing GSM network without adding extra capacity. The concept is illustrated in Table 8.2, which shows that for any given number of carriers there exists an associated amount of Erlang capacity available for a 'best effort' GPRS where blocking probability can be disregarded.

In this case, is possible to implement 'best effort' GPRS with no additional resources in base-station TRXs in the absence of a requirement on packet data service level and performance indicators.

Apparently, when there is a need to support a defined amount of traffic and targeted performance level, additional capacity resources need to be introduced in all network nodes. The dimensioning inputs are: traffic volume, type and performance

Table 8.2 The relationship between carriers, time slots, voice traffic and Erlang capacity [2].

Number of carriers	Number of time slots	Signalling time slots	Traffic time slots	Offered voice traffic (Erlang), 1% GOS	Capacity available for GPRS (Erlang)
1	8	1	7	2.5	4.5
2	16	1	15	8.1	6.9
3	24	2	22	13.7	8.3
4	32	2	30	20.3	9.7
5	40	2	38	27.3	10.7
6	48	3	45	33.4	11.6
7	56	3	53	40.6	12.4
8	64	3	61	47.9	13.1
9	72	3	69	55.2	13.8
10	80	4	76	61.7	14.3

requirements. The dimensioning output is the required amount of traffic-dependent hardware and the associated software configurations.

The primary technique for dividing resources between the circuit-switched (CS) and packet ((E)GPRS) traffic is known as the Territory Method. In this, timeslots within a cell are dynamically divided into the CS and (E)GPRS territories. This means that a certain number of consecutive traffic timeslots are reserved for CS GSM calls with the remainder being available for the (E)GPRS traffic.

The dynamic variation of the territory boundary (and hence the number of timeslots in each territory) are controlled by territory parameters. This enables the system to adapt to different load levels and traffic proportions, thus offering optimized performance under a variety of load conditions. The Figure 8.13 shows an example of how a traffic resource within a cell (2 TRX in this case) can be divided into CS and (E)GPRS territories.

With a service requirement for dedicated (E)GPRS capacity, the number of timeslots are allocated on a permanent basis to (E)GPRS. These timeslots are always configured for (E)GPRS and cannot be used by the circuit-switched traffic. This ensures that the (E)GPRS capacity is always available in a cell. The drawback with this approach is that, for a given cell configuration, blocking levels for the CS traffic will increase in the case of a respective increase in the voice traffic load above the committed level.

The decision on whether to assign the dedicated (E)GPRS territory is a trade-off between providing a minimum level of (E)GPRS service and increasing the blocking for CS services. This decision needs to take into account operator priorities, network performance and predicted (E)GPRS usage levels. The commitment by the mobile operator to support a certain class of multislot terminals will determine the minimum number of time slots allocated for dedicated EGPRS territory.

In addition to dedicated EGPRS territory, which may or may not exist, a default (E)GPRS capacity can be allocated, as shown in Figure 8.13. The default (E)GPRS territory is an area that will always be included in the instantaneous (E)GPRS territory provided that the current CS traffic levels permit. With the exception of the dedicated (E)GPRS area, CS services always take priority over (E)GPRS services and so, if circuit-switched traffic levels dictate, the circuit-switched traffic will occupy as much (E)GPRS default territory as is needed. If the CS level decreases, while previously occupying some of the (E)GPRS default territory, these timeslots will automatically be re-allocated back to (E)GPRS irrespective of the actual (E)GPRS load. In this approach, allocation to (E)GPRS will only occur if the (E)GPRS load reaches a predefined (see later).

The outcome of the EGPRS base-station dimensioning could be a supported mean data rate per cell. Throughput performance is of the EGPRS cell is affected by many factors such as an optimization of link adaptation, application of frequency hopping, coding schemes supported and so on. Some details can be found in [3]. The accepted measure of cell design performance is a mean cell data rate that can be reasonably estimated to be in the range of 30 kbps per time slot averaged over the cell area.

Once the average throughput per time slot has been defined in kbps, the corresponding EGPRS Erlang value can be obtained by dividing the total GPRS traffic load (kbps) by the average throughput per time slot (kbps). This will give the total channel usage per unit of time; that is, 'carried traffic' in Erlangs.

Dimensioning of the PS core network is far more complicated. In addition to the limitations of radio interface that acts more as a data pipe, PS core dimensioning should take into account various overheads for various protocols associated with different traffic types, such as non-real-time versus real-time traffic and respective quality indicators. Dimensioning is based on the internet activity model and queuing theory [2].

8.7 Summary

A summary of main features of the GSM system:

- Network architecture is built to provide Circuit Switch (CS) and Packet Switch (PS) services.
- CS is handled by the Mobile Switch (MSC), a soft switch with a logically and physically separate control call server and switch functionality (IP switch).
- MSC sets up and releases the CS call.
- PS is served by two nodes: SGSN and GGSN.
- SGSN handles the control plane: control signalling for PS call setup and mobility management for the attached MS.
- GGSN handles the user plane; that is, it creates PDP context, terminates mobile PS traffic and routes to a fixed network.
- Control and user plane are not separated on interfaces. This is a clear disadvantage.
- The GSM spectrum is arranged in frequency clusters. The clusters are spatially separated and reused.
- Each cell has a partial set of frequency channels available in the frequency cluster.
- The cell is a logical object: a Base Station that must have a set of Common Control channels mapped to available frequency and time slots.
- GSM based on the FDMA/TDMA principle. Users are separated in different time slots or in frequency if allocated the same time slot.
- Users do not interfere with each other in same cell, neither in downlink nor in uplink, but there is co-channel interference from cells belonging the nearest frequency cluster.

- Normally cells are arranged in sectors (typically 120°); cell, sector and base station are the synonyms.
- The Base-Station Site/Site is a physical base station with a HW that supports several sectors. The site must have an antenna system, transmission equipment, power supply and a back up.
- The GSM air interface is encrypted. For a CS call, encryption/decryption takes place in the Base station. The PS call encryption extends up to SGSN.
- GSM mitigates channel distortion with the following:
  channel coding, interleaving, equalizer, frequency hopping and uplink diversity (space or polarization).
- As an additional tool to combat channel introduced signal distortions, EDGE deploys:
  1) A link adaptation mechanism that allows change to the coding scheme (related to a gross transmission rate) depending on channel condition and
  2) An ARQ protocol that indicates the TFI of erroneous radio blocks of user data to be retransmitted over the air interface.
- GSM supports intercell mobility with Mobile Assisted Handover (MAHO). MAHO means that a mobile station reports the downlink channel status from current and neighbour cells as well as acquiring system information from neighbour cells. The handover itself is commanded by the MSC via BSC and BS.

References

General Packet Radio Service (GPRS); Overall description of the GPRS radio interface; Stage 2; 3GPP TS 43.064 version 12.2.0 Release 12.

Sanders, G., Thorens, L., Reisky, M., Rulik O. and Deylitz, S., *GPRS Networks*, John Wiley & Sons, Ltd, 2003.

Halonen, T., Romero, J. and Melero, J., *GSM, GPRS, and EDGE Performance: Evolution Towards 3G/UMTS*, John Wiley & Sons, Ltd, 2003.

9

Third Generation Network (3G), UMTS

After GSM/EDGE, the third generation (3G) of mobile network system first appeared in 1999 under the name of the Universal Mobile Telecommunication System (UMTS) according to the European approach to 3G standardization. The 3GPP specification makes UMTS backwards compatible with GSM. In addition, GSM and UMTS networks are able to inter-operate between them. The UMTS system has an overall network structure similar to GSM, as shown in Figure 9.1.

The 3G system mobile terminal is called the User Equipment, UE. Physically most terminals are dual-mode multiband terminals capable of both 2G and 3G communications. The general term for the radio access part of the network is RAN (Radio Access Network) that refers to both 2G and 3G. The radio access for WCDMA UMTS is called Universal Terrestrial Radio Access (UTRA) or UTRAN. The base-station controller in UMTS is called the RNC (Radio Network Controller).

The switching system can be common to both GSM and UMTS. Nonetheless, in the lower layers within the UTRAN and the Core Network (CN), UMTS introduces a set of new protocols that require different GSM hardware, software and interfaces.

The important difference between UMTS and GSM is that UMTS separates the user plane from the control plane, the radio network from the transport network and the access network from the core network. Such separation between the radio subsystem and the network subsystem allows the network subsystem to be used with other Radio Access Technologies (RATs). The Core Network (CN) structure is adopted from GSM and consists of two-user traffic-dependent domains:

- circuit-switched traffic in the CS domain;
- packet-switched traffic in the PS domain.

Both traffic-dependent domains use the functions of the remaining network entities: the Home Location Register (HLR) together with the Authentication Centre (AuC) or the Equipment Identity Register (EIR) for subscriber management, mobile station roaming and identification and handling different services. Thus, the HLR contains GSM, GPRS and UMTS subscriber information.

Two domains handle their respective traffic types at the same time for both the GSM and the UMTS access networks. The CS domain handles all circuit-switched traffic for the GSM as well as for the UMTS access network; similarly, the PS domain takes care of all packet-switched traffic for both the access networks.

The UMTS also is composed of a modular structure separating the protocol stack and involved network nodes in terms of support for information flow and connectivity

Introduction to Mobile Network Engineering: GSM, 3G-WCDMA, LTE and the Road to 5G,
First Edition. Alexander Kukushkin.
© 2018 John Wiley & Sons Ltd. Published 2018 by John Wiley & Sons Ltd.

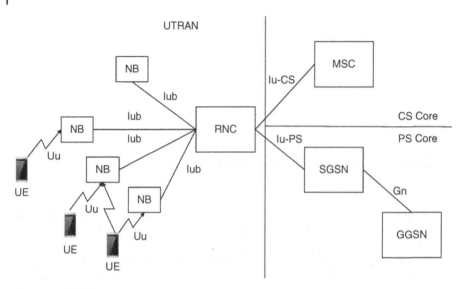

Figure 9.1 UMTS architecture.

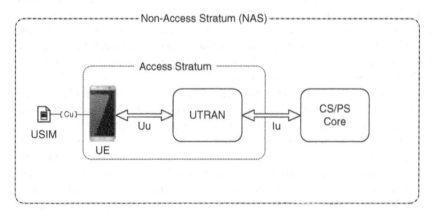

Figure 9.2 Modular functionality split in the UMTS.

and mobility functions. These modules define UMTS in another domain structure that includes Access Stratum (UE and UTRAN) and the Non-Access Stratum containing the USIM (Universal Subscriber Identity Module), serving the Core Network and Access Stratum, as shown in Figure 9.2.

With that definition, the Non-Access Stratum (NAS) can be described as a functional layer in the UMTS protocol stacks between the core network and User Equipment at application layer, while the Access Stratum is a functionality supporting the protocol stack between the mobile and radio access network. NAS supports dialogue between the mobile and core network Nodes (e.g. MSC, SGSN) that is passed transparently through the radio network. Functionalities supported by NAS include the following:

- Identity management
- Communication session establishment, maintaining and terminating.

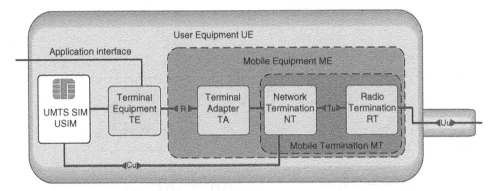

Figure 9.3 Modular architecture of UE [2].

- Call control
- Mobility management

The NAS protocol stack is specified in 3GPP TS 24.301 [1].

The User Equipment (UE) in the UMTS replaces the 2G Mobile Station (MS). The UE has a modular design that is composed of several elements, see Figure 9.3:

- *Mobile Termination* (MT) module terminates the radio interface in the UE.
- *Terminal Adapter* module that terminates application-specific protocols.
- *USIM* is a user subscription module containing all relevant user and network data to enable access to the subscribed network. In contrary to a GSM SIM, the USIM is downloadable, can be accessed via the air interface and can be modified by the network. The USIM is a Universal Integrated Circuit Card (UICC), which has much more capacity than a GSM SIM. It can store profiles containing user management and user rights and can also store Java applications.

9.1 The WCDMA Concept

The main radio technology deployed in the UMTS is WCDMA (Wideband Code Division Multiple-Access) whose variants FDD (Frequency Division Duplex) and Time-Division Duplex (TDD) were selected by the European Telecommunications Institute (ETSI) in 1998. Like 2G CDMA (IS-95), the spread spectrum forms the underlying technique for WCDMA. Compared with IS-95, implementation of spread spectrum techniques is different in 3G WCDMA, with different control channels and signalling, enhanced call control and link performance management.

Fundamental concepts utilized in the system WCDMA are as follows:

- Channelization and scrambling,
- Channel coding,
- Power control and
- Handover.

The channelization used in WCDMA is a spread spectrum technique that involves transmission of a radio signal over a frequency range much greater than the message

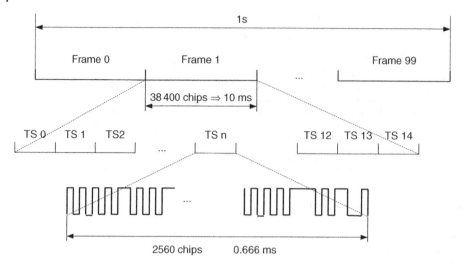

Figure 9.4 WCDMA timing arrangement.

bandwidth. Spreading of the signal spectrum in WCDMA is achieved by filling each information symbol with pseudonoise such as a spreading sequence of '0' and '1' (chips) at a much higher rate compared to the symbol rate. Though symbol rate is varied, the chip rate is always constant, 3.84 Mcps, thus resulting in a varied number of chips per symbol. The information is transmitted in time slots mapped to radio frames. The WCDMA air-interface time arrangement is shown in Figure 9.4. The spreading coder generates 3.84 Mega chips per second (Mcps). This chip stream is divided into 100 of 10 ms radio frames, each radio frame contains 15 slots leaving 2560 chips per time slot and 38 400 chips per radio frame.

9.1.1 Spreading (Channelization)

Figure 9.5 shows the basic operations of spreading and despreading for a WCDMA system. User data bits have a rate R assuming the values of ± 1. The spreading operation, in this example, is the multiplication of each user data bit with a sequence of four code bits called chips. The resulting spread data is at a rate of $4 \times R$ and has the same random (pseudonoise-like) appearance as the spreading code. In this case, we would say that we used a spreading factor (SF) of 4. This wideband signal would then be further processed and transmitted across a wireless channel to the receiving end.

During despreading, the spread user data/chip sequence is multiplied, symbol by symbol, with the very same four coded chip sequence used in the spreading process of these symbols. As shown, the original user symbol sequence is then recovered, provided there is synchronization between the spread signal and the (de)spreading code.

The increase of the signalling rate by a factor of 4 corresponds to a widening (by a factor of 4) of the occupied spectrum bandwidth of the spread user data signal. Due to this virtue, CDMA systems are more generally called spread spectrum systems. Despreading restores a bandwidth proportional to intended rate R for the signal. In addition to widening the signal spectrum, spreading sequence is coded in specific way thus introducing an *orthogonality factor* between different data streams (channels). Within the cell, the

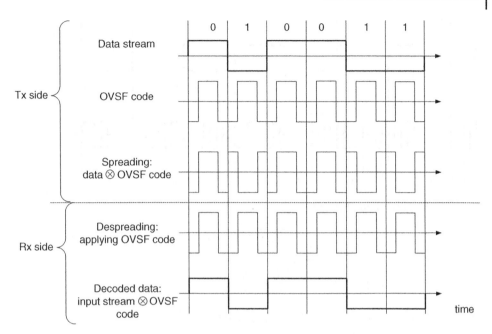

Figure 9.5 Combining data and spreading sequences with SF = 4.

different channels are separated by a *channelization code*, also called the *Orthogonal Variable Spreading Factor* (OVSF) code. The OVSF handles the signal spreading, as illustrated in Figure 9.5. The OVSF possesses two important features:

- an orthogonality of the codes with same length
- and the fact that orthogonality is conserved between OVSFs of variable lengths.

The OVSF orthogonality property ensures that different users of the same cell do not interfere with each other. If a signal coded with a given OVSF is decoded with a different OVSF, the resulting signal is an average null signal producing an equal number of 1s (−1) and 0s (+1).

The length of the OVSF (also called the *Spreading Factor*, SF) refers to the number of chips for a single input bit/symbol: a bit coded with OVSF length 256 would be represented by 256 chips, while a bit coded with OVSF length of 4 would be represented by four chips. Using a long OVSF has the advantage of adding redundancy to the transmitted information. The impact of this redundancy is seen in the spreading gain; that is, the ratio of user bits to transmitted chips.

The bit rate of the user signal represented in the chip sequence is related to the spreading factor. The orthogonal code families containing code sequence of different lengths are used to implement variable data rates. This is necessary even for a single UMTS user since a multibearer service could be available to a single user at the same time. Also, this is necessary to distinguish between control and traffic channels on the uplink from the same terminal.

The principle of implementation of different transmission rate is illustrated in Figure 9.6. This involves the 'multiplication' of every bit of the each data stream by a spreading code with a respective number of chips equal to the SF.

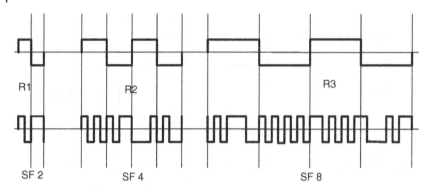

R1

R2

R3

SF 2

SF 4

SF 8

Figure 9.6 Spreading with OSVF codes.

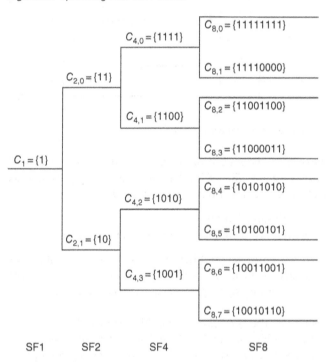

$C_{8,0} = \{11111111\}$

$C_{4,0} = \{1111\}$

$C_{8,1} = \{11110000\}$

$C_{2,0} = \{11\}$

$C_{8,2} = \{11001100\}$

$C_{4,1} = \{1100\}$

$C_{8,3} = \{11000011\}$

$C_1 = \{1\}$

$C_{8,4} = \{10101010\}$

$C_{4,2} = \{1010\}$

$C_{8,5} = \{10100101\}$

$C_{2,1} = \{10\}$

$C_{8,6} = \{10011001\}$

$C_{4,3} = \{1001\}$

$C_{8,7} = \{10010110\}$

SF1   SF2   SF4   SF8

Figure 9.7 OVSF code tree.

   The different bit streams that are to be transmitted simultaneously are multiplied by different OVSF codes and then added together. The receiver that receives the sum of all chip streams must be in position to reconstruct each of the transmitted bit streams. This is only possible if the code sequences of different chip streams are orthogonal to each other. The OVSF codes can be created through the use of code tree, as illustrated in Figure 9.7. Each node of the tree has exactly two branches, each representing a double-length code. The codes of the same level (of the same SF value) have same length. Each code with a spreading factor N is created from a code with spreading factor $N/2$.

Consequently, a set of 2^k spreading codes with a length of 2^k chips are available at the kth level. Availability of OVSFs of a specific length is determined by the number of OVSFs of same length or shorter that are used, as well as by the number of longer OVSFs used.

An OVSF code is basically created though multiplication of a code of the next lower level of a code tree. The code being multiplied is called the *mother code*. Exactly two next level double-length codes are created from a mother code by chaining the two copies of the mother code or chaining the mother code with a copy multiplied by -1. Based on the rules of creating the OVSF, the code of shorter length may be found in a longer one. This means that two codes of different levels of the code tree are orthogonal to each other as long as one of the two codes is not the mother code of other one. Because of this limitation, the number of simultaneously usable codes depends on the bit rate and spreading factor.

The definition for the same code tree means that for transmission from a single source, from either a terminal or a base station, one code tree is used (with one scrambling code on top of the tree). This means that different terminals and different base stations may operate their code trees totally independently of each other; there is no need to coordinate the code tree resource usage between different base stations or terminals.

9.1.2 Scrambling

The aggregate spread signal is then scrambled, as shown in Figure 9.8. The scrambling occurs by a chip-by-chip multiplication of the aggregate signals by a scrambling code. The scrambling code performs on the top of the spreading and, therefore, has the same rate as a chip rate. Scrambling codes are basically transmitter specific, each base station sector or mobile terminal has its own code. For a cell (sector), the downlink scrambling code, named a *Primary Scrambling Code* (PSC), is unique for every cell. The downlink scrambling facilitates an essential requirement to differentiate among different cells in order to establish and keep communication link between the mobile and base station. In the TDMA/FDMA system, this is achieved by using a different frequency for each cell. In the WCDMA system, cells are discriminated by using Primary Scrambling Codes.

The purpose of channelization and scrambling is apparently different. A channelization spreading code differentiates user data channels while scrambling code differentiates transmitters; that is, network cells on the downlink or user terminals on the uplink.

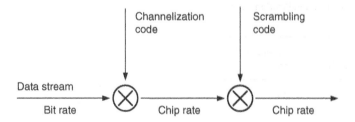

Figure 9.8 Spreading and scrambling for a single data stream.

9.1.3 Multiservice Capacity

The various user services (voice, packet services with different data rate and quality of service) have to be mapped to Radio Access Bearers (RABs), the latter could be regarded as the kind of traffic channels used in GSM, see Section 9.7. In addition, one user can have multiple services simultaneously. Multiple services share a single radio resource available in the transceiver (base station or mobile station); that is, an RF carrier. The resource sharing becomes possible by means of using channelization codes of different lengths depending on service requirements. The channelization (OVSF) code length directly relates to a data service bit rate. In WCDMA, RF channel conditions have rather little and indirect impact on the choice of OVSF length. The requirements for minimum spreading factor (SF) applied to specific service are provided in 3GPP recommendations 3GPP TR 25.993 [3].

Table 9.1 lists a sample of services and recommended OVSF lengths as well as the number of OVSFs available for each service in the cell for downlink. Apparently, the number of OVSF codes is limited. On the downlink, the number of OVSFs available for each dedicated channel is reduced, because some common control channels must be supported in the cell independently of traffic channels. Figure 9.9 summarizes the mandatory downlink common control channels and the mandatory values of their Spreading Factors. It also shows optional downlink common channels. On the uplink, each UE has its own code tree, so the code tree is not a limiting factor in that direction.

In the OVSF code tree structure, one PS 384 connection uses the same resources as four PS 64 connections or six voice connections. However, in terms of the SF, the probability of having $SF = 8$ free channels is not just four (or 16) times less than the probability of having one $SF = 32$ (or $SF = 128$) free, because the equivalent $SF = 32$ (or $SF = 128$) free channels must be contiguous and start at a specific position. Therefore, the availability of an OVSF of a specific length is determined by the number of OVSFs of same length or shorter that are used, as well as by the number of longer OVSFs used. The OVSF allocation algorithm at the NodeB normally manages the availability of consecutive OVSFs. This algorithm also allocates and optimizes the code tree to maximize the availability of shorter OVSFs.

When the system is allocating codes, each code reserves all the codes above and below the same branch as discussed in the previous section. As in the example in Figure 9.7, if the code $C_{8,3}$ is used, the codes above it ($C_{4,1}$, $C_{2,0}$, and C_1) and the codes below it

Table 9.1 Example of allocation of channelization codes in the cell [4].

Service	Minimum SF	Number of OVSF codes available
Voice, AMR	128	125
CS 64	32	31
PS 64	32	31
PS 128	16	15
PS 384	6	7

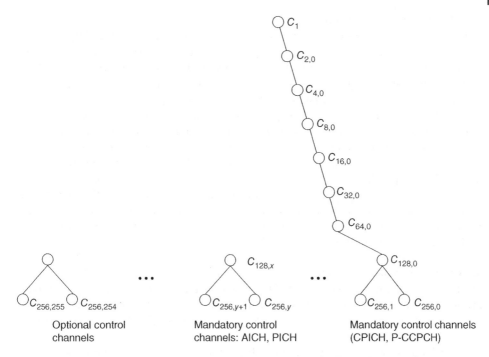

Figure 9.9 Channelization code tree example with control channels allocated in the cell.

($C_{16,6}$, $C_{16,7}$, $C_{32,12}$, $C_{32,13}$ etc.) cannot be used. In WCDMA, the RNC allocates the channelization codes and tries to optimize the usage of the codes.

9.1.4 Power Control

The signal variations (caused by fast fading) decrease with a respective increase in signal bandwidth. This phenomenon is due to the loss of coherence in different multipath components. This is the same effect as the frequency diversity that can be used when the system bandwidth becomes larger than the channel coherence (correlation) bandwidth. The power regulation is somewhat simplified when the spread bandwidth is much larger than the correlation bandwidth. This is applicable to a typical WCDMA system radio environment. Nonetheless, power control is instrumental for WCDMA operation and used basically to:

- estimate minimum transmit power (target Signal-to-Interference Ratio) for acceptable radio link performance and
- maintain the level of Signal-to-Interference Ratio during the call.

Power control is implemented at two levels of control speed, relatively slow (open and outer) and rather fast (closed loop). Altogether, three power control mechanisms are deployed in 3G WCDMA:

- Open-loop power control
- Outer-loop power control
- Closed-loop power control.

9.1.4.1 Open-Loop Power Control

The open-loop power control mechanism is applied in both downlink and uplink communications. In relation to the uplink, it provides a coarse initial power setting of the mobile station for the uplink channel. The Frequency Division Duplex (FDD) WCDMA system cannot use the signal level estimation of the downlink pilot channel received by a mobile, as, for instance, is possible in GSM. The duplex separation in WCDMA is too large to the extent that fading in the uplink and downlink become uncorrelated.

The open-loop power control function is located both in the mobile station and in the network (base station and radio network controller). In the uplink direction, it sets the initial power for the random-access channel and for the uplink traffic channel. In the downlink, the open-loop power control sets the power for the downlink channels. The power levels are based on the mobile station measurements reported to the base station in terms of frame reliability indicators (FRI) related, in turn, to block error rate (BLER). For each level of BLER of FRI there is a respective set point for Signal-to-Interference Ratio (SIR) or output power level.

9.1.4.2 Outer-Loop Power Control

The radio link performance is impacted by mobile speed, slow shadowing and multipath profile. This means that the target SIR may need to be adjusted over time as the mobile environment changes. The changes in SIR target are controlled by outer-loop power control, as shown in Figure 9.10.

Outer-loop power control function resides in the Radio Network Controller (RNC). The RNC monitors the received quality at the mobile station and the base station. If there is a difference in the BLER and the targeted BLER with current SIR, the RNC may change the respective SIR target for either the MS or BS. An additional reason for placement of outer power control in the RNC is a soft handover in WCDMA. The outer control is always performed after soft handover combining in RNC.

Closed-Loop Power Control Closed-loop power control keeps the received signal level of the uplink signal from every mobile at the same specified level in order to minimize the interference level within the cell. Figure 9.11 shows an example of uplink closed-loop

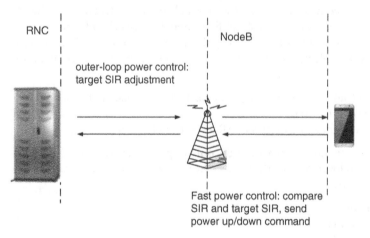

Figure 9.10 Outer-loop power control.

power control. Mobile stations UE 1 and UE 2 operate at the same carrier frequency, separable at the base station only by their respective spreading codes thus producing a noise like interference to each other. Apparently, the path loss for UE 2 is significantly higher than for UE 1. In the absence of a power control mechanism, the uplink signal from UE 1 can easily block reception of the signal from UE 1. This blocking is known as the *near-far problem* in CDMA systems. The optimum strategy to ensure minimum interference and maximum possible capacity is to continuously equalize the received power to *Pmin* of all mobile stations, see Figure 9.11.

With the closed-loop power control in the uplink, the base station performs frequent estimates of the received Signal-to-Interference Ratio (SIR) and compares it to a target SIR. If the measured SIR is higher than the target SIR, the base station will command the mobile station to lower the power or increase the power in opposite case. The measurement cycle is executed at a rate of 1.5 kHz for each mobile station and hence operates faster than the speed of fast Rayleigh fading in car or the pedestrian environment. The power control can be implemented in steps of 1 or 2 dB.

Figure 9.12 illustrates how uplink closed-loop power control should work on a fading channel at low speed. Based on measured SIR, the base station sends a closed-loop power control command to the mobile station to reduce/increase power proportional to the inverse of the SIR. In a nearly perfect situation, the received signal in the base station is essentially non-fading, as shown in the upper section of Figure 9.12.

The positive effect of increased transmit power for mitigating the fading at the cell edge or in a deep fade may produce a negative effect of increased inter-cell interference [5]. This means that the maximum output power of the mobile should also be optimized in order to minimize inter-cell interference. The same closed-loop power control technique is also used on the downlink. In addition to the task of keeping reasonable link performance around the SIR target, the base station tends to keep the overall transmit power to a minimum in order to create some marginal reserve in power of the RF carrier to ensure a trade-off between serving the mobile stations at the cell edge and reducing inter-cell interference.

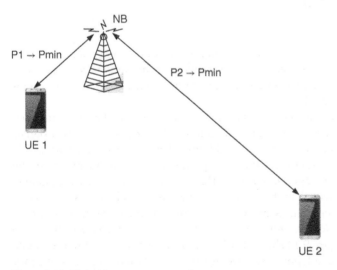

Figure 9.11 Closed-loop power control.

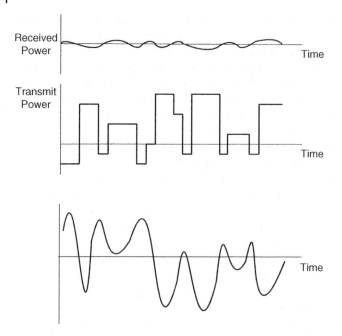

Figure 9.12 Closed-loop power control compensates for a fading channel.

9.1.5 Handover

In WCDMA there are several types of handovers: softer handover, soft handover, inter-frequency handover and intersystem handover. Handovers support mobility and performed when the mobile station moves from the coverage area of one cell to the coverage area of another cell. The intersystem handover (ISHO) is needed when the mobile operator uses both WCDMA and GSM networks together. Another name for such a handover is inter-RAT handover.

9.1.5.1 Softer Handover

Softer handover takes place when a mobile station moves into an overlapping area of radio coverage from two adjacent sectors (base stations) of the same base station site. Two air-interface links between the same mobile and different two sectors are established in this case, see Figure 9.13.

A mobile station can distinguish downlink signals from two base stations since each sector uses a different scrambling code and different OVSF code tree. The two signals are received in the mobile station by means of additional RAKE processing, with different respective codes for each sector for the appropriate despreading operation. In the uplink direction, a similar process takes place at the base station site: the code channel of the mobile station is received in each sector, then routed to a baseband RAKE receiver and finally a maximal ratio combining mechanism is applied. The softer handover processing is confined to the NodeB. *During softer handover only one power control loop per connection is active.* Figure 9.14 shows a soft handover. In a soft handover a mobile station is positioned in the overlapping RF coverage area from two sectors. But contrary to softer handovers, those two sectors belong to different NodeBs, respectively.

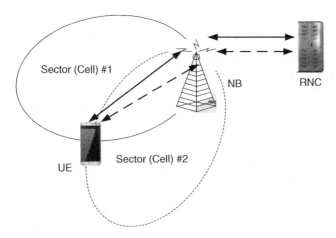

Figure 9.13 Softer handover.

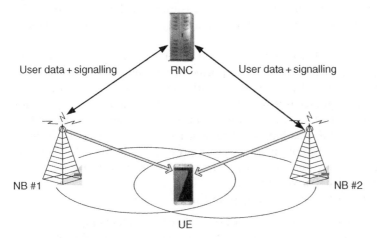

Figure 9.14 Soft handover.

From the mobile station receiver point-of-view there is no principal difference in signal processing compared with a softer handover. As is the case for the softer handover, the downlink communication between mobile station and base station takes place concurrently via two air-interface channels from each base station with separate scrambling and channelization codes. Both channels are received at the mobile station using a maximal ratio combining RAKE processing.

However, the uplink in a soft handover is drastically different: the uplink signal from MS is received by two NodeBs, then two data streams are routed to the RNC for combining. The attached frame reliability indicator as provided for outer loop power control is used to select the better frame between the two possible candidates within the RNC. This selection takes place after each interleaving period; that is, every 10–80 ms. During soft handover *two outer power control loops per connection are active, one for each base station.*

In a general case, both soft and softer handover can take place in combination with each other.

9.1.5.2   Other Handovers

In addition to soft/softer handover, WCDMA supports hard handover types:

- Inter-frequency hard handovers that used to hand a mobile over from one WCDMA frequency carrier to another. These may take place inside the same NodeB with several RF carriers or between different NodeBs.
- Inter-System Hard Handovers (ISHO) that take place between the WCDMA FDD system and another system, such as WCDMA TDD or GSM.

9.1.5.3   Compressed Mode

There is an important difference between CDMA and TDMA systems related to handover measurements. In TDMA the terminal is only active during one slot in a frame. Then, during the rest of the slots, the receiver can be tuned to other frequencies and perform measurements at the neighbour frequencies. With CDMA, the UE transmits and receives continuously during all the frames and slots. The intra-frequency neighbours can be measured simultaneously with normal transmission by the UE using a RAKE receiver. Inter-frequency and inter-system measurements, however, require the UE to measure on a different frequency, which is managed using the *compressed mode* of operation.

In compressed mode, the UE monitors cells on other FDD frequencies or other Radio Access Technologies (RAT) that are supported in multimode UE, such as TDD WCDMA and GSM. In principle, it is possible with multiple receivers incorporated into UE. The alternative approach is to stop the normal transmission and reception for a certain time period enabling the UE to measure on the other frequency. To achieve this gap and not lose any information, the data sent have to be *compressed in time*, that is, the transmission and reception enter in Compressed Mode. The RNC controls the compressed mode parameters in which frames are compressed and sends the information both to the NodeB and to the UE.

There are three methods to generate the gaps in compressed mode by:

1) reducing the data rate used in the upper layers (higher-layer scheduling),
2) reducing the symbol rate used in the physical layer (rate matching and/or puncturing),
3) spreading factor splitting (halving the spreading factor doubles the available symbol rate).

Figure 9.15 shows the spreading factor reduction approach meaning that if the channelization code C_{SF} is used in normal frames, then the code $C_{SF/2}$ is used in compressed frames.

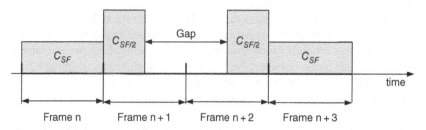

Figure 9.15 Discontinuous transmission during compressed mode.

In this way, the instantaneous bit rate is doubled in certain slots while other slots are idle. The average bit rate over radio frame is kept unchanged. The reduction in the spreading factor requires a higher power in the slots carrying the data with a doubled bit rate in order to maintain coverage and QoS. Such an increase in output power may cause noise rise to other terminals.

9.1.6 RAKE Reception

The multipath propagation channel in WCDMA system can be regarded as a wideband channel; that is, a channel with frequency selective fading and inter-symbol interference. Considering relative parameters of the system and propagation channel, one may note the following:

The chip duration at 3.84 Mcps is 0.26 µs. On the other hand, the delay spread in urban multipath channel is about $1 \div 2$ µs that exceeds the chip duration. All multipath components with a time difference no less than chip duration, that is 0.26 µs, can be resolved in the WCDMA receiver. Using the orthogonality of the chip sequence, the receiver should be able to separate those multipath components and combine them coherently to obtain multipath diversity. This could be possible only if multipath components are uncorrelated when their relative propagation delay exceeds a chip period.

The WCDMA uses RAKE receiver that was patented by Robert Price and Paul Green in 1956. The block diagram of RAKE receiver is shown in Figure 9.16. The receiver employs a code tracking receiver and several digital data receivers/correlators that act as fingers of a RAKE separating and tracking particular multipath components. The basic operation principle is as follows:

1) Capture strongest multipath components by identifying the time-delay positions at which significant energy arrives and allocate correlation receivers, that is RAKE fingers, to those peaks. As a result, each correlator detects a time-shifted version of the original WCDMA transmission and each finger of the RAKE correlates to a portion of the signal, which is delayed by at least one chip in time from the other fingers.
2) Within each correlation receiver, estimate characteristics of time-variant channel using the pilot symbols available via control channels transmitted by the cell

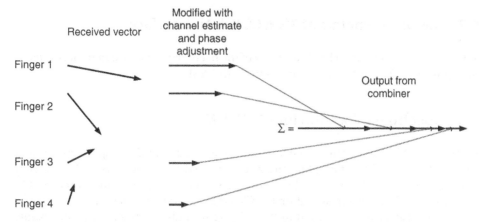

Figure 9.16 Channel estimation and recovery of the multipath component of the signal.

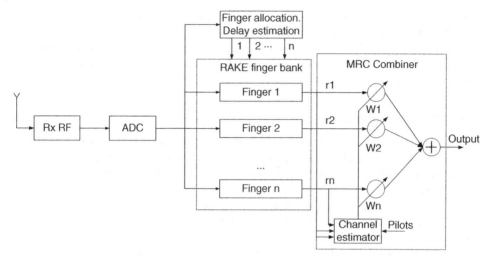

Figure 9.17 RAKE receiver with MRC [7].

transceivers. The estimated channel response is then used to remove distortion in amplitude and phase of each replica of received signal. The estimation is done separately for each finger. The principle is illustrated in Figure 9.16.

3) The symbols from allocated fingers are maximal-ratio-combined to construct the 'combined' symbol. The output symbols from different fingers are multiplied with complex conjugate of the channel estimate and the result of multiplication is summed together into the 'combined' symbol. As shown in Figure 9.17, the receiver uses known pilot symbols that are used to sound the channel and provide an instant estimate of the channel state (value of the weighted phasor) for a particular finger. Then the received symbol is rotated back, so as to undo the phase rotation caused by the channel. Such channel-compensated symbols can then be simply summed together to recover the energy across all delay positions. This processing is called Maximal Ratio Combining (MRC) [6].

9.2 Major Parameters of 3G WCDMA Air Interface

Tables 9.2 and 9.3 provide a brief summary of the major parameters in radio access technology of WCDMA and some comparisons with GSM.

9.3 Spectrum Allocation for 3G WCDMA

Most of the WCDMA deployments use the identified IMT-2000 spectrum around 2 GHz: 1920–1980 MHz for uplink and 2110–2170 MHz for downlink. This spectrum is a major band in IMT-2000 used in Europe, Asia (including Japan and Korea) and in Brazil. The 3GPP spectrum allocation for WCDMA is given in Table 9.4.

WCDMA can also be deployed in the existing second-generation frequency bands initially used by GSM or CDMA, provided it is solid and sufficient for the 3G carrier

Table 9.2 Main parameters of WCDMA.

Access method	Direct Sequence CDMA
Duplex method	FDD/TDD
Synchronization in radio access	Asynchronous operation
Chip rate	3.84 Mcps
Frame length	10 ms
Multiservice capability	Multiple services with different QoS in concurrent operation
Detection method	Coherent detection using pilot symbols

Table 9.3 Comparison of GSM and WCDMA in air-interface technology.

	GSM	WCDMA
Carrier separation	200 kHz	5 MHz
Frequency reuse factor	1–18	1
Power control rate	2 Hz	1.5 kHz
Frequency diversity method	Frequency hopping	5 MHz wide bandwidth ensures frequency selective fading
Downlink transmit diversity	Supported in standards	Supported

Table 9.4 Spectrum allocation for 3G FDD WCDMA.

FDD bands	Uplink (MHz)	Downlink (MHz)
I	1920–1980	2110–2170
II	1850–1910	1930–1980
III	1710–1785	1805–1880
IV	1710–1755	2110–2155
V	824–849	869–894
VI	830–840	875–885
VII	2500–2570	2620–2690
VIII	880–915	925–960
IX	1749.9–1784.9	1844.9–1879.9
X	1710–1770	2110–2170

accommodation block of the spectrum. That approach is called re-farming. WCDMA deployment in the USA started by re-farming WCDMA to the existing cellular bands at 850 MHz and to the PCS band at 1900 MHz, since there were no new frequencies available for WCDMA deployment.

9.4 3G Services

The GSM system is efficient in delivery of voice services over a circuit switch network. The design objective of UMTS is a delivery of any type of service with defined quality alone or in combination to the user over all IP networks, at least starting from Release 5 of UMTS. To meet the challenge, WCDMA air-interface technology has to bring advanced capabilities:

- High bit rates, initially up to 2 Mbps, and beyond 10 Mbps later releases of the system;
- Packet round trip times less than 200 ms;
- Seamless mobility also for packet data applications;
- QoS differentiation across the whole network;
- Simultaneous voice and data capability;
- Interworking with existing GSM/GPRS networks.

9.4.1 Bearer Service and QoS

There is a one-to-one relation between Bearer Services and QoS in UMTS networks. Other than in 2G systems where a bearer was a traffic channel, in 3G the bearer represents a selected QoS for a specific service. Only from the point-of-view of the physical layer can the bearer be regarded as a type of channel. A Bearer Service is a service that guarantees a QoS between two endpoints of communication. Several parameters will have to be defined from operators. A Bearer Service is classified by a set of values for these parameters:

- Traffic class.
- Maximum bit-rate.
- Guaranteed bit-rate.
- Delivery order.
- Maximum SDU (Service Data Unit) size.
- SDU format information.
- SDU error ratio.
- Residual bit-error ratio.
- Delivery of erroneous SDUs.
- Transfer delay.
- Traffic handling priority.
- Allocation/retention priority.

From a user point-of-view, services (applications, such as speech, web browsing, video streaming etc.) are considered end-to-end; that is, from a Terminal Equipment (TE) to another TE. An End-to-End Service may have a certain QoS that is provided for the user through the different networks. In UMTS, it is the UMTS Bearer Service that provides the requested QoS through the use of different QoS classes as defined in 3GPP TS 23.107 [8]. Basic characteristics for main traffic classes are explained in Table 9.5 [3].

The UMTS Bearer Service consists of two parts; the Radio Access Bearer (RAB) Service and the Core Network Bearer Service. The Radio Access Bearer Service is realized by a Radio Bearer (RB) Service and an Iu-Bearer Service. The relationship between the services is illustrated in Figure 9.18.

Table 9.5 Traffic classes.

Traffic class	Characteristics	Example of application
Conversational (real time)	Conversational pattern (stringent, low delay)	speech, video, games…
Streaming (real time)	Preserve time relation between information entities of the stream (synchronize video and speech)	streaming audio and video
Interactive (best effort	Preserve response pattern. Preserve payload content	web browsing
Background (best effort)	Preserve payload content. No restriction on latency	background download, email

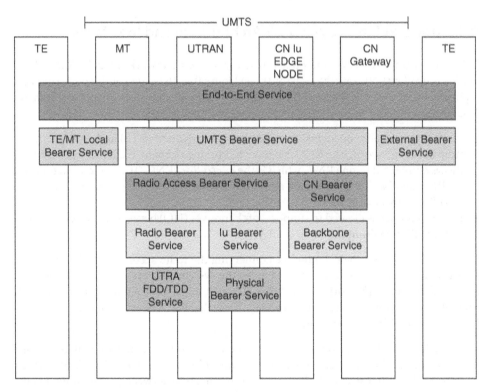

Figure 9.18 UMTS QoS architecture [8].

The layered architecture of a UMTS bearer service is depicted in Figure 9.18; each bearer service on a specific layer offers its individual services using those provided by the layers below. The QoS parameters are given by the core network to the radio network in a radio access bearer setup. Table 9.6 shows an example of mapping the service (application) belonging to a certain QoS traffic class to the Radio Access Bearer at the air interface. The RAB is specified by the CS/PS domain and UL/DL user data rates.

Table 9.6 Mapping traffic class to RAB.

Applications	Domain	Traffic class	RAB (UL/DL) kbps
speech	CS	conversational	CS AMR 12.2/12.2 or HSPA
speech (VoIP)	PS	streaming/interactive	R99/HSDPA
video telephony	CS	conversational	CS UDI 64/64
audio streaming	PS	streaming	PS S 16/64 + PS I/B 8/8
mobile TV	PS	streaming	PS S 16/128 + PS I/B 8/8
web browsing	PS	interactive/background	PS I/B 64/384
file download		interactive/background	PS I/B 64/384
online ticketing	PS	interactive	PS I/B 64/64
Email	PS	interactive/background	PS I/B 64/384

9.5 UMTS Reference Network Architecture and Interfaces

WCDMA introduces two new nodes in Radio Access Network, namely, Radio Network Controller (RNC) and NodeB (NB), see Figure 9.1. These two nodes perform tasks equivalent to the GSM's BSC and BTS, respectively. Given new network protocols in 3G WCDMA, a number of new interfaces are defined between new network nodes:

- The Iub interface is the logical interface between aNodeB (BTS) and an RNC in the radio access network. The RNC is connected to one or many NodeBs. The NodeB is always connected to one RNC, which is the *controlling* RNC. The Iub interface allows the RNC and NodeB to negotiate about radio resources and to transport uplink and downlink transport frames.
- The Iur interface is created to support soft handover (HO) between RNCs connecting multiple RNCs within the same UTRAN.

The Iu interface is a logical interface connecting the radio access network to the core network. Iu is an open interface that divides the system into the radio access network and core network. The core network handles switching, routing and service control. The interface is implemented in accordance with the 3GPP interface specifications. Iu has two major different instances:

1) Iu-CS between the radio access network and the circuit-switched domain in the core network. The interface carries the communication between an RNC, MSC Server and an MGW. It also carries the direct communication between user equipment and the MSC, which is transparent to the RNC.
2) Iu-PS between the radio access network and the packet-switched domain in the core network. The interface carries the communication between an RNC and an SGSN. It also carries the direct communication between user equipment and the SGSN, which is transparent to the RNC.

The main functions of the Iu interface are:

- establishing, maintaining and releasing radio access bearers

- performing intra-system and inter-system handovers as well as serving RNC relocations
- general procedures, not related to specific user equipment, such as reporting general errors
- separating each UE on the protocol level for user-specific signalling management.

In practical networks, there are other interfaces in the operator network used for value added services and network management.

9.5.1 The NodeB (Base Station) Functions in the 3G Network

The main function of the NodeB is to perform the air-interface physical layer processing (channel coding and interleaving, rate adaptation, spreading etc.). It also performs some basic Radio Resource Management (RRM) operations such as the closed (inner) loop power control. It logically corresponds to the GSM Base Station.

9.5.2 Role of the RNC in 3G Network

The RNC is responsible for the following:

1) Call admission control. The RNC must calculate the current traffic load for each individual cell. On the basis of this information, Call Admission Control (CAC) functionality then decides whether the interference level is acceptable and, if necessary, reject the call.
2) RRM: The RNC manages the radio resources in all attached cells. This includes calculating interference and utilization levels and priority control.
3) Radio bearer setup and release. The setup of radio bearer is about the establishment of logical data connection and does not indicate whether PS or CS data transmitted over the radio bearer.
4) Code allocation. The RNC allocate part of code tree to each mobile station and can also change the allocation during the session.
5) Power control. The actual fast power control takes place in NodeB but target control values are established in the RNC.
6) Packet scheduling. Several mobile stations share the same resource. The RNC cyclically allocates transmission capacity to the individual MS, at the same time taking into account negotiated QoS.
7) Handover management. Based on the measurements supplied by NodeB and MS, the RNC decides on a handover, takes responsibility of signalling with new cell and informs the MS about the new channel.
8) Encryption of CS services. The mobile terminated call is encrypted in RNC before transmitting over the air interface.

In order to support soft handover, three types of logical RNC are defined in the 3G network in relation to the link between NodeB and RNC. The first is a Controlling RNC (CRNC) that terminates the Iub interface towards NodeB. The CRNC actually performs the load and congestion control of all connected cells, and also executes the admission control and code allocation for new radio links to be established in those cells. Figure 9.19 illustrates the concept of Serving and Drift RNC.

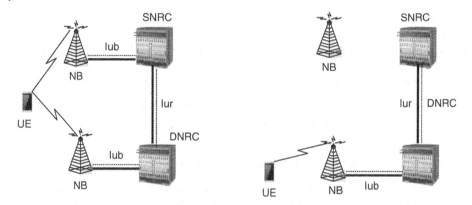

Figure 9.19 Logical role of the RNC for one UE UTRAN connection.

In case one mobile–UTRAN connection uses resources from more than one RNS, the RNCs involved have two separate logical roles (with respect to this mobile–UTRAN connection):

- Serving RNC (SRNC). The SRNC for one mobile is the RNC that terminates both the Iu link for the transport of user data and the corresponding signalling to/from the core network. The SRNC also terminates the Radio Resource Control (RRC) Signalling; that is, the signalling protocol between the UE and UTRAN.
- Basic RRM operations, such as the mapping of Radio Access Bearer parameters into air-interface transport channel parameters, the handover decision and outer loop power control, are executed in the SRNC. The SRNC may also (but not always) be the CRNC of some NodeB used by the mobile for connection with UTRAN. One UE connected to the UTRAN has one and only one SRNC.
- Drift RNC (DRNC). The DRNC is any RNC, other than the SRNC, that controls cells used by the mobile. If needed, the DRNC may perform macrodiversity combining and splitting. The DRNC routes the data transparently between the Iub and Iur interfaces. One UE may have zero, one or more DRNCs.

The left-hand scenario in Figure 9.19 shows a case when UE is in the inter-RNC soft handover (combining is performed in the SRNC) stage. The right-hand scenario represents one UE connected to one NodeB only, controlled by the DRNC. The solid and dash lines correspond to user and control plane connections, respectively. Note that the physical RNC contains all the CRNC, SRNC and DRNC functionalities.

9.6 Air-Interface Architecture and Processing

The Uu interface is the air interface between a mobile station and a radio access network. The Uu interface consists of three layers:

- Physical layer (Layer 1)
- Data-link layer (Layer 2) contains sublayers
  - Medium access control (MAC)
  - Radio link control (RLC)
- Network layer (Layer 3)

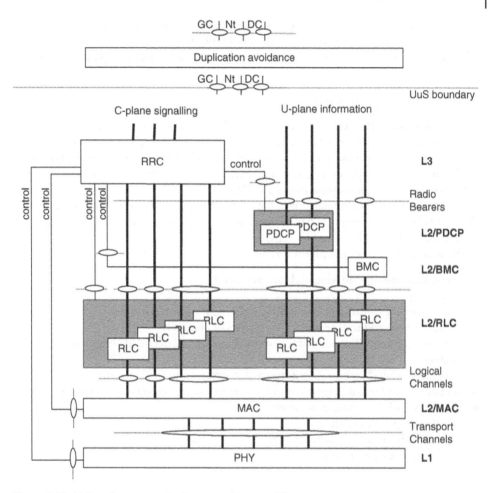

Figure 9.20 Air-interface protocol reference architecture [9].

The protocol architecture is also divided into the control plane responsible for transmission of signalling information and the user plane with dedicated user data transfer. Figure 9.20 shows the reference architecture for a radio interface protocol stack.

The control plane includes application protocols and signalling bearers for transporting the application protocol messages. The application protocols include:

- Radio resource control (RRC)
- Radio link control (RLC)
- Medium access control (MAC)

The user plane consists of transport channels and physical channels allocated for the connection and the data carried over the connection. Essential protocols are:

- Packet data convergence protocol (PDCP), which is used with the packet data.
- Broadcast/multicast control (BMC), which is used for broadcast services.
- Radio link control (RLC).
- Medium access control (MAC).

The layered structure of air interface implies message transfer services between layers. Such a transfer of information requires the definition of adequate interfaces between adjacent layers specifying the path that information follows depending on its nature.

9.6.1 Physical Layer (Layer 1)

The physical layer offers transport channels to the higher layers. The physical layer supports information transfer to the MAC and higher layers through transport channels. A radio link is a bi-directional connection between a mobile station and a base station cell. Each radio link is comprised of one or more traffic channels that are associated with the same physical layer control channel (DPCCH).

The physical layer is controlled by a data-link layer that has two sublayers: MAC and RLC. MAC realizes transport channel management and RLC realizes flow control. RRC manages the physical layer and its activities.

The main functions of the physical layer are:

- channel coding, interleaving and rate matching
- measuring
- macrodiversity splitting or combining and executing soft handovers
- handling cyclic redundancy check on transport channels
- multiplexing/demultiplexing of transport channels and of coded composite transport channels
- mapping of coded composite transport channels on physical channels
- modulating and spreading, or demodulating and despreading of physical channels
- frequency and time synchronization
- closed-loop power control
- power weighting and combining of physical channels
- radio frequency processing

9.6.2 Medium Access Control (MAC) on Layer 2

The MAC layer maps the logical channels to the transport channels. The layer also selects an appropriate transport format for each transport channel depending on the instantaneous source rates of the logical channels. The transport format is selected with respect to a *transport format combination set* (TFCS) that is defined by the admission control for each connection. Some of the main functions of MAC are:

- transferring data and providing unacknowledged transfer of MAC Service Data Units (SDU) between peer MAC entities
- reporting measurements: local measurements like traffic volume, quality indication and so on to RRC
- selecting appropriate transport format from the TFCS for each transport channel depending on instantaneous source rate
- priority handling, between data flows of a MS and between users, by means of dynamic scheduling
- multiplexing or demultiplexing higher-layer *protocol data units* (PDU) into/from transport blocks delivered to/from the physical layer on common transport channels

- multiplexing or demultiplexing higher-layer PDUs into/from transport-block sets delivered to/from the physical layer on dedicated transport channels
- mapping between logical channels and transport channels, monitoring traffic volume and dynamic transport channel type switching.

9.6.3   Radio Link Control (RLC) on Layer 2

The RLC protocol provides segmentation and retransmission services for both the user and control data to the higher layers. In the control plane, the RLC provides a service called the *signalling* radio bearer. In the user plane to circuit-switched core network, the RLC provides a service called the radio bearer (PDCP protocol is not used by the service). In the user plane to the packet-switched core network, PDCP above the RLC provides the radio bearer.

Each radio link control instance is configured by the RRC to operate in one of three modes:

1) Transparent mode (TM): no protocol overhead is added to higher-layer data and erroneous data can be discarded or marked.
2) Unacknowledged mode (UM): no retransmission protocol is used or data delivery guaranteed. Erroneous data is discarded or marked.
3) Acknowledged mode (AM): automatic repeat request is used for error correction, notification is sent in case data is not delivered to the destination.

The main functions of RLC are:

- establishing and releasing data-link layer connections
- segmenting and reassembling packet data units, concatenation and padding
- transferring user data, which is controlled by QoS setting
- error correction, detecting and recovering protocol errors
- sequenced delivery of higher-layer packet data units, controlling data flow
- ciphering.

9.6.4   RRC on Layer 3 in the Control Plane

Control signalling consists mostly of RRC messages in the Uu interface. The control interfaces between the RRC and all the lower layer protocols are used for:

- configuring characteristics of the lower layer protocol entities, including parameters for the physical, transport and logical channels
- reporting measurement results and errors to the RRC.

The RRC layer handles the main part of control signalling between the Mobile Switching System and RAN. The main functions of RRC are:

- broadcasting system information
- paging
- initial cell selection and reselection in idle mode based on system information broadcasted by RNC and on measurements from mobile station (MS)
- establishing, maintaining and releasing RRC connections between mobile station and radio access network

- controlling radio bearers, transport channels and physical channels
- controlling security functions (ciphering and integrity protection)
- RRC connection mobility functions, such as cell or UTRAN registration area (URA) updates and handovers
- supporting SRNS relocation (information to the target SRNS)
- performing downlink outer loop power control and open-loop power control
- higher-layer signalling with core network (direct transfer).

9.7   Channels on the Air Interface

Any service provided to the MS user (e.g. a voice service, web browsing service etc.) relates to a Radio Bearer that specifies the configuration and the parameters of RLC, MAC and the physical layer depending on the characteristics of the service being provided. The information flow associated to a radio bearer (RAB) is mapped into different types of channels at different protocol layers, respectively. Those types are the logical, transport and physical channels. The respective position of different type of channels to a layer in the air-interface protocol stack is shown later in Section 9.8.1 (and Figure 9.23 later).

9.7.1   Logical Channels

Logical channels created at the RLC layer and then mapped onto transport channels in the MAC layer. There are two types of logical channels providing bearers for control plane and user plane, namely:

- Control channels for signalling information and
- Traffic channels to carry user data.

Table 9.7 lists both traffic and control logical channels in UTRA-FDD.

9.7.2   Transport Channels

Compared with GSM, the UMTS transport channel is a new type of channel. This concept is introduced to accommodate different but concurrent services (with different transmission constraints) at the common radio interface. Transport channels are defined between MAC and PHY layers and they specify how the information from logical channels is adapted to the radio transmission medium. Transport channels define the transmission parameters, such as channel coding, interleaving or bit rate. Contrary to logical channels that differentiate based on control or traffic information type, the transport channel does not take into account the nature transmitted information. As a consequence, the transport channel may carry information from different logical channels if their transmission parameters are similar.

Two classes of transport channels are defined in UMTS; namely, dedicated and common transport channels.

Table 9.7 UTRA logical channels.

Control logical channels	Use	Direction
BCCH (Broadcast Control Channel)	Cell broadcast of System Information message that contains cell identifier, code sequences, timers etc.	DL
PCCH (Paging Control Channel)	Notify the users in location area of incoming calls or other messages.	DL
DCCH (Dedicated Control Channel)	Bi-directional signalling channel for every user with RRC connection to RNC. It transmits measurements report, RRC control messages.	DL/UL
CCCH (Common Control Channel)	Connection set up, channel allocation, cell reselection	UL and DL
Traffic logical channels		
DTCH (Dedicated Traffic Channel)	Transfers information for specific service dedicated to a single user. One user may have multiple services provided on different co-existing DTCHs simultaneously.	DL and UL
CTCH (Common Traffic Channel)	Point to multipoint channel carrying information for group of users, such as SMS cell broadcast message.	DL

9.7.2.1 Dedicated Transport Channel (DCH)

DCHs transport user data and control information coming from the upper layers for a single user. This information may include speech frames or packet data, as well as upper layer control data such as handover commands or measurement reports. There is no need for a UE identifying parameter.

One user may have multiple services carrier by multiple DCHs simultaneously. Even in the case of single service, two DCHs have to be allocated: one for transfer of a traffic logical channel (DTCH) and another, DCCH, for a signalling associated with traffic channel. The physical resource allocated to DCH is available for the whole duration of the call.

9.7.2.2 Common Transport Channels

Common transport channels are shared between users in the cell. When a message to a specific user is transmitted over the common transport channel, it should contain explicit UE identification. Table 9.8 lists common transport channels.

When a *single user is allocated multiple transport channels*, then all channels are multiplexed together over the same physical resources forming a Coded Composite Transport Channel (CCTrCH). One physical control channel and one or more physical data channels form a single Coded Composite Transport Channel. There can be more than one CCTrCH on a given connection but only one physical-layer control channel is transmitted in such a case.

Table 9.8 Common transport channels.

Common transport channels	Use	Direction
BCH (Broadcast Channel)	Provides transport for the BCCH logical channel.	DL
PCH (Paging Channel)	Provides transport for the PCCH logical channel	DL
RACH (Random Access Channel)	Short signalling information during the initial access to the system, before traffic channel can be allocated	UL
FACH (Forward Access Channel)	Carries a logical channel for a specific UE. Response to RACH.	DL
CPCH (Common Packet Channel)	Extension to the RACH channel for the transmission of longer data packets.	UL
DSCH (Downlink Shared Channel)	Pool of physical resources allocated on a TTI basis to different users according to packet scheduling policy, similar to GPRS.	DL
	The users that transmits through the DSCH channel must also have an associated bi-directional DCH channel through which control information is sent. This control information indicates the TTI when the DSCH is allocated to the specific UE as well as power control commands for closed-loop power control for that UE.	

9.7.3 Physical Channels and Physical Signals

Physical channels are defined by a combination of carrier frequency, scrambling code, channelization code, time start and stop (giving a duration) and, on the uplink, relative phase (0 or $\pi/2$).

Time durations are defined by start and stop instants and measured in integer multiples of chips. The main time reference is the radio frame with a length of 38 400 chips and 10 ms in duration. The radio frame processing duration consists of 15 slots of 0.666 ms each, with 2560 chips/slot. In contrary to GSM or any other TDMA system, the time slot structure has nothing to do with separation of the users in the time domain. Instead it defines a closed-loop power control cycle for a transmission to a given user with rate 15 periods over the radio frame; that is, $15/10$ ms $= 1500$ Hz.

Physical signals are entities with the same basic on-air attributes as physical channels but do not have transport channels or indicators mapped to them. Physical signals may be associated with physical channels in order to support the function of physical channels. As in the case of transport channels, physical channels are classified as dedicated and common physical channels. Table 9.9 lists some of the physical channels in UTRA-FDD.

9.7.4 Parameters of the Transport Channel

Multiplexing of transport channels onto physical channel is performed at the physical layer. Appropriate configuration of the physical channel is defined by a set of channel parameters, TTI, TB, TF and TFC, listed in Table 9.10.

Table 9.9 Different types of physical channels in UTRA-FDD.

Name	Use
Dedicated Physical Data Channel (DPDCH)	Transmission of user data and higher layer signalling (RRC, NAS) in the uplink direction coming from higher layers.
Dedicated Physical Control Channel (DPCCH)	Transmission of radio control information in uplink direction. This channel exists only once per radio connection.
Dedicated Physical Channel (DPCH)	Transmission of user data and control information in downlink direction. Both types of information will be mapped onto the DPCH
Synchronization Channel (SCH)	Cell search and synchronization of the UE to the Node B signal. Subdivided into Primary Synchronization Channel (P-SCH) and Secondary Synchronization Channel (S-SCH)
Common Control Physical Channel (CCPCH)	Transmission of common information and is divided into Primary Common Control Physical Channel (P-CCPCH) and Secondary Common Control Physical Channel (S-CCPCH). P-CCPCH transmits the broadcast channel (BCH) and S-CCPCH transports the Forward Access Channel (FACH) and the Paging Channel (PCH). FACHs and PCH can be mapped to the same or to separate S-CCPCHs.
Common Pilot Channel (CPICH)	Supports channel estimation and allows estimations in terms of power control. It is subdivided into Primary Common Pilot Channel (P-CPICH) and Secondary Common Pilot Channel (S-CPICH), which differ in scrambling code and availability within a cell.
Physical Random Access Channel (PRACH)	Transmission of the Random Access Channel (RACH), which is used for the random access of UE and for transmission of a small amount of data in the uplink direction.
Physical Common Packet Channel (PCPCH)	Common data transmission using the collision detection CSMA/CD method.
Paging Indicator Channel (PICH)	Transmission of the Page Indicator (PI) to realize the paging in the downlink direction. One PICH is always related to an S-CCPCH, which transports the PCH.
Acquisition Indicator Channel (AICH)	Transmits the positive acknowledgment of a random access of a UE via PRACH or PCPCH. *Phys signal*
Physical Downlink Shared Channel (PDSCH)	Common transmission of data in downlink direction. Parallel UEs will have different codes assigned.

The TFC is fixed during TTI but can be changed at the next TTI. One of the possible changes may be 'no transmission' during a given TTI. The TFC change may involve alteration to physical layer parameters, such as spreading factor. As a consequence, each transport channel is accompanied by the Transport Format Indicator (TFI) at each TTI. The physical layer combines the TFI information from different transport channels to the Transport Format Combination Indicator (TFCI). The details of the processing are given in Section 9.8.1.

Figure 9.21 shows a simplified example for grouping services/applications according to QoS and mapping the logical, transport and physical channels.

Table 9.10 Transport channel parameters.

Transmission Time Interval (TTI)	Time interval for transmission of transport channel that indicates how often data arrives from upper layers to the physical layer. The TTI value is a multiple of the frame time 10 ms, e.g. 10, 20, 40,…ms.
Transport Block (TB)	Data part +MAC+RLC headers. Total number of bits is denoted as transport block size (TBS).
Transport Format (TF)	Number of transport blocks that are transmitted in the corresponding TTI, which consequently defines the instantaneous bit rate or equivalently the spreading factor that should be used in the physical layer. Additionally, TF defines channel coding type (e.g. convolutional or turbo-code), code rate and the number of CRC bits.
Transport Format Combination (TFC)	Used exclusively when several transport channels are multiplexed onto a CCTrCH. TFC defines the number of transport blocks that are transmitted from each of the multiplexed transport channels in a given TTI.

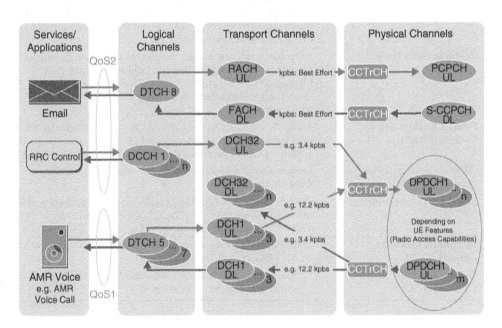

Figure 9.21 Example of mapping services to radio interface channels [10].

9.8 Physical-Layer Procedures

The physical layer at the transmitter side receives transport channels from the MAC layer via Transport Blocks (TBs). These transport blocks may belong either to one or to several transport channels that are simultaneously multiplexed. The physical layer performs procedures and processing targeted to adopt transport channels for transmission over radio interface. The first step in processing TBs for multiplexing different transport channels and introducing the required level of redundancy into the outgoing bit stream.

9.8.1   Processing of Transport Blocks

In each Transmission Time Interval (TTI), the MAC delivers to the physical layer a given number of transport blocks for each of the transport channels multiplexed together according to a Transport Format Combination (TFC). The physical layer executes a set of processes to map the transport blocks onto the available physical resources. Figure 9.22 shows the interface between physical and MAC layer and simplified multiplexing of two transport channels. Such a minimal combination of two DCHs normally includes the logical traffic channel DTCH and logical control channel DCCH. The transport channels may have a different number of TBs to transmit and are not necessarily active all the time. The time periods, TTIs, for transmitting either channel could also be different.

A process of mapping transport to physical channel is performed at the physical layer in several steps as shown in Figure 9.23 for the uplink [11]:

- Adding CRC to each transport block.
- Transport-block concatenation and code block segmentation. Concatenation is required to form a block code of a fixed size. The coded block size value depends on the channel coding type applied, 504 bit for convolutional or 5114 bits, for turbo coding. The *segmentation* of the transport block is required when total number of bits exceeds the maximum size of block code. The *concatenation* is used in the opposite case.
- Channel coding.
- Radio frame equalization. The equalization consists of padding required number of bits after the channel coding procedure in order to ensure that the resulting number of bits fits into multiple of the number of 10 ms frames in a TTI. This procedure ensures that equal-sized blocks are transmitted per frame.
- First interleaving. The first interleaving is executed as inter-frame interleaving with an interleaver length 20, 40 or 80 ms that is directly related to the Transmission Time Interval (TTI). First interleaving is applied to each transport channel separately but start positions of the TTIs for different transport channels multiplexed together for a single connection have to be time aligned.
- Radio frame segmentation. The frame segmentation will distribute the data coming from the first interleaving over two, four or eight consecutive frames in line with the interleaving length.

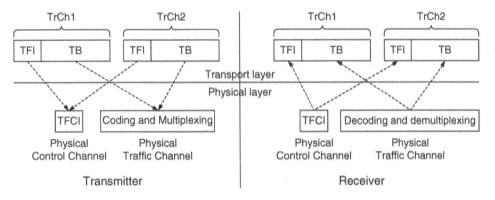

Figure 9.22 Multiplexing the transport channels.

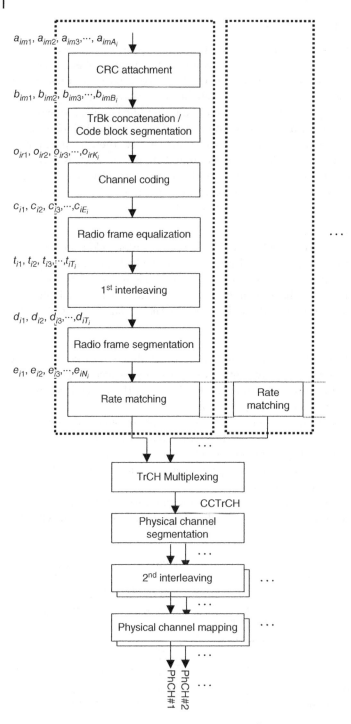

Figure 9.23 Transport channel multiplexing structure for the uplink [11].

- Rate matching. The rate matching is used to ensure that resulting number of bits per radio frame fits into one of the available bit rates of the physical channel. That depends on the spreading factor and on the number of parallel code sequences being used. Rate matching is achieved by either bit puncturing or bit repetition depending on whether the bit rate should be decreased or increased. Higher layers assign a rate-matching attribute for each transport channel. The rate-matching attribute commands the number of bits to be repeated or punctured.
- Multiplexing of transport channels. Every 10 ms, one radio frame from each TrCH is delivered to the TrCH multiplexing. These radio frames are serially multiplexed into a coded composite transport channel (CCTrCH).
- Physical channel segmentation. This procedure is executed whenever the resulting number of bits in one radio frame exceeds a maximum number per single physical channel (i.e. a spreading code sequence). In this case, the bit flow is segmented into blocks of equal size and each of them will be mapped into a different physical channel.
- The second interleaving performs intra-frame interleaving over a 10 ms radio frame over each physical channel segment separately, if applicable.

Downlink mapping is similar in procedures to uplink with a major difference in terms of insertion of DTX (Discontinuous Transmission) indication bits. The objective of DTX is to indicate the frame when there is no transmission for given transport channel. The DTX indication bits are not transmitted over the air interface; they indicate to the transmitter at which bit positions the transmission should be turned off. The use of DTX in the downlink direction allows for two possibilities when multiplexing several transport channels: 'fixed' and 'flexible' positions inside the radio frame at which channels are mapped. Detailed discussion of both options can be found in [12]; the major conclusion is that implementation of DTX bits in the 'flexible' position is more effective in terms of minimal puncturing bits required. The result of transport channel mapping can be illustrated in Table 9.11 using an example from [12].

Here, transport channel TrCH1 carries the logical traffic channel DTCH. It is multiplexed together with the dedicated control channel DCCH = TrCH2. It is assumed that TFC = (TF2, TF1) is selected in a given TTI (2 transport blocks: TTBs) are sent for the interactive service and 1 TB for the signalling).

Table 9.11 shows results of applied procedures at each step of mapping for both transport channels. After CRC attachment and TB concatenation, the code block sizes 688 bits and 164 bits for TrCH1 and TrCH2, respectively. In both cases, no segmentation is required since coded blocks sizes are below the maximum values for the corresponding channel coding. After equalization and segmentation into the radio frame, TrCH1 is mapped onto two frames of 1038 bits/frame and TrCH2 onto four frames of 129 bits/frame.

The rate-matching procedure should adjust the total number of bits per frame to one of the possible values of a DPDCH physical channel in the uplink direction. The closest value is 1200 bits, corresponding to a single physical channel with spreading factor 32. As a result, $1200 - 1038 - 129 = 33$ bits should be repeated. When the rate-matching attribute is equal for both transport channels, this leads to repeating 29 bits from TrCH1 and 4 bits from TrCH2. So, finally, a segment of 1067 bits from TrCH1 will be transmitted in each frame together with a segment of 133 bits from TrCH2.

Table 9.11 Multiplexing the DTCH and DCCH onto a single DPDCH [12].

	TrCh1	TrCh2
TTI	20 ms	40 ms
TFC	TF2 = 2 TBs	TF1 = 1 TB
TB size	336 bits	148 bits
CRC attachment	336 + 16 = 352 bits	148 + 16 = 164 bits
TB concatenation	2 × (336 + 16) = 688 bits	1 × (148 + 16) = 164 bits
Channel coding	Turbo r = 1/3 – > 3 × (688 + 4) = 2076 bits	Conv r = 1/3 – > 3(164 + 8) = 516 bits
Radio frame size equalization	2076 × 10/20 = 1038 bit/frame	526 × 10/40 = 129 bit/frame
Radio frame segmentation	1038 bit/frame	129 bit/frame
Rate matching	1039 + 29 = 1067 bit/frame	129 + 4 = 133 bit/frame
CCTrCH	1067 + 133 = 1200 bit/frame = 120kbps	

9.8.2 Spreading and Modulation

Spreading is applied to the physical channels. As discussed in Section 9.7.3, this consists of two operations. The first is the channelization operation, which transforms every data symbol into a number of chips, thus increasing the bandwidth of the signal, with the number of chips per bit determining the spreading factor (SF). The second operation is the scrambling operation, where a scrambling code is applied to the spread signal. Channelization coding is applied to data symbols on both the I and Q branches in a way that I and Q channels are independently multiplied with an OVSF code C_c, and the resulting chips from both components are combined to form complex symbols prior to the scrambling process, as shown in Figure 9.24.

Spreading is applied to composition of the physical channels. The channelized complex signal is then multiplied by complex-valued scrambling code C_{scr}. At the receiver side, the received sequence is de-scrambled by multiplying it with the complex conjugate of the scrambling code sequence.

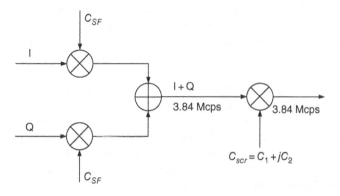

Figure 9.24 Complex-valued scrambling for the physical channel.

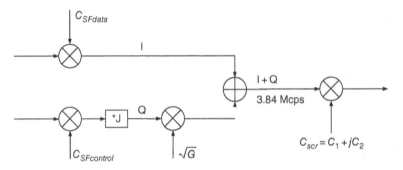

Figure 9.25 I/Q multiplexing with complex scrambling [5].

In WCDMA, I and Q components of the input signal before scrambling may carry different physical channels with different spreading factors and different amplitudes. After modulation, this would result in signals with a high crest factor (i.e. the peak-to-average power ratio) thus leading to distortion of the amplified signal. Reduction in a crest factor is achieved using complex scrambling with a weighted component. An example of this configuration is given in Figure 9.25 where two parallel physical channels DPDCH and DPCCH have different amplitudes with a relative strength difference G. Two channels are mapped at I and Q components with a weight $\sqrt{G}$ as shown in Figure 9.25. As a result, the amplitudes of I and Q components are equalized and the *crest factor* of the modulated signal is reduced.

Figure 9.26 further illustrates spreading for the uplink composition of control and traffic channels. The DPCCH is spread to the chip rate by the channelization code C_c. The n-th DPDCH called *DPDCH$_n$* is spread to the chip rate by the channelization code $C_{c,n}$.

After channelization, the real-valued spread signals are weighted by gain factors, β_c, for DPCCH, β_d for all DPDCHs. The β_c and β_d values are signalled by higher layers or derived as described in [13]. At every instant in time, at least one of the values β_c and β_d has the amplitude 1.0. The β_c and β_d values are quantized into 4 bit words; that is, 16 different values [14]. At a given point in time, the gain value for either DPDCH or DPCCH is set to 1 and then for the other channel a value between 0 and 1 is applied to reflect the desired power difference between the channels. The power differences can have 15 different values between -23.5 dB and 0.0 dB.Finally, a scrambling code is applied as shown in Figure 9.25. A total of $2^{18} - 1 = 262\,143$ scrambling codes, numbered $0\dots262\,142$ can be generated. However, not all the scrambling codes are used. Each scrambling code contains a sequence of 38 400 coded chips that fits a 10 ms radio frame. The scrambling codes are divided into 512 sets each of a primary scrambling code and 15 secondary scrambling codes. Every cell is allocated one and only one primary scrambling code, which carries the CPICH and the P-CCPCH physical channels. The other channels in the cell can use secondary scrambling codes. In each cell, the UE may be configured simultaneously with at most two scrambling codes for uplink transmission.

9.8.3 Modulation Scheme in UTRAN FDD

The modulation is embedded into spreading process with use of modulation mapper [14]. In case of single traffic channel, a fixed set of binary symbols is selected according

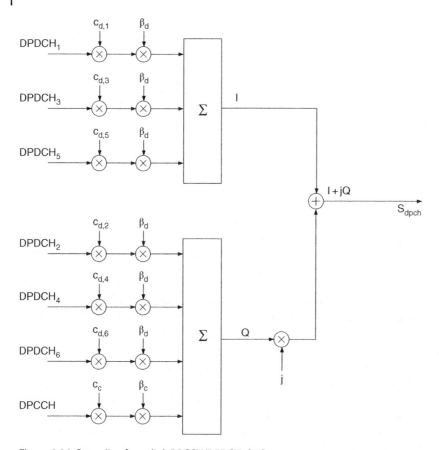

Figure 9.26 Spreading for uplink DPCCH/DPDCHs [14].

to modulation format, from two symbols for QPSK and six for 64QAM, respectively, and then converted to I and Q constellation components by the modulation mapper [14]. The I and Q branches are then both spread to the chip rate by the same real-valued channelization code. The channelization code sequence is aligned in time with the symbol boundary.

The spread complex value baseband signal is then passed to a pulse shaping filter and modulated at carrier frequency to create physical signal that can be transmitted via antenna, see Figure 9.27. The pulse shaping filter additionally spread spectrum mask of the signal in the frequency domain. As a result, the 3.84 MHz baseband signal spreads into 4.68 MHz at the output of the pulse shaping filter.

After modulation, signals are amplified until they reach the desired transmitted power level, which depends on the power control mechanism. For the mobile terminal, different power classes are defined, ranging from 21 to 33 dBm, and the minimum power level is 50 dBm. The transmitted power must be adjusted in steps of 1, 2 or 3 dB at the mobile terminal, while the base station should be able to adjust the power level in steps of 1 dB (mandatory) and 0.5 dB (optional).

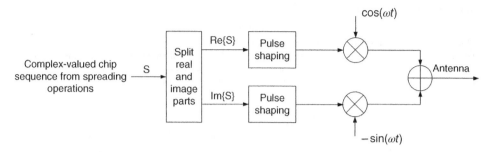

Figure 9.27 Modulation process [14].

9.8.4 Composition of the Physical Channels

9.8.4.1 Dedicated Physical Channel

The information from DCH transport channels is mapped onto the Dedicated Physical Data Channel (DPDCH), which is multiplexed together with the Dedicated Physical Control Channel (DPCCH) that carries physical layer control information. Such multiplexing is done differently in the uplink and downlink cases.

Uplink The composition of the dedicated physical channel in the uplink direction is shown in Figure 9.28. The DPDCH and DPCCH are multiplexed in I and Q components, respectively.

The uplink DPCCH is used to carry control information generated at the physical layer. This information consists of

- Known pilot bits to support channel estimation for coherent detection,

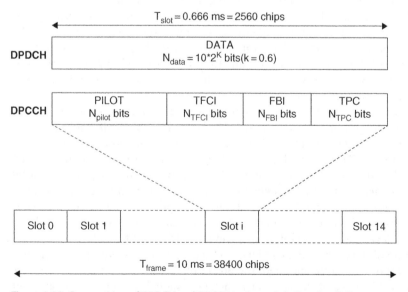

Figure 9.28 Composition of DPDCH and DPCCH in the uplink direction [15].

Table 9.12 DPDCH fields [15].

Slot Format #i	Channel Bit Rate (kbps)	Channel Symbol Rate (ksps)	SF	Bits/Frame	Bits/Slot	N_{data}
0	15	15	256	150	10	10
1	30	30	128	300	20	20
2	60	60	64	600	40	40
3	120	120	32	1200	80	80
4	240	240	16	2400	160	160
5	480	480	8	4800	320	320
6	960	960	4	9600	640	640

- Transmit power control (TPC) commands,
- Feedback information (FBI), and
- An optional transport format combination indicator (TFCI). The transport format combination indicator informs the receiver about the instantaneous transport format combination of the transport channels mapped to the simultaneously transmitted uplink DPDCH radio frame.

There is only one uplink DPCCH on each radio link.

The parameter k in Figure 9.28 determines the number of bits per uplink DPDCH slot. It is related to the SF of the DPDCH as $SF = 256/2^k$. The DPDCH spreading factor may range from 256 down to 4. The spreading factor of the uplink DPCCH is always equal to 256; that is, there are 10 bits per uplink slot. For DPDCH there are seven time-slot formats defined with channel symbol rate varying from 15 to 960 kbps.

Transmissions at variable bit rate can be obtained in the DPDCH channel simply by modifying the slot format or equivalently the spreading factor, as shown in Table 9.12, whenever the transport format is changed at the MAC layer. The physical channel allows the performance of this operation from frame to frame, although in practice the time between variations will be given by the TTI of the corresponding transport channel, which is an integer number of frames.

Variable transmission rate for DPDCH is accompanied by outer loop power control. One of the QoS requirements is specified bit-error-rate, which depends on value of energy per bit-to-noise ratio, Eb/No, for a given modulation format. On the other hand, Eb/No relates to the bit rate and signal-to-noise ratio SIR by equation Eb/No = (W/Rb)*SIR, where W is a bandwidth of the transmitted signal, Rb is a bit rate. As a consequence, the requirement to fix Eb/No, leads to power adjustment of DPDCH when bit rate is changed. Figure 9.29 shows Tx power variations with variable bit-rate transmission. It should be noted that the control channel is transmitted at a constant power level with a constant bit rate and spread factor, SF = 256.

Demodulation of uplink user data on DDDCH is performed coherently; that is, similar to the downlink. The difference with the downlink transmitter is that the users are distributed over cell area and, consequently, each user should generate a separate pilot signal for channel estimate. As shown in Figure 9.28, the pilot signal is embedded in

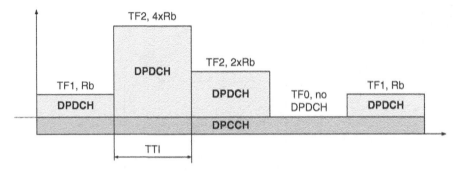

Figure 9.29 Power control and variable bit-rate transmission in the uplink direction.

uplink DPCCH. The number of pilot bits N_{pilot} varies from 3 to 8 bits depending on the slot format, which is configured by higher layers depending on transport channel composition [11].

A scrambling in uplink is user-specific and performed on the top of channelization of user data. Due to the fact that WCDMA is an asynchronous system, the uplink signals from different users are not time synchronized in the NodeB receiver. Therefore, channelization code is only used to separate different channels from the same UE. Due to the lack of time synchronization, orthogonality of OVSF uplink codes between different users cannot be achieved, therefore, all users in the cell can use the same set of channelization codes thus producing noise interference to each other. The user-specific scrambling effectively performs 'whitening' of the intra-cell interference on the uplink.

Downlink Channelization code spreading is applied to all physical channels on the downlink with the exception of SCH. Both I and Q branches are spread to the chip rate by the same real-valued channelization code $C_{ch,SF,m}$; that is, the output for each input symbol on the I and the Q branches presents a sequence of SF chips multiplied by the real-valued symbol. The channelization code sequence is aligned in time with the symbol boundary. Application of the same OVSF code for I and Q branches is possible due to orthogonality of I and Q signal components by itself. In the example in Figure 9.30, the real-valued chip sequence on the Q branch is complex multiplied with j and summed with the corresponding real-valued chip sequence on the I branch, thus resulting in a single complex-valued chip sequence that is further scrambled by a complex-valued scrambling code $S_{dl,n}$.

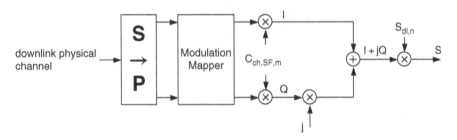

Figure 9.30 Spreading for all downlink physical channels except SCH [14].

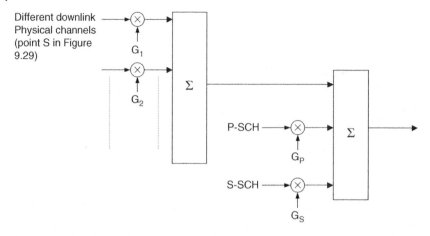

Figure 9.31 Combining downlink physical channels [14].

Different downlink channels are then combined together before channel shaping and RF modulation at the carrier frequency, as shown in Figure 9.31. Each complex-valued spread channel is separately weighted by a weight factor Gi. The complex-valued synchronization channels, primary P-SCH and secondary S-SCH, may also be separately weighted by corresponding weight factors G_p and G_s. All downlink physical channels are then combined using complex addition.

9.8.4.2 Common Downlink Physical Channels

Common Pilot Channel (CPICH) The CPICH is a fixed rate (30 kbps, SF = 256) downlink physical channel that carries a predefined bit sequence. The CPICH is not mapped to a transport channel. With a predefined bit pattern the CPICH provides the estimate of phase reference for the other control channels on downlink. Figure 9.32 shows the frame structure of the CPICH.

There are two types of common pilot channels; Primary and Secondary CPICH. They differ in their use and the limitations placed on their physical features.

Primary Common Pilot Channel (P-CPICH) The P-CPICH has the following characteristics. There is only one P-CPICH per cell, which is always transmitted with the same

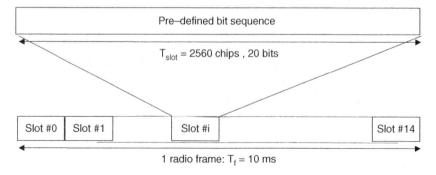

Figure 9.32 Frame structure for the common pilot channel [15].

channelization code, $C_{ch,256}$, and is scrambled by the primary scrambling code. The mobile user can obtain the cell primary scrambling code during the synchronization procedure. The P-CPICH provides a phase reference for other downlink control channels, as well as a channel reference for cell selection and handover management. The allocated power of the P-CPICH is broadcasted in the system message of the BCH channel, thus providing the means to a mobile terminal to estimate the path loss in given location within the cell. The neighbour cells with a measured CPICH power level comprise the *Monitored Set*, the one with highest level is normally selected as a *serving cell*. During soft handover, another set of neighbours is formed, namely the *Active Set*, comprised of cells to which the mobile has established a connection over the air interface.

Secondary Common Pilot Channel (S-CPICH)   The secondary pilot appeared in the cell only in special cases when the transmit diversity or multiple narrow beam antennas are configured in the cell. In the case of transmit diversity in the MIMO mode and of beam forming with different antennas, there may be several S-CPICH per cell transmitted with an arbitrary channelization code of SF=256. The S-CPICH can be scrambled by either the primary or a secondary scrambling code. The S-CPICH can be used as a phase reference for the second, third or fourth transmit antenna by UEs configured in MIMO mode or in MIMO mode with four transmit antennas.

Synchronization Channel (SCH)   The Synchronization Channel (SCH) is a downlink signal used for cell search. The SCH consists of two subchannels, the Primary and Secondary SCH. The 10 ms radio frames of the Primary and Secondary SCH are divided into 15 slots, each with a length of 2560 chips. Figure 9.33 illustrates the structure of the SCH radio frame.

The Primary SCH transmits the Primary Synchronization Code (PSC) once in every slot in a frame, as shown in Figure 9.33. The Primary Synchronization Code (PSC) is denoted c_p in Figure 9.33. The PSC code is the same for every cell in the system and is therefore known to the UE and has a length of 256 complex chips with identical real and imaginary values. The primary SCH is a first channel in the cell which mobile terminal has to acquire in order to be time synchronized with the base station. The Primary Synchronization Code sequence c_p has an autocorrelation correlation function with a sharp peak at zero time shift that ease PSC detection with a matched filter. Since the Primary PCH repeats the PSC sequence as shown in Figure 9.33, the correlator output

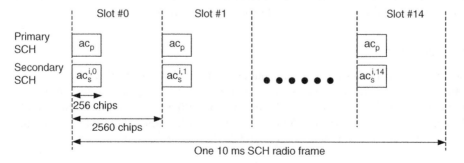

Figure 9.33 Structure of the synchronization channel [16].

will contain peak replicas at every time slot in SCH radio frames. As a result, by reading the Primary SCH, the mobile terminal *will acquire both time slot and chip level synchronization with the base station transceiver.*

The Secondary Synchronization Codes (SSC) are transmitted in parallel with the Primary SCH. The S-SCH contains a sequence of 15 modulated code words $c_s^{i,k}$ each of length 256 chips, where $i = 0, 1, \ldots, 63$ is the number of the scrambling code group and $k = 0, 1, \ldots, 14$ is the slot number. The autocorrelation properties of code word $c_s^{i,k}$ are similar to PSC. Each SSC sequence is chosen from a set of 16 different codes $c_{s,j}$ ($j = 0, \ldots, 16$) of length 256. Total 64 code groups S_i ($I = 0,\ldots, 63$) is constructed by composition of 16 codes $c_{s,j}$. Each code S_i of the group of 64 codes identifies to which scrambling code groups the base station belongs. The code S_i presents a sequence of 15 code words $c_s^{i,k}$, one per time slot as shown in Figure 9.16. As an example, with $I = 17$, the SSC sequence may look like [14]:

$$S_{17} = \{c_s^{17,0}, c_s^{17,1}, \ldots, c_s^{17,14}\}$$
$$= \{c_{s,1}, c_{s,11}, c_{s,14}, c_{s,4}, c_{s,13}, c_{s,2}, c_{s,9}, c_{s,10}, c_{s,12}, c_{s,16}, c_{s,8}, c_{s,5}, c_{s,3}, c_{s,15}, c_{s,6}\} \quad (9.1)$$

It is important that the SSC code S_i is unique such that none of the 64 codes S_i can be generated by cyclical rotation of another code S_n, with $n \neq i$. This property allows synchronization at the frame level. When a mobile detects the SSC sequence, it knows the position of the first time slot in the radio frame; therefore, frame synchronization is achieved.

The detection of SSC code S_i determines one of the 64 code group that the base station belongs to. A total of 512 scrambling codes are defined for the WCDMA system, they are grouped in 64 groups associated with 64 SSC codes. Each group contains eight scrambling codes. Once the UE detects the sequence S_i, it knows the specific group of eight scrambling codes. In the next step, the mobile has to attempt to descramble the Primary Common Control Physical Channel (P-CCPCH) with eight possible scrambling codes and choose one with a high correlation. Knowing the scrambling code mobile can then read the system information message by decoding the Broadcast Transport Channel transmitted at P-CCPCH.

9.9 RRC States

There are two modes in the RRC state machine: RRC Idle Mode and RRC Connected Mode. Figure 9.34 illustrates the RRC states and interstate transitions.

9.9.1 Idle Mode

When a UE is switched on, it selects a public land mobile network (PLMN), normally the home PLMN, based on information stored in the user SIM card, and then the UE searches for a suitable cell of that PLMN to camp on. When the UE chooses the suitable cell to provide available services and tunes to its control channel, it is able to synchronize to the cell and read the broadcast channel. This choosing is known as 'camping on

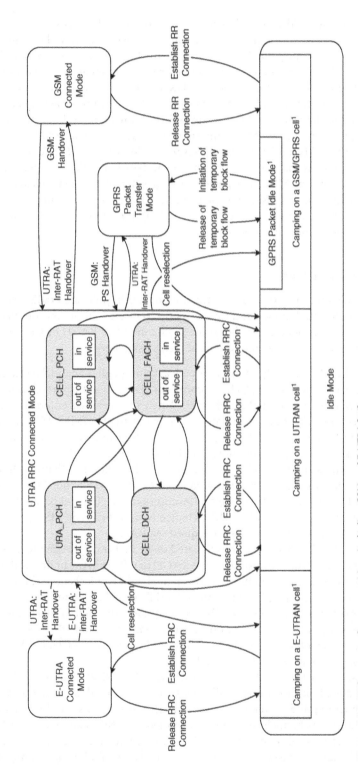

Figure 9.34 RRC states and state transitions including GSM and E-UTRA [16].

the cell'. The terminal in idle mode has no connection with the access stratum of the network. Any transfer of information from mobile to network is allowed by entering in the Connected Mode via an Initial Message sent via the RACH transport channel. While listening, the broadcast channel the mobile compares the broadcasted registration area identity (URA) with one stored in the SIM. If URAs are the same, there is no need for the location registration procedure. In a case where the URA broadcasted is different to the one stored in the SIM, the MS goes into RRC connected mode for the location registration request sent via the RACH and then goes back to the idle mode.

In idle mode, the UE is not known to UTRAN but identified by NAS identities such as IMSI, TMSI. It means that the network knows the position of the terminal on a registration area basis and the MS can be paged. As long as the mobile remains in the same registration area, it is not necessary for the terminal to start an RRC connection. However, when the mobile enters a new registration area, it must enter into RRC connected mode to perform a location area update.

9.9.2 RRC Connected Mode

In RRC connected mode, the MS is able to transmit data using allocated radio resources. Each mobile in connected mode has a RNTI (Radio Network Temporary Identity). The RNTI identifies the UE identifier in the common channels like RACH or CPCH.

The RRC connected mode has four different states listed here.

1) Cell_PCH. The MS can only monitor broadcast and paging channels and cannot transmit either dedicated (DTCH) or common (DCCH) transport channel. The mobile location is known at the cell level. When the terminal enters a new cell, a cell reselection procedure is initiated by moving to the Cell_FACH state and transmitting through the RACH. After the cell reselection procedure, the terminal switches back to Cell_PCH, if no more activity is detected.

2) URA_PCH state is similar to Cell_PCH state with the difference that the mobile is known to the network at registration area level. Neither DTCH nor DCCH can be allocated to the MS in this state. The MS monitors the paging and broadcast channels and perform LR update when URA change is identified. The MS moves to the Cell_FACH state and transmits the LR request via RACH.

3) Cell_FACH state. The MS cannot be allocated dedicated transport channel DCH in this state. The MS can only transmit via the RACH or CPCH in the uplink and receive through the FACH in the downlink. However, the DCCH logical channel is available for the transfer of signalling information and a DTCH logical channel for the transfer of data may also be configured. The MS can enter the Cell_FACH state from Idle mode setting the RRC connection. In RRC connected mode the Cell_FACH is reachable from any other states of RRC connected mode, Cell_PCH, URA_PCH or Cell_DCH states, as shown in Figure 9.34.

4) Cell_DCH state. In Cell_DCH state, the dedicated physical channel both in the uplink and downlink directions is allocated to the terminal, so that the MS can transmit signalling and/or data information. The MS can use a combination of dedicated transport channels and control channels. The MS is known on a cell level according to its current active set. The MS can initiate soft and hard handovers. During soft handover, the network changes the active set and location of mobile.

The MS can enter the Cell_DCH state either from Cell_FACH by means of establishing a dedicated physical channel or from idle mode directly going through RRC connection establishment procedure.

It is important to note that the MS can send and receive data in the Cell_FACH state using RACH and FACH channels, a transition to Cell_DCH is based on a trade-off between amount of data to be sent by mobile or network, service delay constraints, signalling load and terminal power consumption. A significant amount of signalling between the MS and SRNC is required for setting up the DCH channel. When data runs out and timers have elapsed the MS goes away from Cell_DCH state, as shown in Figure 9.34. The RRC state transition can be initiated either by the mobile or network. Two paging states Cell_PCH and URA-PCH consume minimum power since the UE sleeps in those states and awakes periodically just to check paging messages.

9.9.3 RRC Connection Procedures

The RRC functions are listed in Section 9.6.4. The main function of the RRC layer is signalling between the UE and the network regarding the RRC connection establishment, maintenance and release. Those include:

- Periodic broadcast of system information messages on the BCCH channel.
- Paging. The network pages the MS in case of network-initiated call establishment, request for cell update procedure, or network-initiated state transition due to downlink packet data activity. The paging message is sent through the PCH channel to mobile in Idle mode or in Cell_PCH and URA_PCH states. For a mobile in Cell_DCH or Cell_FACH, the DCCH channel is used for paging.
- RRC connection establishment and release. The MS originated RRC connection establishment is initiated on request from higher layers (NAS) on the UE side. In the UTRAN originated case, the network sends a paging message first.

The message flow for RRC connection establishment is shown in Figure 9.35. The MS transmits an RRC CONNECTION REQUEST message on the uplink CCCH mapped to RACH. As with any request from NAS, the UTRAN needs to setup a Signalling Radio Bearer (SRB) between the UTRAN and the UE. The mobile identifies itself by means of 'Initial UE ID', that is an NAS identifier, like the IMSI or the TMSI. The Initial UE ID is included in RRC CONNECTION REQUEST content.

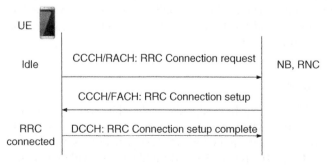

Figure 9.35 RRC connection establishment.

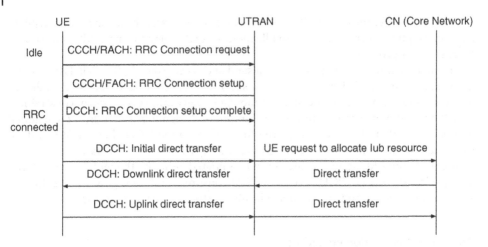

Figure 9.36 Service establishment process between the UE and core network.

Upon receiving an RRC CONNECTION REQUEST message, the UTRAN transmits an ***RRC CONNECTION SETUP*** message on the downlink CCCH mapped at FACH or transmit an ***RRC CONNECTION REJECT*** message on the downlink CCCH directing the UE to another UTRA carrier or to another system.

The RRC Connection Setup message may include a dedicated physical channel assignment for the UE (transition to Cell_DCH state), or it can command the UE to use common channels (transition to Cell_FACH state). In the latter case, a Radio Network Temporary Identity (RNTI) to be used as UE identity on common transport channels is allocated to the UE.

In the case of acceptance of an RRC CONNECTION REQUEST the RNC sends the RRC Connection Setup message that includes the RNTI for the mobile and the command for state transition; that is, move to Cell_DCH or to Cell_FACH. It also includes the characterization of the allocated dedicated radio channel in terms of code sequence and TFCs in both the uplink and downlink direction when the user is moved to the Cell_DCH. In any case, the mobile terminal is now in connected mode and there is a DCCH logical channel allocated to the Signalling Radio Bearer.

The RRC connection setup procedure leads to establishment of NAS signalling between terminal and core network. This allows the direct transfer of the upper layer signalling messages between the UE and the core network through the RRC layer either in the uplink or downlink direction. In UTRAN the signalling messages are mapped on DCCH established between the UE and the RNC [17]. A simplified RRC connection and direct transfer is illustrated in Figure 9.36.

9.9.4 RRC State Transition Cases

In RRC connected mode, the UE stays in the Cell_DCH state while transmitting and receiving data over the dedicated traffic channel. As soon as the data runs out and timers have elapsed, the UE will be moved away from the Cell_DCH state to other states of RRC connected mode or to Idle mode and the respective dedicated channel will be released together with the signalling link and all radio bearers between the UE and UTRAN.

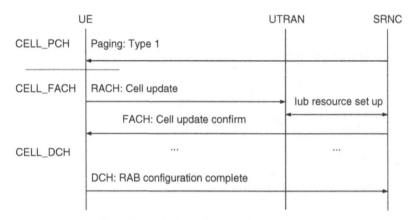

Figure 9.37 Network-initiated state transition.

Moving back to the Cell_DCH state will require signalling between UE and UTRAN and allocation of radio bearers and traffic channels. The use of Cell_DCH or Cell_FACH state is always a trade-off.

Depending on amount of application data to be transmitted, service delay, signalling load and network resource utilization, the UE can be allowed to use either the Cell_DCH or Cell_FACH state. The Cell_FACH state does not imply allocation of the dedicated traffic channel using RACH instead for data transmission. The network-initiated RRC state change occurs when there is a large amount of data to be transmitted that requires a dedicated transport channel. The network first transmits the paging message to the UE sleeping in CELL_PCH state. Upon reception of the paging message, the UE moves to the Cell_FACH state and initiates signalling on the RACH, as illustrated in Figure 9.37. After radio access bearer reconfiguration, the UE sends the reconfiguration complete message being ready to receive data from network.

9.10   RRM Functions

The major objective of RRM is to ensure QoS, maintain the planned radio coverage and maximize capacity of the network in the Radio Access part. The RRM is implemented by means of executing RRM software algorithms in network nodes. The RRM algorithms include handover control, power control, admission control, load control and packet scheduling [18]. On the network side, all RRM algorithms reside in the RNC with the exception of the inner loop (fast) power control located in the NodeB.

9.10.1   Admission Control Principle

The admission control is executed when the existing bearer is to be modified or a new radio bearer is setup. The admission control estimates the load and interference increase ΔI that a new bearer would produce in the network. The scheme is shown in Figure 9.38. A new UE is admitted by the uplink admission control algorithm if the new resulting total interference level $I_o + \Delta I$ is lower than the threshold value $I_{threshold}$:

$$I_o + \Delta I < I_{threshold} \tag{9.2}$$

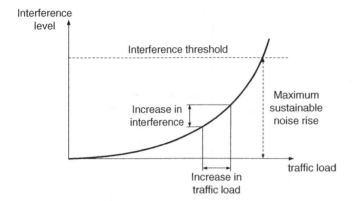

Figure 9.38 Estimation of load increase caused by the admission of additional traffic.

9.10.2   Load/Congestion Control

A maximum load level is planned in the network. Load control algorithm has to ensure that the system is not overloaded. The network load is generally produced by a combination of RT (Real-Time) and NRT (Non-Real-Time) service transmissions. If the RT traffic load itself is below the planned value and overload is caused by NRT users, then congestion may be alleviated by reducing the transmission rates of a number of NRT users. If this is still not sufficient then other actions may be applied, such as:

- Reduce the throughput of packet data traffic.
- Handover to another WCDMA carrier.
- Handover to GSM.
- Decrease bit rates of real-time UEs, for example, an AMR speech codec.
- Drop low priority calls in a controlled fashion.

9.10.3   Code Management

Since the number of OVSF codes inside the code tree is a limited resource, it is important to be able to allocate/reallocate the channelization codes in the downlink with an efficient method in order to prevent 'code blocking'. Code blocking could be caused by existing code assignment when spare capacity is not available for a new call because of the 'mother code' induced loss of orthogonality. This means that code assignment should be rather dynamic and allow code reallocation in order to eliminate code blocking. Execution of this strategy may require 'code handover'; that is, a call using a given code is forced to use a different code belonging to the same layer.

9.10.4   Packet Scheduling

All four UMTS traffic classes, from background to conversational, can be supported by WCDMA radio networks. The conversational class traffic belongs to real-time services and is transmitted on dedicated channel DCH, no scheduling is applied. Background and interactive traffic classes do not require any guaranteed minimum bit rate and they are transmitted through the packet scheduler. The video streaming class requires a minimum guaranteed bit rate but tolerates some delay and packet scheduling can be utilized for a streaming.

The WCDMA packet scheduler is located in RNC. The base station provides the air-interface load measurements and the mobile provides uplink traffic volume measurements for the packet scheduler. There are two types of packet scheduling in WCDMA; user-specific and cell-specific scheduling. The user-specific scheduling controls the utilization of RRC states, transport channels and their bit rates according to the traffic volume. The cell-specific part controls the sharing of the radio resources between the simultaneous users.

- In user-specific packet scheduling, WCDMA supports three types of transport channel that can be used to transmit packet data: common, dedicated and shared transport channels. Important to note that for small data traffic volume common channels, FACH, RACH and CPCH can be utilized in packet scheduler within RRC state Cell_FACH. The state transitions can be controlled by the traffic volume threshold from Cell_FACH to Cell_DCH and by the inactivity timer from Cell_DCH to Cell_FACH. If the amount of data in the mobile uplink buffer or in the RNC downlink buffer exceeds the traffic volume threshold, DCH allocation takes place. When DCH is allocated, either for uplink or for downlink, it must be allocated for the other direction at the same time as well, since DCH is a bi-directional channel. The DCH bit rate can be different in the uplink and downlink depending on which direction triggers the DCH allocation.
- The cell-specific packet scheduler divides the non-real-time capacity between simultaneous users. The cell-specific scheduler operates periodically. This period is a configuration parameter and its value typically ranges from 100 ms to 1 s. If the load exceeds the target, the packet scheduler can decrease the load by decreasing the bit rates of packet bearers; if the load is less than the target, it can increase the load by allocating higher bit rates.

9.11   Initial Access to the Network

The initial UE radio access procedure includes a cell search to connect to. In the cell-search procedure the UE has to acquire cell identity and time synchronization in order to read system information messages on the broadcast control channel. The access procedure involves several steps as shown in Figure 9.39:

1) The UE reads the Primary Synchronization Channel, which is not scrambled and spread by a predefined spreading code (SF=256). By reading this, the UE become time synchronized with the NodeB at the time slot level, since P-SCH is transmitted every slot during the radio frame. Also, the UE matches the received P-SCH sequence with one of three sequences associated with cell-identity group. As a result, the UE knows the cell-identity group to which Cell ID belongs but not the Cell ID itself.
2) The UE reads the S-SCH, which is also not scrambled and transmitted in parallel with P-SCH. The S-SCH will transmit a unique 15 code word sequence that is mapped to time slots in predefined order [14]. By reading these values, the UE will be frame synchronized and able to identify the Cell ID and scrambling group the actual NodeB is using.

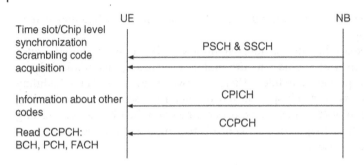

Figure 9.39 Initial UE radio access.

3) The UE can now read the Common Pilot Channel, which is scrambled with one of eight primary scrambling codes of the scrambling group. It is a matter of trial and error to find the correct code. The Pilot Channel will contain further information about other necessary codes and about the downlink diversity synchronization pattern.

4) The UE will read the Common Control Physical Channel, which uses the same scrambling code as the CPICH, to get detailed information about UTRAN and the CN, to allow the P-CCPCH to transport the BCH, and to be able to get paged, and to allow the S-CCPCH to transport PCH. The system information in the BCH will also indicate the secondary scrambling code of the actual NodeB for further data transmission on the DCHs.

9.12 Summary

The main features of WCDMA can be summarized as follows:

- 3G WCDMA Technology is based on a spread spectrum concept.
- Different channels are separated by the Orthogonal Variable Spreading Factor (OVSF) codes, also called channelization codes.
- OVSF produces a spreading chip sequence at the rate of 3.84 Mcps. The information bit is then mapped to the chip sequence.
- The OVSF spreading factor, SF, refers to a number of chips per information symbol; that is, defines SF length from 1 to 512.
- The orthogonal codes with a variable SF is used to form a family of usable (see limitation by 'mother code') codes to transmit the number bit streams with variable bit rates on the same frequency without interference to each other.
- Spread aggregate signal is then scrambled. The scrambling code performs on the top of the spreading, has the same rate as a chip rate.
- Scrambling codes are transmitter specific, each base station sector or mobile terminal has its own code.
- Scrambling code is not orthogonal, but has the properties of random noise. The scrambling results in randomization of the inter-cell interference.
- It is important that the scrambling code is a complex one with I and Q components weighted to reduce the peak-to-average ratio of the signal.

- On uplink, each user runs its own OVSF code tree, as well as a scrambling code. Users are not required to be synchronized.
- WCDMA (like CDMA) cannot work without fast power control for uplink. This is because of all users on uplink interfere to each other and the goal of power control is to *minimize intra-cell interference* by equalizing the received power per bit of all mobile stations at all times.
- Frequency reuse is once in WCDMA: that is, all cells use the same carrier frequency and frequency reuse principle; as such, this is not applicable in WCDMA.
- Soft/softer handover is used in WCMA in both uplink and downlink. This allows effective space diversity mechanism to mitigate large-scale fading, it can be regarded as effective as frequency hopping in GSM.
- During soft/softer handover, two close power control loops are active in order to handle near-far problem in both cells.
- The means for fading and multipath mitigations include channel coding, interleaving and are specific for a CDMA-RAKE receiver that deals with the multipath at chip level.
- Concept of WCDMA time slot is different from GSM. The 10 ms radio frame is subdivided into 15 time slots. Each time slot corresponds to a power control cycle with a rate of 1500 Hz.
- Network architecture is similar to GSM.
- The principal difference from GSM network system architecture is a separation of control and user planes.

References

1 Technical Specification Group Core Network and Terminals; Non-Access-Stratum (NAS) protocol for Evolved Packet System (EPS); Stage 3 (Release 8); 3GPP TS 24.301 V8.1.0 (2009–03).

2 Kreher, R. and Gaenger, K., *LTE Signalling, Troubleshooting and Optimization*, John Wiley & Sons, Ltd, 2011.

3 Technical Specification Group Radio Access Network; Typical examples of Radio Access Bearers (RABs) and Radio Bearers (RBs) supported by Universal Terrestrial Radio Access (UTRA); 3GPP TR 25.993 version 12.0.0 Release 12.

4 Chevallier, C., Brunner, C., Garavaglia, A., Murray, K.P. and Baker, K.R. (eds); *WCDMA (UMTS) Deployment Handbook: Planning and Optimization Aspects*, John Wiley & Sons, Ltd, 2006.

5 Holma, H. and Toskala, A. (eds); *WCDMA for UMTS*, John Wiley & Sons, Ltd, 2004.

6 Brennan, D.G. Linear diversity combining techniques, *Proc. IRE*, vol. 47, pp. 1075–1102, June 1959.

7 Heikkilä, T. and Järvelä, M. RAKE training slides by Nokia, 03.04.2001, Oulu.

8 Technical Specification Group Services and System Aspects; Quality of Service (QoS) concept and architecture (Release 14); 3GPP TS 23.107 V14.0.0 (2017–03).

9 Radio Interface Protocol Architecture (Release 14); 3GPP TS 25.301 V14.0.0 (2017–03).

10 Kreher, R. and Rudebusch, T. *UMTS Signalling*, John Wiley & Sons, Ltd, 2007.

11 Multiplexing and channel coding (FDD) (Release 14); 3GPP TS 25.212 V14.0.0 (2016–12).

12 Perez-Romero, J., Sallent, O., Agustí, R. and Díaz-Guerra, M.A. *Radio Resource Management Strategies in UMTS*, John Wiley & Sons, Ltd, 2005.

13 Physical-layer procedures (FDD) (Release 14); 3GPP TS 25.214 V14.1.0 (2017–03).

14 Spreading and modulation (FDD) (Release 14); 3GPP TS 25.213 V14.0.0 (2017–03).

15 Physical channels and mapping of transport channels onto physical channels (FDD) (Release 14); 3GPP TS 25.211 V14.0.0 (2017–03).

16 Radio Resource Control (RRC); Protocol specification (Release 14); 3GPP TS 25.331 V14.3.0 (2017–06).

17 Interlayer Procedures in Connected Mode (Release 1999); 3GPP TS 25.303 V3.3.0 (2000–03).

18 Technical Specification Group Radio Access Network; Radio resource management strategies (Release 7); 3GPP TR 25.922 V7.1.0 (2007–03).

10

High-Speed Packet Data Access (HSPA)

Successful introduction of WCDMA UMTS in the first Release 99 assisted smooth evolution from GSM to a third generation of mobile communications system. The WCDMA brought tangible benefits to operators in terms of network quality, voice capacity and new-data service capabilities considerably better that provided with EDGE. The next step in WCDMA standards evolution was the introduction of High-Speed Downlink Packet Access (HSDPA) and High-Speed Uplink Packet Access (HSUPA) in Releases 5 and 6 of UMTS. The combination of HSDPA and HSUPA was called HSPA; that is, High-Speed Packet data Access. The HSPA technology supports much higher data rates, higher capacity in both uplink and downlink directions and significantly reduces latency compared with the initial WCDMA system. The major features of Release 99 that drastically changed with HSPA introduction were about channel structure for packet services and centralized radio resource control.

Conversational, streaming or high data volume services in Release 99 are supported with dedicated channels (DCHs) or shared (DSCHs) channels with attached priorities identifiers. Use of DCH for packet services leads to radio resource limitations, such resources as channel element, power and code cannot be shared with other users even when the application is not sending or receiving data. The shared-channel DSCH was inefficient for high throughput-low-latency services because of slow scheduling functionality in Release 99.

The packet scheduling function in WCDMA Release 99 is centralized in the RNC. The RNC makes decisions based on the measurement reports sent by the NodeB in intervals of order 100 ms. In practice, variation in data traffic in the cell is often much shorter than packet scheduler's operating period. This results in low efficiency in cell throughput and long latency for individual application.

The HSPA addresses these issues with the introduction of new shared data channels and transition of some RRM functions including the packet scheduler into NodeB.

10.1 HSDPA, High-Speed Downlink Packet Data Access

In the HSDPA system, downlink radio resources such as channelization codes and transmission power belong to a common pool of resources that is dynamically shared between users in the time domain. This constitutes a concept of shared-channel transmission that is implemented via the newly introduced *High-Speed Downlink Shared Channel* (HS-DSCH). This channel operates on a three timeslot TTI frame structure

Introduction to Mobile Network Engineering: GSM, 3G-WCDMA, LTE and the Road to 5G,
First Edition. Alexander Kukushkin.

with a total of 2 ms duration, 2 ms = 3·0.667µs, meaning that HSDPA introduced a new Transmission Time Interval, TTI = 2 ms.

The HSDPA uses set of channelization codes, a maximum of 15, with a fixed spreading factor SF = 16. A specific number of codes available for HS-DSCH transmission is configurable between 1 and 15 and can be dynamically allocated to a single user, as shown in Figure 10.1 and Figure 10.2. In support of the assigned set of the channelization codes a certain transmitter power is allocated for HS-DSCH transmission. The HS-DSCH power level is constant. There is no power control in HSDPA. Instead, HSDPA is rate controlled through channel dependent scheduling.

With the introduction of HSDPA, the overall RRM architecture has changed since the scheduling functionality has moved to the NodeB. The HSDPA transmission is not supported by soft and softer handovers. Therefore, HSDPA users connected to a single RNC, and the only single cell supports HSDPA service at the time. Each HSDPA user could have assigned a dedicated uplink DCH associated with downlink HS-DSCH. The associated DCH is a Release 99 type of channel and could be in a soft handover controlled by SNRC.

The HSDPA RRM functions transferred to the NodeB include:

- Channel-dependent packet scheduling
- Dynamic resource allocation
- QoS provision
- Load and overload control
- Physical-layer retransmission with soft combining, HARQ

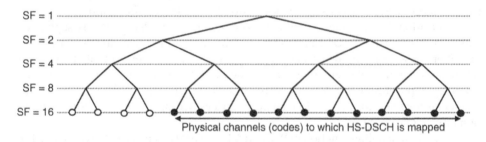

SF_{HSDPA} = 16 (example)
Number of codes to which HSDPA transmission is mapped: 12 (example)

Figure 10.1 HSDPA mapping to physical channels with a fixed spreading factor [1].

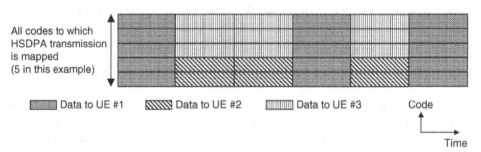

Figure 10.2 Sharing by means of time multiplex as well as code multiplex [1].

The changes in RRM functions of the network nodes led to additional modifications of HSDPA air interface including introduction of higher-order modulation (16QAM, 32QAM) and coding schemes, modification of the MAC protocol architecture for enabling the fast scheduling in the BTS over 2 ms TTI, introduction of Adaptive Modulation and Coding (AMC) and new error-correction mechanism in the MAC layer (HARQ).

10.2 HSPA RRM Functions

10.2.1 Channel-Dependent Scheduling for HS-DSCH

The scheduling mechanism controls to which user the shared-channel transmission is directed at a given time instant. In each TTI, the scheduler decides to which user(s) the HS-DSCH should be allocated. The NB also dynamically applies AMC mechanism that defines at what data rate transmission is optimal.

The principle of channel-dependent scheduling is based on fundamental assumption of independent channel conditions for different users at each time instance. Such an assumption is truly satisfied with Raleigh fading. Independence of signal level fading for different users leads to high likelihood that there is a user with the best possible radio channel conditions (at fading peak) at each time instance, as shown in Figure 10.3. A radio link to this user is likely to be of good quality, therefore, a high data rate channel can be configured for a selected user in that TTI.

The highest overall throughput of the scheduler is achieved in the case of statistically independent fading on radio links to different users. In that way, the scheduling algorithm just exploits Rayleigh fading properties rather that mitigate its impact.

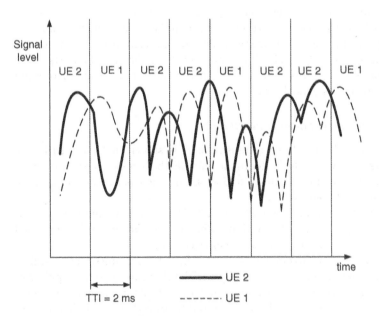

Figure 10.3 Channel-dependent scheduling.

10.2.2 Rate Control, Dynamic Resource Allocation, Adaptive Modulation and Coding

The data rate control in HSDPA is implemented by dynamically adjusting the channel coding rate as well as dynamically selecting between QPSK and M-QAM modulation. Higher-order modulation such as QAM allows for higher bandwidth utilization than QPSK, but requires a higher received E_b/N_0. Consequently, M-QAM (16QAM, 32QAM, 64QAM) is mainly useful in good channel conditions often at shorter ranges compared with QPSK. The data rate is selected independently for each 2 ms TTI by the NodeB and the rate control mechanism can therefore track rapid channel variations.

The AMC is sensitive to measurement error and delay. In order to select the appropriate modulation, the scheduler must be aware of the channel quality. Errors in the channel estimate will cause the scheduler to select the wrong modulation/coding scheme and respective transmission power allocated for HS_DSCH. The total power of HSDPA carrier is shared between Release 99 channels, which are power controlled and HS-DSCH, which is rate controlled. The HS-DSCH data rate is then selected to match the radio conditions and the amount of power instantaneously available for HS-DSCH transmission.

The feedback to NodeB for selection of the HS-DSCH transport format is supported in two ways:

1) The UE estimates the downlink channel quality and calculate a suitable transport format that is reported to the NodeB.
2) The NodeB may determine the transport format based on power control gain of the associated dedicated physical channel.

10.2.3 Hybrid-ARQ with Soft Combining, HARQ

HARQ is a link adaptation technique that works along with AMC by reducing the sensitivity to channel measurement and respective prediction error and traffic fluctuations. With HARQ, link layer acknowledgements are used for retransmission decisions. The terminal attempts to decode received transport block, buffers it and reports to the NodeB its success or failure in 5 ms intervals. A relatively short interval allows for rapid retransmissions of erroneous data blocks.

10.2.4 Retransmission Mechanism in the NodeB

The principle of HSPA physical-layer retransmission is shown in Figure 10.4. The HSDPA packet is first received in the buffer in the NodeB. The NodeB keeps the number of packets in the buffer even after sending it to the UE. In case of packet decoding error, the NodeB retransmits buffered packet.

The UE buffers the received data as well and then is able to combine both (or all) received replicas of the packet. As observed, the RNC is not involved in the described retransmission process that is performed at the physical layer.

Should physical-layer retransmission fail, then RNC based retransmission may still be applied on top, as shown in Figure 10.4. Typically, RLC level retransmission is applied in RNC due to signalling errors or in connection to serving cell change in mobility operation. The HSDPA RLC retransmission is normally 100 times less probable than NodeB retransmission.

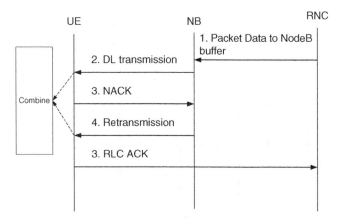

Figure 10.4 Packet retransmission principle in NodeB.

10.2.5   Impact to Protocol Architecture

The transfer of RRM functions, such as fast scheduling and packet retransmission, to the NodeB leads to the changes in protocol architecture. The specific HSDPA additions to user-plane protocol architecture are shown in Figure 10.5. In particular, the new protocol entity MAC-hs (high-speed) is placed in BTS. The MAC-hs handles the scheduling and user packet priority. The RNC retains part of MAC-d ('d' stands for dedicated) functionality, specifically, transport channel switching as all other functionalities while scheduling and priority handling are moved to MAC-hs in BTS. The higher layers, starting from RLC, have no impact from HSPA.

With HSUPA (E-DCH) there is also a new MAC entity added to the BTS, MAC-e, as shown in Figure 10.6.

A new HSUPA related protocol entity is also introduced in the terminal. This is a new MAC entity, MAC-es/e. The MAC-e resides in both the UE and NodeB and is responsible for fast HARQ retransmissions, scheduling and demultiplexing. The

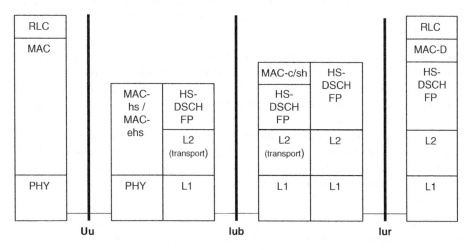

Figure 10.5 Protocol architecture of HS-DSCH, configuration with MAC-c/sh [2].

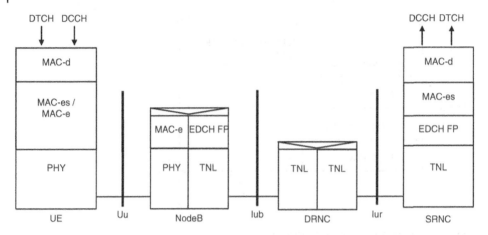

Figure 10.6 Protocol architecture of E-DCH [3].

MAC-es resides in RNC and the UE and responsible for in-sequence delivery of retransmitted packets (reordering) and enables uplink soft combining for data when UE is involved in inter-NodeB soft handover.

10.2.6 HARQ Schemes

There are many schemes for implementing HARQ: chase combining, rate compatible punctured turbo codes and incremental redundancy [1]. All these methods are based on retransmission of erroneous transport blocks, while some retransmit the same block, the other retransmit erroneous block with addition reductant information (parity bits, different puncturing pattern). The decoder at the receiver combines these multiple copies of the transmitted packet weighted by the received SNR.

The hybrid-ARQ functionality spans both the MAC-hs and the physical layer. As the MAC-hs is located in the NodeB, erroneous transport blocks can be rapidly retransmitted. Hybrid-ARQ retransmissions are therefore significantly less costly in terms of delay compared to RLC-based retransmissions that involve RNC. There are two fundamental reasons for this difference:

1) There is no need for signalling between the NodeB and the RNC for the hybrid-ARQ retransmission. Consequently, any Iub/Iur delays are avoided for retransmissions. Handling retransmission in the NodeB is also beneficial from a pure Iub/Iur capacity perspective; hybrid-ARQ retransmission comes at no cost in terms of transport-network capacity.
2) In order to reduce signalling over Iub/Iur interface, the RLC protocol is configured with a status report once per several TTIs, therefore it is slow compared with HARQ. In HSDPA, the hybrid-ARQ operates per transport block or, equivalently, per TTI. That is, whenever the HS-DSCH CRC indicates an error, a retransmission of transport block is then requested.

Incremental redundancy is the basic scheme for soft combining; that is, retransmissions may consist of a different set of coded bits than the original transmission. Different redundancy versions, that is, different sets of coded bits, are generated as part of the rate-matching mechanism, as described in Section 10.3. The rate matcher uses puncturing (or repetition) to match the number of code bits to the number of physical channel bits available. By using different puncturing patterns, different sets of coded bits, that is different redundancy versions, result. Chase combining is a special case of incremental redundancy. The NodeB decides whether to use incremental redundancy or chase combining by selecting the appropriate puncturing pattern for the retransmission.

The UE receives the coded bits and attempts to decode them. In case the decoding attempts fail, the UE buffers the received soft bits and requests a retransmission by sending a NAK. Once the retransmission occurs, the UE combines the buffered soft bits with the received soft bits from the retransmission and tries to decode the combination as shown in Figure 10.7.

In order to distinguish between new and retransmitted (old) data for soft combining, the downlink control signalling includes a new-data indicator. To operate properly, the UE needs to know whether the transmission is a retransmission of previously transmitted data or whether it is transmission of new data. For this purpose, the downlink control signalling includes a new-data indicator. When the retransmitted block arrives, the soft combining takes place, as shown in Figure 10.7 for block 2.

In practice, HSPA uses multiple HARQ processes carried for several UEs in parallel. Each of multiple HARQ uses 'stop-and-wait' protocol. There are two different methods for N-channel HARQ [1]:

1) either signal the sub-channel number explicitly (fully asynchronous), or
2) tie the sub-channel number to, for example, frame timing (partially asynchronous).

Method (1) is illustrated in the Figure 10.8, which shows an example sequence of events when packets 1–7 are being transmitted to UE1 and packet 1 to UE2 using the N-channel HARQ with N = 4 and 3 slot TTI. The UE1 is scheduled during four consecutive TTIs and the UE2 is allocated one TTI. Scheduled packets are transmitted using

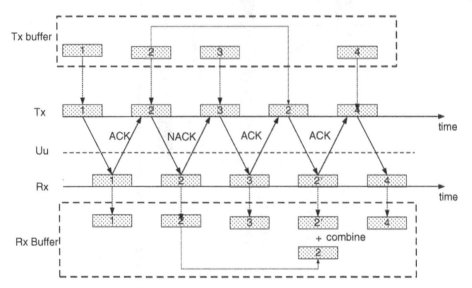

Figure 10.7 HARQ retransmission and soft combining.

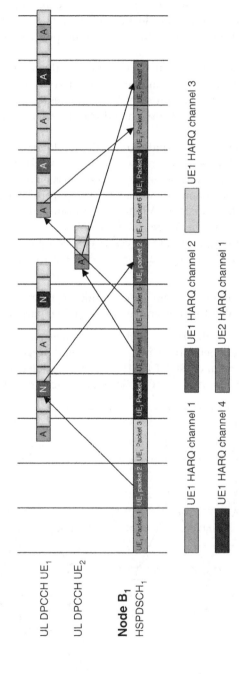

Figure 10.8 Principle N-channel stop-and-wait HARQ (N = 4) [1].

four parallel ARQ processes for UE1 and one ARQ process for UE2, respectively, each using stop-and-wait principle. Each packet is acknowledged during the transmission of other packets so that the downlink channel can be kept occupied all the time if there are packets to transmit. The receiver needs to know which HARQ process the packet belongs to that can be explicitly signalled on the HSDPA control channel.

The asynchronous feature of N-channel HARQ is also shown in Figure 10.8: after four packets to UE1, a packet is transmitted to UE2 and the transmission to UE1 is delayed by one TTI. Also, there are five packets to UE1 between packets to UE2.

10.3   MAC-hs and Physical-Layer Processing

The MAC-hs is comprised of four different functional entities: flow control, scheduling, hybrid-ARQ mechanism protocol and transport format selection (rate control) (see Figure 10.9).

1) *Flow Control*

   This is a flow control function providing a controlled data flow between the MAC-d and MAC-hs. The flow control takes dynamically the transmission capabilities of the air interface into account. This function assists in reducing discarded and retransmitted data as a result of HS-DSCH congestion and minimizes layer 2

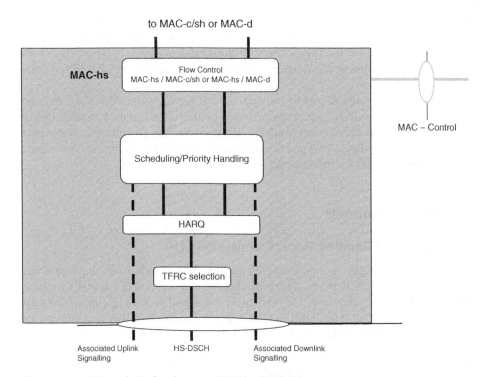

Figure 10.9 UTRAN side MAC architecture/MAC-hs details [2].

signalling. Flow control is provided independently per priority class for each MAC-d flow.

2) *Scheduling/Priority Handling*

This function manages HS-DSCH resources between HARQ entities and data flows according to their priority class. Based on status reports from associated uplink signalling either new transmission or retransmission is determined when operating in CELL_DCH state. When operating in CELL_FACH, CELL_PCH and URA_PCH state of the FDD system, the MAC-hs can perform retransmission without uplink signalling. Further it sets the priority class identifier and Transmission Sequence Number (TSN) for each new data block being serviced.

The TSN is unique to each priority class within a HS-DSCH, and is incremented for each new data block.

3) *HARQ*

One HARQ entity handles the hybrid-ARQ functionality for one user. One HARQ entity is capable of supporting multiple instances (HARQ process) of stop-and-wait HARQ protocols.

4) *TFRC selection*

Selection of an appropriate transport format and resource combination (TFRC) for the data to be transmitted on HS-DSCH.

The HS-DSCH physical-layer processing is straightforward. A 24-bit CRC is attached to each transport block. The CRC is used by the UE to detect errors in the received transport block. HS-DSCH supports multiple modulation schemes from QPSK to high order QAM.

The fundamental coding scheme in HSDPA is rate-*1/3 Turbo coding*. The rate matching in terms of puncturing or repetition is utilized in order to match physical channel data rate selected by MAC-hs. The rate matching mechanism is used in conjunction with HARQ to generate different redundancy versions for soft combining with incremental redundancy.

To obtain the code rate selected by the rate-control mechanism in the MAC-hs, rate matching, that is, puncturing or repetition, is used to match the number of coded bits to the number of physical-channel bits available. The rate-matching mechanism is also part of the physical-layer hybrid-ARQ and is used to generate different redundancy versions of retransmitted packets for incremental redundancy.

10.4 HSDPA Channels

10.4.1 High-Speed Downlink Shared Channel (HS-DSCH)

Instrumental for HSDPA is a new shared packet data channel for the downlink, HS-DSCH. The HS-DSCH is the transport channel that carries the actual user data with HSDPA. In the physical layer the HS-DSCH is mapped on the high-speed physical downlink shared channel (HS-PDSCH). The key features of HS-DSCH and differences from the Release 99 DCH-based packet data operation are as follows:

- The HS-DSCH has no power control. Instead, link adaptation selects the suitable combination of codes, coding rates and modulation to be used.

- Support of higher-order modulation than the DCH.
- Operates on short TTI = 2 ms.
- Use of physical-layer retransmissions and retransmission combining, while with DCH if retransmissions are used they are based on RLC level retransmissions.
- There is no soft handover mechanism applied to HS-DSCH. Data are sent from one serving HS-DSCH cell only.
- Control signalling for HS-PDSCH is carried instead on the separate control channel, HS-SCCH for HSDPA use and on the associated DCH (uplink power control etc.)
- The HS-DSCH corresponds to a set of channelization codes, each with a fixed spreading factor of 16. Each such channelization code is also known as an HS-PDSCH (High-Speed Physical Downlink Shared Channel). Up to 15 channelization codes can be configured for HS-DSCH.
- With HSDPA only turbo coding is used, while with the DCH convolutional coding may also be used.
- There is no discontinuous transmission (DTX) on the slot level. The HS-PDSCH is either fully transmitted or not transmitted at all during the 2-ms TTI (3 slots).

The terminal may also receive the concurrent DCH that carries services like circuit-switched AMR speech or video.

10.4.2   HSDPA Control Channels

The HS-DSCH is accompanied by the *High-Speed Shared Control Channel* (HS-SCCH) that carries *downlink control signalling*. The HS-SCCH uses a separate channelization code. The HS-SCCH is a shared channel, received by all UEs scheduled at HS-DSCH. The control information on HS-SCCH contains the identities of the UE(s) currently being scheduled as well as the physical resource (the channelization codes) and transport format used for transmission to that UE. Several HS-SSCHs can be configured in a single HSDPA cell depending on traffic needs and capability of scheduler.

HSDPA operation needs an uplink control channel to support channel-dependent scheduling, rate control and HARQ. In each TTI, the UE has to confirm HS-DSCH decoding with ACK or NAK sent on uplink. This data is carried on the uplink. The control channel on uplink the uplink *High-Speed Dedicated Physical Control Channel* (HS-DPCCH). Each HSDPA user has associated HS-DPCCH setup in the cell.

Since uplink data transfer in HSDPA is supported with the Release 99 DCH channel, the HSDPA UE need power control information from NodeB in support of the fast closed-loop power control. The power control commands are mapped to new downlink control channel, Fractional DPCH (F-DPCH). The F-DPCH contains power control commands for up to 10 UEs. In that way, all users share a single downlink channelization code.

Table 10.1 provides a summary of HSPDA channels.

None of the downlink HSDPA channels (HS-DSCH, HS-SSCH, F-DPCH) can be in any kind of handover, since they are scheduled from a single cell. However, *the uplink channels*, as well as any Rel99 dedicated downlink channels, *can be in soft handover*. As these channels are not subject to channel-dependent scheduling, macrodiversity provides a direct coverage benefit.

Table 10.1 HSPDA channels.

Channel name	Purpose	Nature of resource	Direction
HS-PDSCH	User data	Shared	DL
HS-SCCH	Signalling associated with HS-DSCH	Shared	DL
F-DPCH	Power control	Dedicated	DL
DPDCH	User data	Dedicated	UL
DPCCH	Signalling associated with DPDCH	Dedicated	UL
HS-DPCCH	Signalling associated with HS-DSCH	Dedicated	UL

10.4.2.1 Fractional Downlink Power Control Channel

As described before, for each UE for which HS-DSCH can be transmitted, there is also an associated downlink DPCH. In principle, if all data transmission, including RRC signalling, is mapped to the HS-DSCH, there is no need to carry any user data on the DPCH. Consequently, there is no need for downlink Transport Format Combination Indicator (TFCI) or dedicated pilots on such a DPCH. In this case, the only use for the downlink DPCH in case of HS-DSCH transmission is to carry power control commands to the UE in order to adjust the uplink transmission power. This fact is exploited by the F-DPCH or *fractional DPCH*, introduced in Release 6 as a means to reduce the amount of downlink channelization codes used for dedicated channels. Instead of allocating one DPCH with spreading factor 256 for the sole purpose of transmitting one power control command per slot, the F-DPCH allows up to 10 UEs to share a single channelization code for this purpose.

In essence, the F-DPCH is a slot format supporting TPC bits only. Two TPC bits (one QPSK symbol) is transmitted in one tenth of a slot, using a spreading factor 256, and the rest of the slot remains unused. By setting the downlink timing of multiple UEs appropriately, up to ten UEs can then share a single channelization code, see Figure 10.10. This can also be seen as time-multiplexing power control commands to several users on one channelization code.

10.4.3 HS-DSCH Link Adaptation

Link adaptation operates at HS-DSCH over 2 ms TTI. As a part of scheduling decision, the MAC-hs in the NodeB decides coding and modulation format for transmission in each TTI. The input information for decision is provided by the UE in form of Channel-Quality Indicators (CQI). Link adaptation switches between the modulation-and-coding levels to maximize the throughput for given values of CIR and CQI. By means of switching from QPSK to 16QAM and changing the coding rates and the number of codes downlink power control dynamic range can be effectively improved without extensive increase of output power. The goal of link adaptation is an optimal section of transmission parameters in order to match output power to a minimum symbol energy/interference ratio required for acceptable error rate with requested service level throughput. Figure 10.11 illustrates the principle of link adaptation based on symbol energy/interference ratio, E_s/N_o.

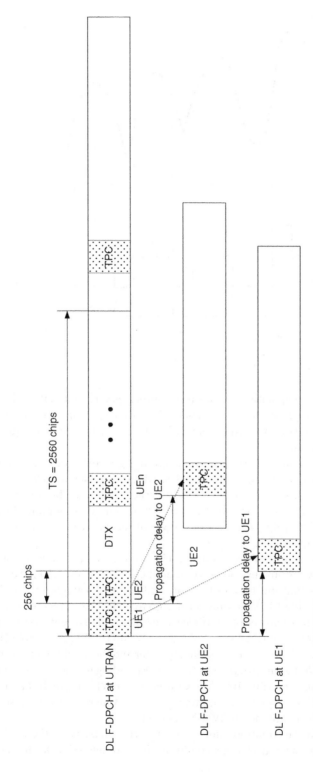

Figure 10.10 Fractional DPCH [4].

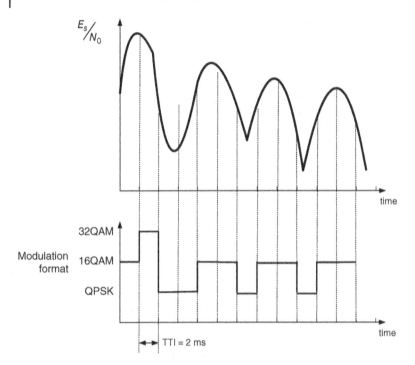

Figure 10.11 Link adaptation based on CIR.

The CQI does not explicitly indicate the channel quality, but rather the data rate supported by the UE given the current channel conditions. More specifically, the CQI directly relates to a recommended transport-block size, that, in turn, is equivalent to a recommended data rate.

Each 5-bit CQI value corresponds to a given transport-block size, modulation scheme and number of channelization codes. Different tables are used for different UE categories as a UE shall not report a CQI exceeding its capabilities. For example, a UE only supporting five codes shall not report a CQI corresponding to 15 codes, while a 15-code UE may do so. Therefore, power offsets are used for channel qualities exceeding the UE capabilities.

A power offset of Δ dB indicates that the UE can receive a certain transport-block size, but at Δ dB lower transmission power than the CQI report was based upon. This is illustrated in Tables 10.2 and 10.3 for two some different UE categories. UEs belonging to a category 1–6 (Table 10.2) can only receive up to 5 HS-DSCH channelization codes and therefore must use a power offset (Reference Power Adjustment, Δ) for the highest CQI values, while category 10 UEs are able to receive up to 15 codes (Table 10.3).

The CQI values listed are sorted in ascending order and the UE shall report the highest CQI for which transmission with parameters corresponding to the CQI result in a block error probability not exceeding 10%. The CQI values are chosen such that an increase in CQI by one step corresponds to approximately 1 dB increase in the instantaneous carrier-to-interference ratio on an AWGN channel.

Measurements on the common pilot (CPICH) form the basis for the CQI. The CQI represents the instantaneous channel conditions in a predefined 3-slot interval ending

Table 10.2 CQI mapping for UE category 1–6 [4].

CQI value	Transport-Block Size	Number of HS-PDSCH codes	Modulation	Reference power adjustment Δ	N_{IR}	X_{rv}
0	N/A			Out of range		
1	137	1	QPSK	0	9600	0
2	173	1	QPSK	0		
3	233	1	QPSK	0		
4	317	1	QPSK	0		
5	377	1	QPSK	0		
6	461	1	QPSK	0		
7	650	2	QPSK	0		
8	792	2	QPSK	0		
9	931	2	QPSK	0		
10	1262	3	QPSK	0		
11	1483	3	QPSK	0		
12	1742	3	QPSK	0		
13	2279	4	QPSK	0		
14	2583	4	QPSK	0		
15	3319	5	QPSK	0		
16	3565	5	16-QAM	0		
17	4189	5	16-QAM	0		
18	4664	5	16-QAM	0		
19	5287	5	16-QAM	0		
20	5887	5	16-QAM	0		
21	6554	5	16-QAM	0		
22	7168	5	16-QAM	0		
23	7168	5	16-QAM	−1		
24	7168	5	16-QAM	−2		
25	7168	5	16-QAM	−3		
26	7168	5	16-QAM	−4		
27	7168	5	16-QAM	−5		
28	7168	5	16-QAM	−6		
29	7168	5	16-QAM	−7		
30	7168	5	16-QAM	−8		

Table 10.3 CQI mapping for UE category 10 [4].

CQI or CQI$_S$ value	Transport-Block Size	Number of HS-PDSCH codes	Modulation	Reference power adjustment Δ	N_{IR}	X_{rv} or X_{rvpb}
0	N/A			Out of range		
1	137	1	QPSK	0	28 800	0
2	173	1	QPSK	0		
3	233	1	QPSK	0		
4	317	1	QPSK	0		
5	377	1	QPSK	0		
6	461	1	QPSK	0		
7	650	2	QPSK	0		
8	792	2	QPSK	0		
9	931	2	QPSK	0		
10	1262	3	QPSK	0		
11	1483	3	QPSK	0		
12	1742	3	QPSK	0		
13	2279	4	QPSK	0		
14	2583	4	QPSK	0		
15	3319	5	QPSK	0		
16	3565	5	16QAM	0		
17	4189	5	16QAM	0		
18	4664	5	16QAM	0		
19	5287	5	16QAM	0		
20	5887	5	16QAM	0		
21	6554	5	16QAM	0		
22	7168	5	16QAM	0		
23	9719	7	16QAM	0		
24	11 418	8	16QAM	0		
25	14 411	10	16QAM	0		
26	17 237	12	16QAM	0		
27	21 754	15	16QAM	0		
28	23 370	15	16QAM	0		
29	24 222	15	16QAM	0		
30	25 558	15	16QAM	0		

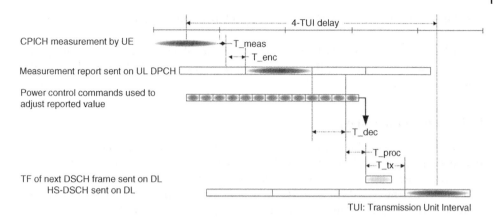

Figure 10.12 Use of power control commands for DL channel-quality report adjustment [1].

one slot prior to the CQI transmission. Specifying which interval the CQI relates to allows the NodeB to track changes in the channel quality between the CQI reports by using the power control commands for the associated downlink (F-)DPCH as described next. The timing of the CQI reports and the earliest possible time the report can be used for scheduling purposes is illustrated in Figure 10.12.

10.5 HSUPA (Enhanced Uplink, E-DCH)

There are two basic technologies at the core of Enhanced Uplink used also for HSDPA – fast scheduling and fast hybrid-ARQ with soft combining. For similar reasons as for HSDPA, Enhanced Uplink also introduces a short 2 ms uplink TTI. These enhancements are implemented in WCDMA by means of a new transport channel, the *Enhanced Dedicated Channel* (E-DCH).

Although the same technologies are used both for HSDPA and Enhanced Uplink, there are fundamental differences between them that affected implementation of the features:

- In the downlink, the shared resources include transmission power and the code space, both are located in *one* central node, the NodeB. In the uplink, the shared resource is the amount of allowed uplink interference, which depends on the transmission power of *multiple distributed* nodes, the UEs.
- The scheduler and the transmission buffers are located in the same node in the downlink, while in the uplink the scheduler is located in the NodeB while the data buffers are distributed across the UEs. Hence, the UEs need to signal the buffer status information to the scheduler.

The WCDMA uplink, also with Enhanced Uplink, is inherently non-orthogonal and subject to interference from uplink transmissions within the same cell. This is in contrast to the downlink, where different transmitted channels are *orthogonal*. Fast power control is therefore essential for the uplink to handle the near-far problem. Power control is applied to the uplink control channel. The E-DCH is transmitted with power offset relative to the power of control channel just follow power control command. The uplink scheduler controls the E-DCH data rate by adjusting the maximum allowed power offset.

Soft handover is supported by the E-DCH just enabling macrodiversity for uplink. Soft handover also implies uplink *power control by multiple cells* in order to limit the amount of interference.

10.5.1 Control Signalling

Downlink control signalling is necessary for the operation of the E-DCH. The downlink, as well as uplink, control channels used for E-DCH support are listed in Table 10.4.

The E-channels listed in Table 10.4 support scheduling decisions and uplink data transmission. The *E-DCH Hybrid-ARQ Indicator Channel* (E-HICH) transmits ACK/NACK messages related to reception of E-DCH channel by NodeB as a part of HARQ procedure. In case of a soft handover, each connected cell sends one E-HICH to the respective UE. The shared *E-DCH Absolute Grant Channel* (E-AGCH) carries scheduling grants to the UE from NodeB scheduler. The grant contains information on timing and data rate for uplink transmission. The E-AGCH is sent from the serving cell and is received by all UEs with an E-DCH configured. In addition, scheduling grant information can also be communicated to the UE through an *E-DCH Relative Grant Channel* (E-RGCH). The E-AGCH is typically used for large changes in the data rate, while the E-RGCH is used for smaller adjustments during an ongoing data transmission.

Given that uplink transmission format is issued in scheduling grant, the serving cell has that knowledge necessary to decode the E-DCH. However, non-serving cell in soft handover needs to be signalled about transmission format for E-DCH. This control signalling is outband and supported by the *E-DCH Dedicated Physical Control Channel* (E-DPCCH). The power control for E-DCH is performed by NodeB, which measures the received signal-to-interference ratio and sends power control commands in the downlink to the UE using DPCH or F-DPCH.

10.5.2 Scheduling

The scheduling framework is based on *scheduling grants* sent by the NodeB to the UE. The Scheduling Grants have the following characteristics:

- Scheduling Grants are only to be used for the E-DCH TFC selection algorithm (i.e. they do not influence the TFC selection for the DCHs);

Table 10.4 Enhanced uplink physical channels.

Channel name	Purpose	Nature of resource	Direction
E-DPCH	User data	Dedicated	UL
E-DPCCH	Signalling associated with E-DPCH	Dedicated	UL
E-AGCH	Absolute grant	Shared	DL
E-RGCH	Relative grant	Dedicated	DL
E-HICH	Hybrid-ARQ ACK/NACK	Dedicated	DL

- Scheduling Grants control the maximum allowed E-DPDCH/DPCCH power ratio of the active processes. For the inactive processes, the power ratio is 0 and the UE is not allowed to transmit scheduled data;
- All grants are deterministic;
- Scheduling Grants can be sent once per TTI or slower.

There are two types of grant:

1) The Absolute Grants provide an absolute limitation of the maximum amount of uplink resources the UE may use. The UE select the data rate or, more precisely, the *E-DCH Transport Format Combination* (E-TFC) within the restrictions set by the scheduler and maintains the *Serving Grant* (SG).
2) The Relative Grants increase or decrease the resource limitation compared to the previously used value.

Serving Grants are sent by the Serving E-DCH cell. In a soft handover, the serving cell takes main responsibility for scheduling based on load and interference level in serving cell. On the other hand, non-serving cell monitors its own load and may request non-serving UEs to reduce data rate on their E-DCHs by transmitting overload indicators in the downlink, as shown in Figure 10.13.

The NodeB can update the serving grant in the UE by sending an *absolute grant* or a *relative grant* to the UE. Absolute grants are transmitted on the shared E-AGCH and are used for absolute changes of the serving grant. Typically, these changes are relatively large; for example to assign the UE a high data rate for an upcoming packet transmission.

Relative grants are transmitted on the E-RGCH and are used for relative changes of the serving grant. Unlike the Absolute Grants, these changes are small; the change in transmission power due to a relative grant is typically in the order of 1 dB. Relative grants can be sent from both serving and, in case of the UE being in soft handover, also from the non-serving cells.

For efficient scheduling, the scheduler needs information from UE about its buffer status and available transmission power. The UE indicates it status by sending the outband 'happy bit' on E-DPCCH. The 'happy bit' means a request for a E-DCH higher data rate stating that the UE has sufficient power available for transmission and substantial amount of data in the buffer.

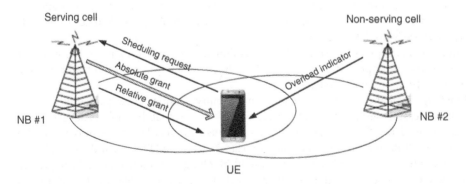

Figure 10.13 Enhanced Uplink scheduling.

10.6   Air-Interface Dimensioning

The design and dimensioning of 3G networks depend significantly on the offered traffic and its mix. Besides the voice application, the main applications of 3G networks require high data rate bearers with symmetric and asymmetric data flow. The grade of service (GOS) required for all these services has to be considered during the coverage and capacity planning phase. Detailed coverage and capacity dimensioning of the WCDMA air interface can be done with help of the SW planning and dimensioning tool. In this section, we discuss both principles of dimensioning and approach to obtaining rough estimates for coverage and capacity.

10.6.1   Input Parameters and Requirements

In 3G mobile networks, a mixture of different traffic types and user profiles have to be taken into account. Because of the close relation between offered traffic and its profile and the provided coverage via the interference situation, an exact knowledge of the expected traffic is mandatory for any accurate planning.

Evaluation of cell range is no different from other technologies, it can be obtained from propagation loss model equation given the estimated MAPL value from link budget. The MAPL values can be different for different bearers and depend on the cell load in own and neighbouring cells.

The quality requirements for the network comprise firstly the coverage requirements (already mentioned) and the quality of service aspects. There are four main QoS classes described in the 3GPP standard – conversational, streaming, interactive and background. Initial network design and dimensioning are mainly based on the best effort approach for PS applications and on the commonly known blocking requirement for CS applications.

Important QoS parameters for traffic modelling in the scope of UTRAN are:

- Radio access bearer rate (peak rate, maximum bit rate, and information rate) In UTRAN, the radio access bearer (RAB) defines the maximum data rate (maximum bit rate attribute allocated during RAB setup) accessible for each connection. In ideal situations, where no transmission errors occur, this data rate can be achieved. This rate is denoted as the information rate for CS services and as the peak rate for PS services.
- Mean rate. This term is mainly used for PS services. It denotes the mean data rate averaged over the session (or RAB life), taking into account possible inactive periods where there is no data transmission.
- Traffic demand per subscriber. The traffic demand per subscriber is often given as mean throughput per subscriber in the busy hour, typically in bps or as traffic volume per busy hour, in bits or Bytes.
- Blocking probability. The blocking probability defines the probability of a non-successful setup of a RAB because of non-availability of resources. It is mainly used for CS services. It occurs because of insufficient resources on the respective links (e.g. air interface, Iub interface) or/and in UTRAN and core network elements (e.g. RNC, MSC). The end-to-end blocking is influenced by all the factors introduced by certain sections of the network. PS services are assumed to be always accepted – however, with reduced data rate in terms of lack of resources.

- Delay (in terms of a time offset to the perfect transmission time). A delay requirement is defined for PS services. This delay denotes the time offset in data transmission, taking into account the RAB rate and the packet or file sizes to be transmitted. Overall end-to-end delay includes the propagation delay on the links, processing delay within network elements (UTRAN and CN), and buffering delay because of insufficient resources on the respective links.
- RAB life and session. Distinction is made between RAB life and DCH session in UTRAN (denoted by RAB session) and the session in the core network. A session in SGSN and GGSN refers to 'attach' and 'primary/secondary' PDP context. In addition, sessions are distinguished for each application. Note that there may be an interruption in RAB sessions because of time-outs. During the RAB Life (continuously, if RAB is mapped on R99 RT radio bearer or discontinuously if NRT R99), the user might have dedicated resources or share common resources (e.g. traffic inactivity of NRT R99).

10.6.2 Traffic Demand Estimation

Speech, voice telephony and PS RT services are characterized by:

- Busy-hour-call-attempts (BHCA)
- Call duration (s)
- Bearer rate (information rate) (kbps)
- Activity (DCH activity)

The traffic demand is calculated by

$$\text{Traffic}_{\text{volume}}[\text{kb}] = \text{No}_{\text{subscribers}} \cdot \text{BHCA} \cdot \text{Call}_{\text{Duration}} \cdot \text{Bearer}_{\text{Rate}} \cdot \text{activity\_factor}$$

$$\text{Throughput [kbps]} = \frac{\text{Traffic\_volume[kb]}}{3600 \text{ s}} \tag{10.1}$$

$$\text{Traffic}_{\text{demand}}[\text{mErl}] = \frac{\text{No\_subscribers} \cdot \text{BHCA} \cdot \text{Call\_Duration} \cdot 1000}{3600}$$

For voice services the DCH activity depends on the interface in the UTRAN to be considered. Default value of activity on-air interface is 0.5.

10.6.2.1 PS Data Services (Release 99)

The traffic demand is calculated using the same formulas (10.1) with activity factor taken for DCH activity. Note that one DCH session per BHCA without Channel Type Switching (CTS) is assumed. In case of channel type switching, the DCH session is interrupted. All parts of the DCH session can be added to a single DCH duration. Channel type switching is included in the signalling traffic demand by the number of CTS events assumed per (PS) subscriber.

10.6.2.2 HSPA Data Services

Relevant parameters for calculating the traffic demand for HSDPA services are:

- Busy-hour-call-attempts (BHCA)
- HS-DSCH (session) duration (s)
- HS-DSCH nominal peak rate (kbps)
- HS-DSCH activity

$$\text{Traffic\_volume [kb]} = \text{No}_{\text{subscribers}} \cdot \text{BHCA} \cdot \text{HS} - \text{SDCH}_{\text{duration}} \cdot \text{Nominal}_{\text{peak}_{\text{rate}}}$$
$$\cdot \text{HS} - \text{SDCH\_activity\_factor}$$

$$\text{Throughput[kbps]} = \frac{\text{Traffic\_volume[kb]}}{3600 \text{ s}} \qquad (10.2)$$

$$\text{Traffic\_demand[mErl]} = \frac{No_{\text{subscribers}} \cdot BHCA \cdot HS - SDCH\_duration \cdot 1000}{3600}$$

One HS-DSCH session per BHCA without channel type switching is assumed. In the case of channel type switching, the HS-DSCH session is interrupted. All parts of the HS-DSCH session can be added up to a single HS-DSCH duration. Channel type switching is included in the signalling traffic demand by the number of CTS events assumed per (PS) subscriber.

Relevant parameters for calculating the HSUPA traffic demand are:

- Busy-hour-call-attempts (BHCA)
- E-DCH (session) duration (s)
- E-DCH nominal peak rate (kbps)
- E-DCH activity.

The UL traffic demand calculates as follows:

$$\text{Traffic}_{\text{volume [kb]}} = \text{No}_{\text{subscribers}} \cdot \text{BHCA} \cdot \text{E} - \text{DCH}_{\text{duration}} \cdot \text{Nominal}_{\text{peak}_{\text{rate}}} \cdot \text{E}$$
$$- \text{DCH\_activity\_factor}$$

$$\text{Throughput [kbps]} = \frac{\text{Traffic}_{\text{volume}}[\text{kb}]}{3600 \text{ s}} \qquad (10.3)$$

$$\text{Traffic}_{\text{demand}}[\text{mErl}] = \frac{\text{No}_{\text{subscribers}} \cdot \text{BHCA} \cdot \text{E} - \text{DCH\_duration} \cdot 1000}{3600}$$

One E-DCH session per BHCA without channel type switching is assumed. In case of channel type switching, the E-DCH session is interrupted. All parts of the E-DCH session can be added to a single E-DCH duration. Channel type switching is included in the signalling traffic demand by the number of CTS events assumed per subscriber (PS). Note that because the HS-DSCH duration and E-DCH duration are identical, UL E-DCH is always accompanied by DL HS-DSCH.

The R99 UL HSUPA traffic demand is calculated as described in the previous section for R99 DCH sessions.

10.6.3 Standard Traffic Model

The purpose of a standard traffic model is to have information about the traffic demand available if no detailed traffic model is provided for network dimensioning. Standard traffic model is defined assuming a 'standard' subscriber, using all accounted services in parallel – the service split is applied on the traffic demand basis (no split on the subscriber basis). Selected default services are:

- Speech with 0.6 BHCA and a traffic demand of 24.5 mErlang per subscriber in the busy hour, which corresponds to a call duration of 147 s, mapped to CS Conversational AMR 12.2 and CS voice over HSPA.

- Video telephony mapped to CS Conversational class, 64/64 kbps RAB, with 0.05 BHCA and a traffic demand of 3.2 mErlang per subscriber in the busy hour, which corresponds to a call duration of 230s.
- Data services mapped to PS Interactive/Background RAB services, differentiated between R99 and HSxPA with 0.34 BHCA and a traffic demand (in DL) of 1430 bps per subscriber in the busy hour. This traffic demand is split into:
  - R99 traffic demand per subscriber: 480 bps
  - HSxPA traffic demand per subscriber: 950 bps

These are applicable assumptions:

- BHCA split between R99 – HSxPA (Release 5/6) of 60% : 40%
- BHCA split between Release 5 UE and Release 6 UE of 25% : 15%

Asymmetry:

- overall (UL:DL): 1:4.54
- R99 (UL:DL): 1:5.1
- HSDPA Release 5 (UL R99 : DL HSDPA): 1:4.3
- HSxPA Release 6 (UL HSUPA : DL HSDPA): 1:4.3

10.6.4   Link Budgets

There are some WCDMA-specific parameters in the link budget that are not used in a TDMA-based radio access system such as GSM. The most important ones are as follows:

- *Interference margin (noise rise):* The interference margin is needed in the link budget because the loading of the cell, the load factor, affects the coverage. The more loading is allowed in the system, the larger the interference margin needed in the uplink and the smaller the coverage area. For coverage-limited cases a smaller interference margin is suggested, while in capacity-limited cases a larger interference margin should be used. In the coverage-limited cases, the cell size is limited by the maximum allowed path loss in the link budget, and the maximum air-interface capacity of the base station site is not used. Typical values for the interference margin in the coverage-limited cases are 1.0–3.0 dB, corresponding to 20–50% loading.
- *Fast fading margin (power control headroom):* Some allowance is needed in the mobile station transmission power for maintaining adequate closed-loop fast power control. This applies especially to slow-moving pedestrian mobiles where fast power control is able to effectively compensate for the fast fading. Typical values for the fast fading margin are 2.0–5.0 dB for slow-moving mobiles.
- *Soft handover gain:* Handovers – soft or hard – give a gain against slow (lognormal) by reducing the required lognormal fading margin. This is because the slow fading is partly uncorrelated between the base stations and by making a handover the mobile can select a better base station. Soft handover gives an additional macrodiversity gain against fast fading by reducing the required E_b/N_0 relative to a single radio link, due to the effect of macrodiversity combining. The total soft handover gain is assumed to be between 2.0 and 3.0 dB in the examples that follow, including the gain against slow and fast fading.

Tables 10.5 and 10.6 summarize typical assumptions for a mobile and a base station, respectively.

Table 10.5 Typical assumptions for the mobile station.

MS Transmission power	23 dBm
MS Antenna gain	0 dBi
Body loss	3 dB

Table 10.6 Typical assumptions for the base station.

Noise figure	5 dB
Antenna gain	18 dBi
Cable loss	3 dB
E_b/N_0 requirements	5.0 dB for voice
	1.5 dB for 144 kbps data
	1 dB for 384 kbps

10.6.4.1 Uplink Load Factor

The theoretical spectral efficiency of a WCDMA cell can be calculated from the load equation whose derivation is shown in equation (10.4). We first define

$$\left(E_b/N_0\right)_j = \text{Processing gain of user}_j \cdot \frac{\text{Signal power of user } j}{\text{Total received power} - \text{power of user } j} \tag{10.4}$$

This can be written:

$$\left(E_b/N_0\right)_j = \frac{W}{v_j R_j} \cdot \frac{P_j}{P_{tot} - P_j} \tag{10.5}$$

where W is the chip rate, Pj is the received signal power from user j, v_j is the activity factor of user j, Rj is the bit rate of user j, and P_{tot} is the total received wideband power including thermal noise power in the base station. Solving for Pj gives

$$P_j = \frac{P_{tot}}{1 + \dfrac{W}{(E_b/N_0)_j v_j R_j}} \tag{10.6}$$

We define $P_j = \eta_j \cdot P_{tot}$ and obtain the load factor η_j of one connection

$$\eta_j = \frac{1}{1 + \dfrac{W}{(E_b/N_0)_j v_j R_j}} \tag{10.7}$$

The total received interference, excluding the thermal noise P_N, can be written as the sum of the received powers from all N users in the same cell

$$P_{tot} - P_N = \sum_{j=1}^{N} P_j = P_{tot} \sum_{j=1}^{N} \eta_j \tag{10.8}$$

The noise rise is defined as the ratio of the total received wideband power to the noise power

$$\text{Noise rise} = \frac{P_{tot}}{P_N} \tag{10.9}$$

and using equation (10.8) for P_{tot} we obtain

$$\text{Noise rise} = \frac{P_{tot}}{P_N} = \frac{1}{1 - \eta_{UL}} \tag{10.10}$$

where we have defined the load factor η_{UL} as

$$\eta_{UL} = \sum_{j=1}^{N} \eta_j \tag{10.11}$$

When η_{UL} becomes close to 1, the corresponding noise rise approaches infinity and the system has reached its pole capacity. Additionally, in the load factor the interference from the other cells must be taken into account by the ratio of other cell to own cell interference, i:

$$i = \frac{\text{other cell interference}}{\text{own cell interference}} \tag{10.12}$$

The uplink load factor can then be written as

$$\eta_{UL} = (1 + i) \sum_{j=1}^{N} \eta_j = (1 + i) \sum_{j=1}^{N} \frac{1}{1 + \dfrac{W}{(E_b/N_0)_j v_j R_j}}. \tag{10.13}$$

The load equation predicts the amount of noise rise over thermal noise due to interference. The noise rise in dB scale defines uplink interference margin

$$M_{UL\ inf} = -10 \cdot \log_{10}(1 - \eta_{UL}) \tag{10.14}$$

The interference margin in the link budget must be equal to the maximum planned noise rise.

10.6.4.2 Downlink Load Factor

The downlink load factor, η_{DL}, can be defined based on a similar principle as for the uplink, although the parameters are slightly different:

$$\eta_{DL} = \sum_{j=1}^{N} v_j \frac{(E_b/N_0)_j}{W/R_j} [(1 - \alpha_j) + i_j] \tag{10.15}$$

The interference margin on downlink

$$M_{DL\_inf} = -10 \cdot \log_{10}(1 - \eta_{DL}) \tag{10.16}$$

is equal to the noise rise over thermal noise due to multiple-access interference. The parameter α_j represents orthogonality factor in the downlink. WCDMA employs orthogonal codes in the downlink to separate users and without any multipath propagation the orthogonality remains when the base station signal is received by the mobile. However, if there is sufficient delay spread in the radio channel, the mobile will see part of the base station signal as multiple-access interference. The orthogonality of 1 corresponds to perfectly orthogonal users. Typically, the orthogonality is between 0.4 and 0.9 in multipath channels.

10.6.4.3 Link Budget for R99 Bearers

Maximum path loss for uplink and downlink are

Uplink:

$$
\begin{aligned}
L_{\max\_UL} = {} & P_{UE} + G_{a,UE} + G_{a,NB} - L_{fedeer,NB} - \text{Information\_Rate} \\
& - \text{Thermal\_noise\_density} - NF_{NB} - {}^{E_b}/_{N_0} - M_{\inf\_UL} \\
& + G_{SHO} - M_{LNF} - M_{fast} - M_{body}
\end{aligned}
\tag{10.16}
$$

Downlink:

$$
\begin{aligned}
L_{\max\_DL} = {} & P_{NB\_per\_user} + G_{a,UE} + G_{a,NB} - L_{feeder,NB} - \text{Information\_rate} \\
& - \text{Thermal\_noise\_density} - NF_{UE} - {}^{E_b}/_{N_0} - M_{Inf\_DL} \\
& + G_{SHO} - M_{LNF} - M_{fast} - M_{body}
\end{aligned}
\tag{10.17}
$$

Parameters in (10.16) and (10.17) are as follows:

In the uplink, the P_{UE} [dBm] is a maximum output power of the user equipment. In the downlink, the total power is divided between signalling and traffic channels. $P_{NB\_per\_user}$ [dBm] is the power of the WCDMA BTS dedicated to one user.

In WCDMA, the base station serves all active users simultaneously. As a result, the total power P_{tot} of the WCDMA BTS must be divided into the power reserved for the signalling and the N_{user} users served

$$
P_{NB\_per\_user} = 10 \log \left(P_{tot} \frac{(1 - \text{sig}(\%))}{N_{user}} \right),
\tag{10.18}
$$

where sig(%) is a fraction of total NodeB transmit power assigned to control channels. The value sig~20% can be used in rough calculation. Power assigned to a single user ($P_{NB\_per\_user}$ [dBm]) cannot exceed a certain limit (Pmax) set by the RNC.

The other parameters in (10.17) and (10.18) are:

- $G_{a,\,NB}$ [*dBi*] Antenna gain of the WCDMA BTS antenna.
- $G_{a,\,UE}$ [*dBi*] Antenna gain of the user equipment antenna.
- $L_{Lfeeder,\,NB}$: feeder loss between the WCDMA BTS and the antenna connector. At the site with the legacy antenna-feeder connection, a tower mounted amplifier (TMA) is usually installed in uplink path. The TMA will compensate for the feeder loss between the receiver antenna and the WCDMA BTS. In case of TMA the $L_{feeder,NB}$ loss could be omitted, but additional *LTMA* − TMA Insertion Loss [dB] should be added to downlink budget.
- Information_Rate = 10 log (*Rb*) [dB/Hz] is the channel bit rate. *Rb* is the bit rate in [bps] of the considered bearer. Please note that, in the link budget formulas presented earlier, the processing gain is taken into account by counting noise only over information rate of the channel; that is, Radio Access Bearer rate.
- Thermal_Noise_Density = −174 dBm/Hz.
- NF_{NB} and NF_{UE} [dB]: noise figure of the NodeB and the UE, respectively.
- E_b/N_0 [dB], E_b/N_0 is the minimum value of energy received per bit relative to the noise.

- $M_{UL,DL\ inf}$ [dB]: interference margins for uplink and downlink given by formulas (10.14) and (10.16). With these margins, the dependency of the cell range on the traffic load in the cell is considered (cell breathing).
- G_{SHO} [dB]: Soft handover gain.
- M_{body} [dB]: body loss.
- M_{LNF} [dB]: shadowing margin (Slow Fade Margin or Lognormal Fade Margin).
- M_{fast} [dB]: Fast fading margin (power control headroom).

10.6.4.4 Link Budget for HSPA

The link budget equation is similar to (10.16) and (10.17) with some adjustments to inclusion of margins and losses.

$$L_{\text{max\_DL}} = P_{NB-per\ user} + G_{a,NB} + G_{a,UE} - L_{feeder,NB} - \text{Information \_Rate}$$
$$- \text{Thermal\_Noise\_Density} - NF_{UE} - {^{E_b}/_{N_0}} \tag{10.19}$$
$$- M_{\text{inf \_DL}} - M_{LNF} - M_{fast} - M_{body}$$

$$L_{\text{max\_UL}} = P_{UE} + G_{a,UE} - L_{feeder,NB} - \text{Information \_Rate}$$
$$- \text{Thermal\_Noise\_Density} - NF_{NB} - {^{E_b}/_{N_0}} - M_{\text{inf \_UL}} \tag{10.20}$$
$$+ G_{SHO} - M_{HS-DPCCH} - M_{LNF} - M_{body}$$

The parameters in (10.19) and (10.20) are as follows:

- $P_{NB_{per}user}$ –HSDPA power (per user in one TTI) [%]. The maximum amount of power that can be dedicated to one HSDPA user is determined by setting relevant RNC data base parameter. Default value of $P_{NB_{per}user}$ is set to 6 dB over the CPICH power level. With a CPICH power of 33 dBm (usually 10% of the total Tx power), the maximum HSDPA power per user can be estimated as a 40% of the total Tx power (~39 dBm). Power per HSDPA user can be used by one or multiple HS-PDSCH codes assigned to this one user. The total power per cell dedicated to HSDPA is normally at maximum of 60% of the total Tx power. The other 40% of the power is reserved for common control channels, HS-SCCH channels and associated DCH channels in the cell.
- $M_{HS-DPCCH}$ is HS-DPCCH overhead (dB). In then UL direction, of HSDPA/Rel99 Link Budget, additional margin for HS-DPCCH channel is considered. HS-DPCCH channel includes the ACK/NACK and CQI. HS-DPCCH Overhead is dependent upon the selected associated DCH (16/64/128/384).

10.6.4.5 Results of Link Budget: Cell Range Calculation, Balancing UL with DL

The main result of the link budget calculation is the estimation of the cell range for different area types and environment types. Cell range estimation is similar to any other radio technology. Cell range is calculated from solving the propagation equation for the maximum allowable path loss MAPL. The cell range estimate depends on the bearer type used in calculation of link budget.

As the links are usually not balanced, the cell ranges calculated for downlink and uplink may differ. The worst (smaller) result should always be taken. In general, changing the load or data rate may balance the links. The process of balancing may be performed as follows:

- Identify the limiting bearer in UL: bearer_UL;
- Identify the limiting bearer in DL: bearer_DL;
- Identify the limiting link: for example, UL.

In order to balance DL with bearer_DL one may change the load (or Data Rate at the cell edge) in the link budget DL to match the UL cell range of the bearer_UL. The result of the link budget calculation can be used as input for the next step of dimensioning – traffic calculation.

10.6.4.6 Link Budget for Common Pilot Channel Signal

The coverage of the pilot signal should be at least as good as the coverage of the traffic channels. To check this, calculate a link budget for the pilot. The way the Interference Margin is calculated in CPICH channel link budget is the same as for the HSDPA DL channel. The pilot range calculated should be greater than or equal to the cell range. The Tx power for the common pilot channel is given as a percentage of the total transmit power of a WCDMA BTS. This is usually equal to around 10% (CPICH power is a part of the overall signalling power where the other signalling channels are taken into account with power values relative to the CPICH power).

10.6.4.7 Link Budget Calculation for the Shared Release 99 and HSDPA Carriers

In the previous sections, the link budget calculations were considered for the case when only R99 or only HSDPA users were present in the cell. In the system, also a mix of R99 and HSDPA users in the same cell is possible. There is no difference in the UL Link Budget calculation. Therefore, the following information concentrates on DL Link Budget calculation.

In a shared scenario, the power of the WCDMA BTS is shared among the common control channels, shared control channels (HS-SCCH), R99 users (DCH channels) and different simultaneously served HSDPA users (shared channels and DCCH channels). With RF planning, several scenarios of power assignment can be considered according to operator requirements. The example of NodeB transmit power sharing is as follows:

- CPICII,BCH,SCH,PCH/FACH– 20%
- Release 99 DCH   – 50%
- HS-SCCH   – 10%
- HS-PDSCH   – 20%

HSDPA channels can only use residual power left after the common control and R99 channels are assigned. Power assigned to HSDPA may change dynamically, depending on the demand from R99 users, specifically for voice that has priority in resource allocation.

The reduction in power required for R99 also reduce the number of R99 users per cell. The power demand per R99 users does not change since the user at the cell border will always require the same power. Also, the interference experienced by the R99 user in DL does not change as it depends on the DL Tx power and not on the type of channel the power is used for. The lower the power available for R99, the lower the R99 load becomes.

At the same time, when HSDPA power increases, HSDPA coverage and maximum achievable HDSPA data rate at the cell border can be higher. HSDPA power per user cannot be increased over a certain predefined level (e.g. 20% of the total power in the example).

10.6.5 Uplink Capacity Estimation

For the uplink, the upper boundary of the capacity *(Npole)* of a WCDMA carrier can be estimated using the standard uplink capacity equation. This widely accepted formula is derived from early Code Division Multiple-Access (CDMA) work and applies to WCDMA if variables are set properly. Equation (10.21) estimates the capacity of a single cell from the spreading bandwidth (W), the radio access bearer (RAB) bit rate (Rb), the required Energy per bit-to-Noise ratio (E_b/N_0), the activity factor (v), and the interference factor (i). To estimate the capacity of a NodeB, sectorization gain must be included. This gain is typically estimated at 2.55 for a three-sector cell.

$$N_{\text{pole}} = \frac{W/R_b}{E_b/N_0 \cdot v \cdot (1+i)}$$

(10.21)

The resultant pole capacity has no practical application; each user transmits at maximum power to overcome the noise, while the uplink coverage is reduced to nothing. Nonetheless, the pole capacity can provide theoretical estimates of the maximum capacity of the WCDMA carrier for different types of traffic bearers, as shown in Table 10.7. From the pole capacity, a practical capacity (N_{user}) for a system can be calculated after the uplink loading (η) operating point has been determined.

$$N_{user} = N_{pole}\eta$$

(10.22)

The maximum uplink loading is selected to ensure that the network remains stable and that coverage is not adversely affected. The uplink link budget is affected by the interference margin (noise rise), which is directly dependent on load. The traffic mix also affects the Uplink loading, mainly when PS data is predominant, because PS data services are asymmetric.

The capacity calculated from (10.22) represents the number of resources available on the uplink radio link and not the capacity in Erlangs. Other resource dimensioning is considered later in the planning process after the uplink and downlink capacity estimations are known. All calculations are based on these limiting factors.

On the downlink, in most cases the code space is a limiting factor. On the uplink, it is assumed that code space can never be limiting factor, because users have their own code spaces to distinguish different channels (Dedicated Physical Control Channel (DPCCH) or Dedicated Physical Data Channel (DPDCH)). In addition, every user is differentiated

Table 10.7 Example of the maximum available channel resources with pole capacity (100% load).

Bearer	AMR 12.2	CS 64
Pole capacity: kbps/carrier	1757	3904
Data rate	12.2	64
Number of users	144	61

by means of scrambling codes. With 2^{24} scrambling codes available, no limitation is expected.

10.6.5.1 Required Bandwidth and Load for Multiple Bearers with GOS Considerations

The required bandwidth is calculated individually per bearer. If GOS requirements are given per bearer, a certain multiplexing gain can be achieved as all bearers share the same resource: the air interface. For the circuit-switched service the *Multidimensional Erlang* (MDE) formula [5] is used.

Therefore, the traffic demand per site, bearer and link must be transformed in terms of number of basic channels. These values and the necessary number of basic channels per bearer are the input for the MDE formula to calculate the needed number of channels (MDE_CH) with bandwidth equal to the bandwidth of reference bearer each. In the absence of simple analytical formulas. the input mix of RABs, CS and PS with respective QoS requirements is handled via the SW dimensioning tool.

The HSPA capacity per cell is limited by capacity of the schedulers, number of codes per user and modulation format.

10.6.5.2 Simplified Estimation of HSDPA Throughput Capacity

Given complex mechanism of HSPA data transfer with link adaptation and HARQ, the theoretical analysis of HSDPA/HSUPA throughput capacity can be performed with help of SW tool based simulations. Nonetheless, Nokia researches proposed a simplified estimate for HDSPA throughput capacity [6]. They introduced 'average HSDPA Signal-to-Interference-and-Noise Ratio (SINR)'. Following that approach, the estimate for average HSDPA is as follows:

$$SINR = 16 \frac{P_{HS-PDSCH}}{P_{tot}(1 - \alpha + 1/G)} \tag{10.23}$$

where $P_{HS-PDSCH}$ is the power of the HS-PDSCH channel; α and G are the orthogonality and the geometry factor; P_{tot} is the total transmit power in the downlink including the HSDPA portion as multipath propagation influences in the same way all downlink channels and value 16 is the fixed spreading factor for HSDPA. The geometry factor G is defined as the ratio of the received power from the serving cell divided by the received power from surrounding cells plus thermal noise, that is:

$$G = \frac{I_{own}}{I_{other} + P_N} \tag{10.24}$$

One can see that if the network is interference limited in the downlink, that is $P_N \ll I_{oth}$, then approximately $G \approx 1/i$. The geometry factor reflects the distance of the UE from the BS antenna. A typical range is ~3 dB for the cell edge.

Extended link-level simulations according to 3GPP specifications have produced mapping tables between the throughput and average SINR. For five parallel codes and by simple second-order curve fitting the following estimate for average HSDPA cell throughput *Thr* was proposed in [6]:

$$Thr[Mbps] = 0.0039 \cdot SINR^2 + 0.0476 \cdot SINR + 0.1421, \quad -5\text{dB} \leq SINR \leq 20 \text{ dB} \tag{10.25}$$

One can notice that the equations (10.23) and (10.25) may also provide rough estimate of NodeB transmit power to be assigned for HSDPA given projected throughput in the planned cell.

10.7   Summary

The main features of HSPA are:

- The HSDPA use Shared-Channel transmission, that is the certain fraction of the total downlink radio resources available within a cell, channelization codes and transmission power in case of WCDMA, is seen as a common resource that is dynamically shared between users in the time domain.
- Spreading factor is fixed, SF = 16, code multiplexing is used to increase user data throughput.
- There is no fast power control as in R99 WCDMA. New, and shorter, transmission time interval is used for HSPA, TTI = 2 ms.
- In order to share downlink resources, the channel depending scheduling is deployed in HSDPA. In each TTI, the scheduler decides to which user(s) the HS-DSCH (downlink shared channel) should be transmitted and, in close cooperation with the rate-control mechanism, at what data rate.
- The channel-dependent scheduling in combination with Layer 2 H-ARQ allow the utilization of the statistical spatial property of fading in different part of the cell for effective share of the common channel resource; that is, allocate most radio resources to the users in the best channel conditions at the each instant TTI. Given the statistical independence of fading in different locations within the cell, a reasonable statistical gain in cell throughput could be achieved, as well as a 'fair' distribution of the channel resources between users.
- HSDPA functions, primarily scheduling and rate control, rely on rapid adaptation of the transmission parameters to the instantaneous channel conditions as experienced by the UE. The NodeB estimates channel conditions using uplink control signalling from the UEs in the form of a Channel-Quality Indicator (CQI). The CQI value corresponds to a given transport-block size, modulation scheme and number of channelization codes.
- For each UE for which HS-DSCH can be transmitted, there is also an associated downlink control channel (DPCH). The only command transmitted on user associated downlink control channel is a power control command to the UE in order to adjust the uplink transmission power. The fractional –DPCH (F-DPCH) is used for this purpose, it carries two bits of TPC command (QPSK modulated) for the user transmitted in one tenth of a slot, using a spreading factor of 256. The F-DPCH channel is a R99 channel that uses TTI = 10 ms. Up to 10 users can then share a single channelization code being multiplexed into F-DPCH.
- The uplink HSPA (HSUPA) called also Enhanced Uplink, E-DCH, like HSDPA, deploys fast scheduling and fast hybrid-ARQ with soft combining and short TTI = 2 ms.
- There are fundamental differences between them, which has affected the detailed implementation of the features:

a) In the downlink, the shared resource is transmission power and the code space, both of which are located in *one* central node, the NodeB. In the uplink, the shared resource is the amount of allowed uplink interference, which depends on the transmission power of *multiple distributed* nodes, the UEs.

b) The scheduler and the transmission buffers are located in the same node in the downlink, while in the uplink the scheduler is located in the NodeB while the data buffers are distributed in the UEs. Hence, the UEs need to signal buffer status information to the scheduler.

- The WCDMA uplink, also with Enhanced Uplink, is inherently non-orthogonal and subject to interference between uplink transmissions within the same cell. This is in contrast to the downlink, where different transmitted channels are *orthogonal*. Fast power control is therefore essential for the uplink to handle the near-far problem. In contrary, HSDPA transmitted power is more or less constant.

- Soft handover is supported by the E-DCH. *Receiving* data from a terminal in multiple cells provides diversity. Soft handover also implies *power control by multiple cells*. On the contrary, HSDPA has no soft handover.

References

1 Technical Specification Group Radio Access Network; Physical-layer aspects of UTRA High-Speed Downlink Packet Access (Release 4); 3GPP TR 25.848 V4.0.0 (2001–03).

2 High-Speed Downlink Packet Access (HSDPA); Overall description; Stage 2 (Release 14); 3GPP TS 25.308 V14.0.0 (2016–12).

3 Technical Specification Group Radio Access Network; FDD Enhanced Uplink; Overall description. Stage 2 (Release 6); 3GPP TS 25.309 V6.6.0 (2006–03).

4 Technical Specification Group Radio Access Network; Physical-layer procedures (FDD) (Release 14); 3GPP TS 25.214 V14.1.0 (2017–03).

5 ITU-D Study Group 2, *Teletraffic Engineering Handbook*, Geneva, 2005. Available online at https://www.itu.int/dms_pub/itu-d/opb/stg/D-STG-SG02.16.2.1-2002-PDF-E.pdf (accessed February 2018).

6 Laiho, J., Wacker A. and Novosad, T. *Radio Network Planning and Optimization for UMTS*, John Wiley & Sons, Ltd, 2006.

11

4G-Long Term Evolution (LTE) System

11.1   Introduction

Widespread acceptance of the 3G/HSPA system led to tremendous growth in usage of mobile data. That was also stimulated by the availability of affordable mobile devices and flat data pricing by the operators. Mobile internet access extended from laptop usage to smart phones, thus facilitating development in mobile network performance towards very high instant peak data rates and very low latencies. Huge growth in mobile users and their traffic to be carried by mobile networks demands a significant increase in system capacity that, in turn, instigates a new technological solution to network design.

When high capacity and high performance at flat pricing are offered to the end customer, then cost per bit becomes a critical issue for the service provider. These three key drivers, capacity, user experience and lower cost per bit, have led to the specification of a Long Term Evolution (LTE) of UTRAN while mobile system core specification is defined as a System Architecture Evolution (SAE), also called Enhanced Packet Core (EPC). LTE together with EPC forms the Evolved Packet System (EPS).

In order to satisfy capacity demands, an additional portion of radio spectrum was released for LTE in the 2.6 GHz to 700 MHz range. New radio technology deployed in LTE delivers high spectrum efficiency and high capacity per site that reduces CAPEX and OPEX for service providers. A major reduction in cost per bit is ensured with flat IP-based LTE network architecture, cost efficient high bandwidth backhaul and transport network.

The 3GPP has set performance targets for an LTE of peak data rates >100 Mbps in DL and >50 Mbps in UL with latency less than 5 ms on the air interface per link. Spectral efficiency of LTE can exceed the one of UMTS Release 6 by a factor of 3–4 in DL and a factor of 2–3 in UL. The access scheme in LTE is OFDMA in downlink and a SC-FDMA in uplink.

OFDM allows for improved interference control, advanced scheduling techniques and ease of implementation of MIMO to improve spectrum efficiency. Further, OFDM enables scaling of user bandwidth very dynamically from very low bit rates required; for example, for control up to very high instantaneous peak data rates above 100 Mbps in downlink and 50 Mbps in uplink. Together with scalable RF bandwidth, OFDM allows for scaling the operator bandwidth from 1.4 or 3 MHz in re-farming scenarios up to 20 MHz for very high capacities. OFDM technology can be used in both FDD and TDD multiple-access schemes, so that both LTE FDD and TDD system are standardized, thus allowing flexibility in implementation.

Introduction to Mobile Network Engineering: GSM, 3G-WCDMA, LTE and the Road to 5G, First Edition. Alexander Kukushkin.
© 2018 John Wiley & Sons Ltd. Published 2018 by John Wiley & Sons Ltd.

11.2 Architecture of an Evolved Packet System

The Evolved Packet System (EPS) is made of the Evolved UTRAN (E-UTRAN), Evolved Packet Core (EPC) and connectivity to legacy 3GPP access and non-3GPP access systems. Figure 11.1 shows the EPS for 3GPP access system built as an evolution to existing 2G/3G architecture.

The EPS architecture has a reduced number of network elements on the data path compared to GPRS/UMTS, RAN functionality supported in a single node, and the separation of the control and user-plane network elements (MME and Serving Gateway).

The new network elements are as follows:

- *Mobility Management Entity* (MME), which is the control plane (C-plane) functional element in EPC. MME manages and stores UE context, generates temporary identities and allocates them to UEs, authenticates the user, manages mobility and bearers and acts as a termination point for Non-Access Stratum (NAS) signalling.
- Serving Gateway (S-GW), which is the user plane (U-plane) gateway to the E-UTRAN. S-GW serves as an anchor point both for inter-eNodeB (eNB) handover and for intra-3GPP mobility (i.e. inter-3GPP access mobility between LTE and 2G or 3G). It is also responsible for packet forwarding, routing and buffering of downlink data for UEs that are in ECM-IDLE state.
- Packet Data Network Gateway (P-GW), which is the U-plane gateway to the PDN (e.g. the Internet or the operator's IP Multimedia Subsystem (IMS)). P-GW is responsible for policy enforcement, charging support and the user's IP address allocation.
- E-UTRAN which is the radio access part of LTE Network.

The legacy network elements interfacing LTE/SAE are as follows:

- Gateway GPRS Support Node (GGSN), which is responsible for terminating the Gi interface towards the PDN for legacy 2G/3G access networks. LTE/SAE interfaces this node only as a part of P-GW functionality and from the perspective of inter-system mobility management.
- Serving GPRS Support Node (SGSN), which is responsible for the transfer of packet data between the Core Network and the legacy 2G/3G RAN. LTE/SAE interfaces the SGSN only in case of inter-system mobility management.
- Home Subscriber Server (HSS) is the IMS Core Network entity responsible for managing user profiles, performing the authentication and authorization of users.

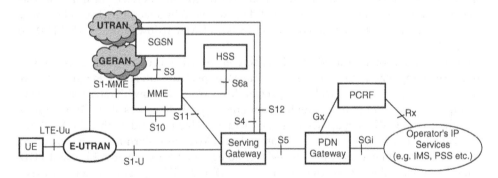

Figure 11.1 EPS architecture for 3GPP accesses.

The user profiles managed by HSS consist of subscription and security information as well as details on the physical location of the user. While IMS is not a mandatory network element, the HSS is a necessary node for operation of the LTE system.

- Policy Charging and Rules Function (PCRF) is responsible for brokering QoS Policy and Charging Policy on a per-flow basis.

11.3   LTE Integration with Existing 2G/3G Network

The LTE system is normally integrated into existing 2G/3G network architecture. Figure 11.2 illustrates combined 2G/3G/LTE network architecture with connectivity to non-3GPP access systems, such as fixed and wireless internet networks. As observed, in addition to 2G/3G network subsystem and nodes, the network includes new logical entity referred as Authorization & Accounting (AAA). The MME functionality and SGSN functionality can be installed as SW logical entities into single physical network node.

The LTE system is normally integrated into existing 2G/3G network architecture. Figure 11.2 illustrates combined 2G/3G/LTE network architecture with connectivity to non-3GPP access systems, such as fixed and wireless internet networks. As observed, in addition to 2G/3G network subsystem and nodes, the network includes new logical entity referred as AAA. The MME functionality and SGSN functionality can be installed as SW logical entities into single physical network node.

The AAA function is responsible for relaying authentication and authorization information to and from non-3GPP access network connected to the EPC.

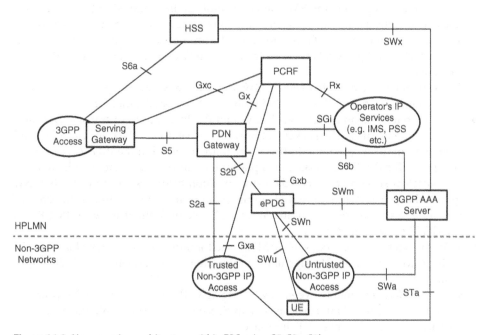

Figure 11.2 Non-roaming architecture within EPS using S5, S2a, S2b.

11.3.1   EPS Reference Points and Interfaces

The EPS reference points as specified in [1], [2] and [3] include:

- S1-MME Control plane reference point between E-UTRAN and MME. This control plane protocol is the S1AP, which is quite similar to UTRAN RANAP.
- S1-U is the user-plane reference point between E-UTRAN and the S GW. S1-U uses tunnelling protocol GTP-U (GPRS Tunnelling Protocol for User plane) for bearer services.
- X2 is the control and user-plane reference point between two E-UTRAN nodes.
- S2a/S2b is PMIPv6 based reference points (S2a also supports Client Mobile IPv4 FA mode) between P-GW and non-3GPP access network (e.g. WLAN, CDMA2000), used for control and mobility support for non-3GPP access interworking.
- S2c is a DSMIPv6 based reference point providing user plane with related control and mobility support between UE and P-GW.
- S3 is a reference point between MME and SGSN, used for user and bearer information exchange for inter-3GPP access network mobility.
- Gn is the reference point between pre-release 8 SGSN and MME/P-GW.
- Gp is the reference point between pre-release 8 SGSN and P-GW in roaming scenario.
- S4 is the reference point between S GW and release 8 SGSN, used for U-plane tunnelling and related mobility support as S GW is an anchor point for 3GPP handover.
- S5 is the reference point between S GW and PGW but not crossing a PLMN boundary, used for U-plane tunnelling and tunnel management and for S GW relocation. S5 includes both GTP and IETF variants. The protocol used at this reference point is GTP for both the control plane and user plane.
- S6a is the reference point between MME and HSS, used for transfer of subscription and authentication data. The S6a reference point can be regarded as the AAA interface between the MME and HSS. The functionality provided by S6a is similar to the one on the Gr interface in 2G/3G, however, the protocol used at this reference point is the DIAMETER protocol inherited from IP standards and modified to 3GPP specifications.
- S6b is the reference point between P-GW and 3GPP AAA Server/proxy for mobility related authentication and retrieval of mobility/QoS related parameters.
- Gx is the reference point between P GW and the PCRF, used to transfer QoS policy and charging rules. The Packet Gateway (P-GW) enforces the rules via Policy and Charging Enforcement Function (PCEF) that are included in the P-GW set of functionalities. The P-GW requests and applies the set of rules associated with charging the particular service flow (user data stream) upon bearer establishment. That may apply volume-based, time-based charging or no charging; the latter in the case of a flat rate. Ga, Gxb are the variants of the GX reference point between PCRF and non-3GPP network nodes.
- S10 is the reference point between MMEs, used for information transfer; for example, during MME relocation. The signalling on this interface is triggered by UE during S1 handover.
- S11 is the reference point between MME and S GW. The S11 uses the GTP-C protocol that is used for control information such as EPS bearer management. The respective user plane is routed across S1-U interface.

- S12 is the reference point between S GW and UTRAN, used for U-plane tunnelling when a Direct Tunnel is established. This is an optional interface that could be established by operator to route user plane from RNC to S-GW.
- SGi is the reference point between P GW and the Packet Data Network (PDN), based on the UMTS Gi.
- STa is the reference point between trusted non-3GPP access and 3GPP AAA Server/Proxy to carry out AAA procedures.
- SWa is the reference point between untrusted non-3GPP access and 3GPP AAA Server/Proxy to carry out AAA procedures.
- SWd is the reference point between 3GPP AAA Server and 3GPP AAA Proxy.
- SWx is the reference point between 3GPP AAA Server and HSS for transfer of authentication data.
- Rx is the reference point between PCRF and Application Function in the PDN, based on the Rx interface of UMTS [TS23.203]. It used for real-time applications and services different from common packet transfer.

There are multiple variants of the Gx reference point. Gx is used for transfer of policy and charging information from PCRF to P-GW. Gxa is used for transfer of policy and charging information from PCRF to trusted non-3GPP access. The new node ePDG (evolved Packet Data Gateway) in Figure 11.2 is a PDN Gateway to non-3GPP network. There are 3GPP (GTP) and IETF (PMIP) variants of the S5 reference points. The protocol over the S1-U will be GTP-U. The protocols over the S3, S4, S5 reference points are based on GTP.

11.4 E-UTRAN Interfaces

E-UTRAN contains only one type of node, eNB, which provides the air interface to UE. eNBs can be connected to each other via the X2 interface and connected to MMEs and S-GWs via the S1 interface. A single eNB can connect to multiple MMEs and multiple S-GWs. This ability provides flexibility and reliability and is referred to as S1-flex. This eNB connection options are illustrated in Figure 11.3.

The eNB is responsible for radio transmission to and reception from UE. As observed, the RNC node is absent in the LTE network. Instead, RNC functionalities reside in the eNB. This involves management of the radio resources (including admission control), radio bearer control, scheduling of user data and control signalling over the air interface. Additionally, the eNB performs ciphering and header compression over the air interface.

eNBs are interconnected via an X2 interface, which is somewhat similar to the Iur interface between RNCs in the WCDMA network. However, the X2 interface will only connect eNodeBs with neighbouring cells since there is no anchor point and functionality in e-UTRAN. The X2 does not have RNC drift-like functionality; instead, it supports relocation functionality with packet forwarding.

eNB is connected to the core network using the S1 interface. The S1 interface is somewhat similar to Iu-ps interface in 3G system. Both the S1 and Iu-ps user planes are transport tunnels based on IP, agnostic to the content of the packet sent. The IP packets of the end user are put into the S1 IP tunnel by the EPC or the eNB and retrieved

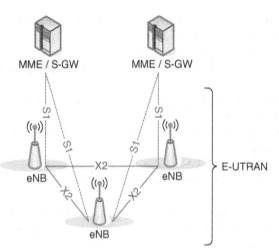

Figure 11.3 E-UTRAN and EPS with S1-flex interface [1].

in the other end (eNB or EPC). The S1 interface is developed to enable connectivity between eNB and multiple MMEs and multiple S-GWs. This feature referred as S1-flex illustrated in Figure 11.2. With S1-flex, if one of the EPC nodes becomes unavailable another EPC node can take over the lost traffic. In addition, this enable dynamic scaling of the network, EPC nodes can be added when needed due to traffic demands and not due to increase in coverage.

11.5 User Equipment

As in UMTS, the LTE mobile station is called User Equipment (UE). It is constructed using a modular architecture that consists of three main components (see Figure 9.3 in Chapter 9).

1) *Mobile Termination*: The MT represents termination of the radio interface. In this entity the RRC signalling is terminated and RRC messages are sent/received.
2) *Terminal Adapter*: The terminal adapter represents the termination of the application-specific service protocols; for example, SIP signalling for VoIP. The terminal adapter could be in fact an external device, such as a modem with a USB interface to connect to a laptop.
3) *Terminal Equipment*: The TE represents termination of the service. In case of USB, the laptop acts as a TE. Otherwise, a smart phone is the TE where the service is terminated in application on mobile device.

11.5.1 LTE UE Category

Table 11.1 shows the LTE UE equipment categories and respective theoretical maximum data rates as comes with modulation capabilities.

Table 11.1 LTE UE equipment categories.

User Equipment Category	Max. L1 data rate downlink (Mbps)	Max. number of DL MIMO layers	Max. L1 data rate uplink (Mbps)	3GPP Release
0	1.0	1	1.0	Rel 12
1	10.3	1	5.2	Rel 8
2	51.0	2	25.5	Rel 8
3	102.0	2	51.0	Rel 8
4	150.8	2	51.0	Rel 8
5	299.6	4	75.4	Rel 8
6	301.5	2 or 4	51.0	Rel 10
7	301.5	2 or 4	102.0	Rel 10
13	391.7	2 or 4	150.8	Rel 12
9	452.2	2 or 4	51.0	Rel 11
10	452.2	2 or 4	102.0	Rel 11
11	603.0	2 or 4	51.0	Rel 11
12	603.0	2 or 4	102.0	Rel 11
15	750	4 or 2	N/A	Rel 12
16	979	4 or 2	N/A	Rel 12
8	2998.6	8	1497.8	Rel 10
14	3917	8	N/A	Rel 12

Note: Maximum data rates shown are for 20 MHz of channel bandwidth. Categories 6 and above include data rates from combining multiple 20 MHz channels. Maximum data rates will be lower if less bandwidth is utilized.

11.6 QoS in LTE

The Quality of Service architecture for the bearer service in LTE is shown in Figure 11.4. It is quite similar to one developed for UMTS, see Figure 9.18 though; of course, the reference points and bearers are different from 3G.

Two types of EPS bearers exist: *default bearers* (similar to primary PDP contexts) and *dedicated bearers* (similar to secondary PDP contexts in 3G but controlled by EPC).

In LTE, different to 2.5 and 3G PS connections, the default bearer with default QoS is already established when the UE attaches to the network. The QoS attributes of default bearer are determined by the subscribed QoS parameters stored in the HSS; that is, the same as in 2.5/3G networks. The default bearer remains active as long as UE is attached to EPC.

Dedicated bearers are created for QoS differentiation purposes: creation of dedicated bearers is controlled by EPC. When the user-plane packet is sent by the UE it is routed to the PCRF. The PCRF analyses the requested end-to-end service and, depending on this service, the PCRF may now trigger a modification of QoS parameters in all the

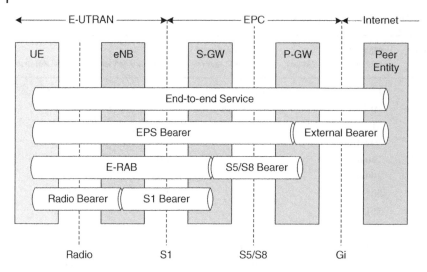

Figure 11.4 LTE bearer services architecture [1].

involved bearers. It means that QoS is managed by network and tied to applications, the subscriber cannot request change to QoS parameters.

A single UE in LTE can have multiple end-to-end services active, each of these services then has its own individual bearer. In principle, 256 individual E-RABs for a single UE can be addressed by E-UTRAN protocols (15 RAB-IDs are defined in UMTS).

To standardize the QoS handling, a set of QoS Class Indicators (QCIs) have been defined by 3GPP. There are four classes with a Guaranteed Bit Rate (GBR) and five classes with a Non-Guaranteed Bit Rate (Non-GBR). Besides the bit rate, the parameter priority, packet delay budget and packet error loss rate are critical factors as given in Table 11.2.

11.7 LTE Security

General 3GPP security architecture, mechanisms for Authentication and Key Agreement (UMTS AKA) and the principles of air-interface protection are described in the standard TS 33.102 [5]. The IP layer security is described in standard TS 33.210 [6] and TS 33.310 [7]. 3GPP standard [2] (and technical report TR 33.821 [8]) specifies the security architecture and mechanisms for SAE/LTE systems. The 3GPP security architecture is depicted in Figure 11.5.

Most of the standards related to security of the TCP/IP protocol suite such as security for various protocols, cryptographic protocols of TCP/IP are specified by other standardization bodies (IETF, ITU-T), but can also be applied to mobile networks. Some security areas such as IP packet filtering and firewalls are mostly not standardized at all (if e.g. no interoperability between vendors is needed).

The drivers for new developments in LTE security are as follows:

- All traffic (except Uu) is based on IP which is very efficient transport, but also susceptible to the threat of attacks
- The user-plane traffic in LTE is IP only (no circuit-switched voice anymore)

Table 11.2 Standardized QCI characteristics [4].

QCI	Priority Level	Packet Delay Budget	Packet Error Loss	Example Services	Resource Type
1	2	100 ms	1.00E-02	Conversational voice	
2	4	150 ms	1.00E-03	Conversational video (live streaming)	
3	3	50 ms	1.00E-03	Real-time gaming, V2X messages	GBR
4	5	300 ms	1.00E-06	Non-conversational video (buffered streaming)	
65	0.7	75 ms	1.00E-02	Mission critical user-plane push to talk voice (e.g. MCPTT)	
66	2	100 ms	1.00E-02	Non-mission-critical user-plane push to talk voice	
75	2.5	50 ms	1.00E-02	V2X messages	
5	1	100 ms	1.00E-06	IMS signalling	
6	6	300 ms	1.00E-06	Video (buffered streaming) TCP-based (e.g. www, e-mail, chat, ftp, p2p file sharing, progressive video, etc.)	
7		100 ms	1.00E-03	Voice, Video (live streaming) interactive gaming	Non-GBR
8	8	300 ms	1.00E-06	Video (buffered streaming) TCP-based (e.g. www, e-mail, chat, ftp, p2p file sharing, progressive video etc.)	
9	9	300 ms	1.00E-06	Video (buffered streaming) TCP-based (e.g. www, e-mail, chat, ftp, p2p file sharing, progressive video etc.)	
69	0.5	60 ms	1.00E-06	Mission critical delay sensitive signalling (e.g. MC-PTT signalling)	
70	5.5	200 ms	1.00E-06	Mission critical data (example services are the same as QCI 6/8/9)	
79	6.5	50 ms	1.00E-02	V2X messages	

- Mobile backhaul on S1, X2 interfaces and also between core sites could be integrated into external IP networks.

Compared with 2G/3G networks, two new security features have been introduced in LTE:

- new ciphering mechanism for NAS signalling and
- encryption of transport on the S1 reference point.

As can be observed from Figure 11.5, with encryption of NAS signalling a double ciphering is applied on the radio interface. The NAS signalling messages between the

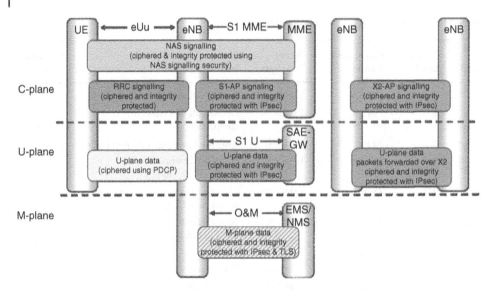

Figure 11.5 3GPP LTE security architecture.

UE and MME are encrypted on top of the protocol stack, the underlying RRC layer is secured by ciphering mechanisms as well, so that the NAS message is ciphered together with its RRC transport message for a second time.

Encryption of transport on S1 and X2 interfaces is supported with IPsec mechanism for all of the control, user and management planes. This feature enables secure eNB control and bulk data communication between the eNB and other eNB and Core Nodes by utilizing IPsec for provision of encryption, integrity protection and communication peer authentication according RFC 4301. Control plane traffic, user-plane traffic and management traffic can be separated with IPsec VPN tunnels from each other and from any other operator's traffic if any part of the transport network is shared. The separation also ensures that flooding attacks at the control/signalling network will have no impact to the separated data paths.

11.8 LTE Mobility

LTE mobility is specified in [1], specific procedures in idle mode are specified in [9]. Mobility management functions are used for keeping track of the current location of UE. The EPS has two main that states, where mobility management needs to be handled:

- ECM (EPS Connection Management) IDLE state and
- ECM-CONNECTED state.

11.8.1 Idle Mode Mobility

Cell selection and reselections is done autonomously by UE and based on UE measurements. The parameters for cell reselection are broadcasted by the network. The UE starts receiving the broadcast channels of that cell and checks if the cell is suitable for camping, which requires that the cell is not barred and that radio quality is sufficient. After cell selection, the UE must register itself to the network. In IDLE state the location of registered is known to a Tracking Area granularity level.

Whenever UE has camped to a cell, it monitors and measures the neighbour cells broadcast channels. The decision for cell reselection is triggered when signal of serving cell is not high enough to satisfy default QoS criteria.

The neighbour cells are indicated in the neighbour cell list broadcasted by serving cell. During the search for cell reselection candidate, the UE ranks the neighbour cells in descending order according the received signal level (in dBm). The reselection occurs when best ranked neighbour cell level exceed the signal level in serving cell to a specified margin (in dB), called hysteresis. This criterion is similar to one used the cell reselection in 2G/3G.

UE is known with an accuracy of Tracking Area (TA) UE is currently registered with. The size of the Tracking Area could be in order of hundreds of eNBs and has to be optimized as a trade-off between paging signalling load and signalling related to a tracking area update. With latter, the objective is to avoid a ping-ponging between neighbour cells of different tracking areas. The UTRAN concept similar to Tracking Area is a Routing Area (RA). The UE in LTE may be registered in both LTE TA and UTRAN RA.

In the ECM-IDLE state no signalling connection exists between UE and the network. UE in ECM-IDLE state is paged in all cells of the Tracking Areas in which it is currently registered with. In IDLE state UE performs the following mobility management functions:

- Tracking area update (TAU) triggered by TA change
- Periodic TAU (Tracking Area Update)
- Intersystem TAU/RAU
- Answer to paging from MME with Service Request.

For tracking area update in IDLE mode, UE enters to ECM-CONNECTED state in order to establish signalling connection between UE and MME.

11.8.2   ECM-CONNECTED Mode Mobility

In the CONNECTED state, the location of UE is known with an accuracy of eNB. The mobility in RRC connected state is entirely managed by handovers. Several mobility scenarios are supported in CONNECTED mode:

- Intra-LTE intra-eNodeB mobility. The handover may take place between cells belonging the same eNB.
- Intra-LTE inter-eNodeB mobility. The handovers happen between adjacent base stations.

3GPP inter-radio access technology (inter-RAT) mobility involves handovers between the Evolved UTRAN and a non-LTE 3GPP access network (UTRAN or GERAN).

In the non-3GPP inter-RAT mobility scenario, the handover takes place between the Evolved UTRAN and a non-3GPP access network, for instance WLAN, WiMAX or 3GPP2 access network.

11.8.3 Mobility Anchor

During mobility, the user-plane data path to the Packet Data Network (PDN) is maintained using a concept called mobility anchoring illustrated in Figure 11.6. The path from UE to the mobility anchor point may change during the handover. However, the path from the anchor point to PDN does not change. EPS includes several mobility anchor points:

- With intra-eNB handover, eNB serves as the anchor point
- With inter-eNB handover, the anchor point is located in the Serving Gateway
- With 3GPP inter-RAT handover, the anchor point is located in the Serving Gateway
- With non-3GPP inter-RAT handover, the anchor point is located in the PDN Gateway

Despite of the location of mobility anchor, the overall session anchor point is always located in the PDN Gateway.

11.8.4 Inter-eNB Handover

Based on measurements received from UE, the source eNB selects a target eNB and initiates the handover. The handover signalling takes place over the X2 interface. If there is no X2 connectivity between the base stations, the signalling must take place via the Mobility Management Entity and via the S1-MME interface. These two alternatives are illustrated in Figure 11.7 and later in 11.9.

The user-plane switching procedure is shown in Figure 11.8. In the preparation phase for handover, the target eNB has to reserve resource and set up the bearers. When preparation phase is successfully completed, the source eNB and UE are notified. The source eNB acts as an anchor and forwards all downlink packets not yet acknowledged by the UE via the X2 interface to the target eNodeB, see Figure 11.7 and Figure 11.8. The

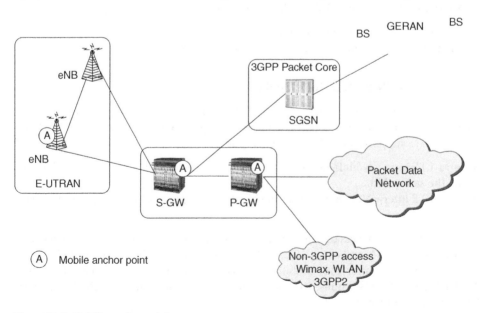

Figure 11.6 Mobility anchor points.

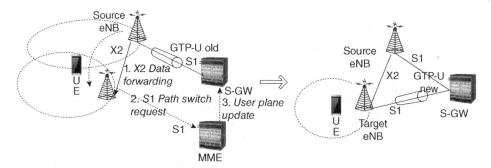

Figure 11.7 Inter-eNB handover with an X2 interface.

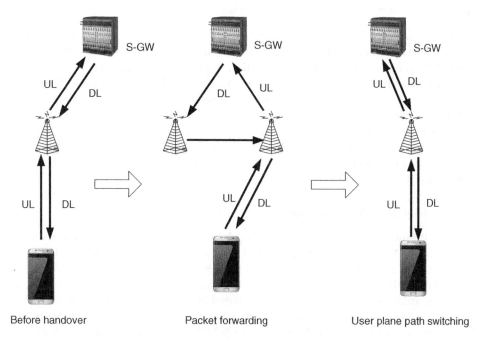

Before handover   Packet forwarding   User plane path switching

Figure 11.8 User-plane switching in handover with an X2 interface.

target eNB then retransmits and prioritizes downlink packets forwarded by the source eNodeB.

In the uplink, after switching to the target cell, the UE will retransmit all packets that have not been acknowledged in sequence before the handover. The reordering and the removal of duplicated packets in the uplink are done in the packet core. At the Serving Gateway, path switching to the target eNB is performed only after HO completion.

In a case when two neighbour eNBs are not connected via the X2 interface, the UE needs to change the serving cell by means of an S1 interface handover. The S1 handover procedure is also used in inter-RAT handover.Figure 11.9 shows S1 handover using the three steps; handover preparation, handover resource allocation and modification of the S1-U bearer.

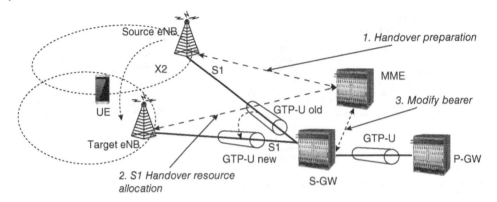

Figure 11.9 Inter-eNB handover without an X2 interface.

The eNB triggers the serving cell change in the stage of handover preparation after receiving measurement reports. The relocation request is sent from eNB to MME. When the MME receives the relocation preparation request of the source eNB, it starts the handover resource allocation procedure in target eNB. After the target eNB sends a handover command message to MME with the required radio interface parameters. The MME forwards this handover command message transparently to the UE that executes the handover. The MME triggers the bearer modification procedure at the S11 interface between MME and S-GW. The S-GW executes the path switch for user plane from S-GW to the target eNB establishing a new GTP tunnel on S1-U to send payload packets in the UL/DL direction. The S-GW acts as an anchor during S1 handover.

11.8.5 3GPP Inter-RAT Handover

With 3GPP inter-RAT handovers, signalling between the access systems always takes place via the MME and SGSN. Inter-RAT handover can be performed in both the E-UTRA RRC connected state as well as in E-UTRA RRC IDLE state.

Like inter-eNB S1 handovers, 3GPP inter-RAT handovers are typically backward handovers meaning that radio resources are prepared in the target access system before UE is commanded by the source access system to switch over to the target access system. The alternative procedure for the 3GPP inter-RAT handover is to force the UE into an idle state and perform a tracking area update (TAU) if the target network is the LTE network, or routing area update (RAU) if the target network is the non-LTE network.

11.8.6 Differences in E-UTRAN and UTRAN Mobility

The idle mode mobility is similar in UTRAN and E-UTRAN. The flat architecture in E-UTRAN brings some differences to the CONNECTED mode mobility. In UTRAN the UE must update the location both to the circuit-switched core network (Location Areas) and to the packet core network (Routing areas) while E-UTRAN only uses Tracking Areas (packet core).

With a dual-mode terminal, the incoming voice call may initiate a Circuit-Switched Fall Back (CSFB) handover. The CS core network can send a paging message to the E-UTRAN UE. In that case the MME maps the Tracking area to the Location Area.

The summary of UTRAN and E-UTRAN mobility features is provided in Table 11.3.

Table 11.3 UTRAN and E-UTRAN differences in mobility [10].

UTRAN	E-UTRAN	Notes
Location area	Not relevant	For CS fallback handover, MME maps tracking area to location area
Routing area	Tracking area	
Soft handover is used for WCDMA uplink and downlink and for HSUPA uplink	No soft handovers	
Cell_FACH, Cell_PCH, URA_PCH, Cell_FACH, Cell_PCH, URA_PCH	No similar RRC states	E-UTRAN always uses handovers for RRC connected users
RNC hides most of mobility	Core network sees every handover	

11.9   LTE Radio Interface

The E-UTRA system uses an Orthogonal Frequency Division Multiplex (OFDM) for the downlink and Single-Carrier Frequency Division Multiple Access (SC-FDMA) for the uplink. OFDM divides the available bandwidth in a number of narrow mutually orthogonal subcarriers as shown in Figure 11.10.

Orthogonality in the frequency domain brings the following OFDM advantages:

- Eliminates intra-cell interference
- Very high spectral efficiency
- Allows for a small guard bands within the nominal bandwidth.

These characteristics enable flexible spectrum usage. LTE FDD supports 1.4 MHz, 3 MHz, 5, MHz, 10 MHz, 15 MHz and 20 MHz carrier bandwidths. Subcarriers are spaced by 15 kHz and there is a maximum of 2048 subcarriers available in LTE.

In the time domain, the OFDM transmitter sends a sequence of OFDM symbols separated by guard time intervals, as shown in Figure 11.10. The guard time interval is filled with a copy of the succeeding symbol tail called a Cyclic Prefix (CP). There are two versions of the CP, one is short with a normal length 5.21 µs and the other is an extended CP with a length 16.67 µs. The receiver uses the CP to mitigate the inter-symbol interference caused by multipath propagation, ease estimation of channel response and respective equalization. The extended CP is normally used in very large cells. The transmission is divided in time into multiples of OFDM symbols with a duration of 66.67 µs. The time slot of 0.5 ms duration comprises seven OFDM symbols with a normal-length CP.

The orthogonality of subcarriers in OFDM technology is prone to a Doppler shift spread in the propagation channel. Since the OFDMA signal is a composition of many orthogonal subcarriers presenting independently modulated data symbols, it reveals the properties of random Gaussian noise with high Peak-to-Average Power Ratio (PAPR) and as a consequence a very large random burst in signal amplitude may occur. In order to avoid OFDM signal quality degradation, the linearized amplifier needs to be deployed

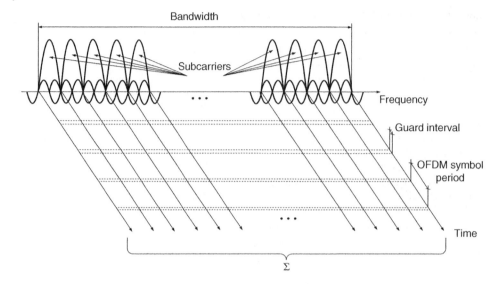

Figure 11.10 Orthogonal Frequency Division Multiple Access concept.

in OFDM transceiver. Normally, an amplifier with a linearization feature is used in the eNB for the downlink. In order to avoid complexity in the UE design, the uplink is built using alternative technology, namely Single-Carrier FDMA, which produces a more even signal.

11.10   Principle of OFDM

The OFDM carrier signal is composed by a very large number of relatively narrowband subcarriers. Each subcarrier is constructed from rectangular pulse in time domain with duration T_s that corresponds to a sinc-square shaped per-subcarrier spectrum with a bandwidth Δf, as shown in Figure 11.11. Then $T_s = 1/_{\Delta f}$.

The composite OFDM carrier is packed with subcarriers separated at $\Delta f = 1/T_s$, see Figure 11.11. The subcarrier pulse duration T_s corresponds to the period of modulated data symbol. The subcarrier spacing is thus equal to the per-subcarrier modulation rate $1/T_s$.

In 3GPP LTE, the basic subcarrier spacing equals 15 kHz. The number of subcarriers depends on the transmission bandwidth. For instance, with 10 MHz spectrum allocation about 600 subcarriers can be packed into OFDM carrier overall transmission bandwidth, as shown in Figure 11.10. In a complex baseband notation, an OFDM signal $s(t)$ during the time interval $mT_s \leq t < (m+1)T_s$ can be presented by composition of modulated subcarriers:

$$s(t) = \sum_{n=0}^{N-1} s_n(t) = \sum_{n=0}^{N-1} a_n^m e^{j2\pi n \Delta f t} \tag{11.1}$$

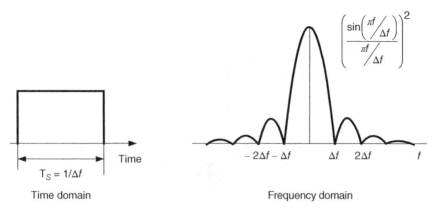

Figure 11.11 Subcarrier shape in time and frequency domains.

where $s_n(t) = a_n^{(m)} \cdot e^{2\pi n \Delta ft}$ is the nth modulated subcarrier with frequency $f_n = n\Delta f$, $a_n^{(m)}$ is the modulation symbol (a complex one in general) applied to the nth subcarrier during the m-th OFDM symbol interval; that is, during the time interval $mT_s \le t < (m+1)T_s$. The subcarrier waveform is presented by the pulse function

$$\xi_n(t) = \begin{cases} e^{2\pi n \Delta ft}, & 0 \le t < T_s \\ 0 & \end{cases} \tag{11.2}$$

The shape $\xi_n(t)$ of single subcarrier ensures mutual orthogonality between subcarriers. The fact that two modulated OFDM subcarriers s_p and s_q are mutually orthogonal over the time interval $T_s \le t < (m+1)T_s$ is supported by basic equation:

$$\int_{mT_s}^{(m+1)T_s} s_p(t)s_q^*(t)dt = a_p a_q^* \int_{mT_s}^{(m+1)T_s} e^{j2\pi p\Delta ft}e^{-j2\pi q\Delta ft}dt = 0, \text{ for } p \ne q \tag{11.3}$$

The orthogonal nature of subcarrier waveform given by equation (11.3) forms the basis of the OFDM principle. OFDM transmission is performed in blocks of N symbols. During each OFDM symbol interval, N modulation symbols are transmitted in parallel. The modulation symbols can be mapped to a modulation alphabet, such as QPSK, 16QAM or 64QAM.

In OFDM, a block of N data symbols is first converted from serial-to-parallel for modulation onto N parallel subcarriers as shown in Figure 11.12. This effectively increases the OFDM symbol duration $a_k^{(m)}$ on each subcarrier approximately by a factor N compared with input symbol duration in the serial stream. The duration of OFDM symbol $s_n(t)$ may be significantly longer than the channel delay spread as intended with the OFDM design. Simplified structure of an OFDM modulator is provided in Figure 11.12. It consists of a bank of N complex modulators, where each modulator corresponds to one OFDM subcarrier.

The OFDM demodulation can be implemented using bank of correlators, one for each subcarrier, as shown in Figure 11.13. In an ideal case, the orthogonality between subcarriers implies that the OFDM subcarriers do not cause any interference to each other after demodulation.

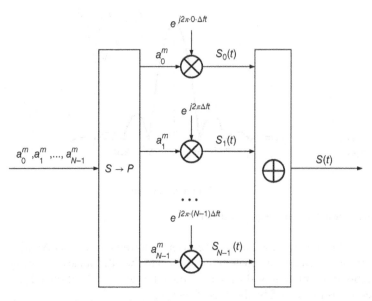

Figure 11.12 OFDM modulator [11].

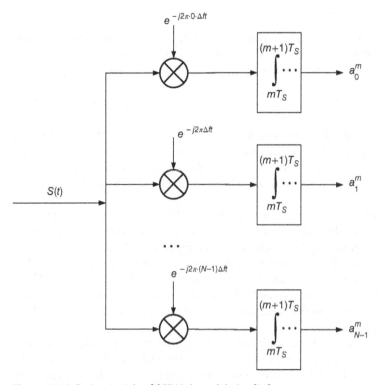

Figure 11.13 Basic principle of OFDM demodulation [11].

11.11 OFDM Implementation using IFFT/FFT Processing

A common practical approach to implementation of OFDM is based on low-cost computationally efficient Fast Fourier Transform (FFT) processing. With FFT implementation the input data stream is sampled at a sampling rate f_s, which is a multiple M of subcarrier spacing Δf; that is, $f_s = M \cdot \Delta f$. The parameter M must exceed the number of subcarriers N with a sufficient margin. On the other hand, the parameter M determines the size of the discrete FFT window, which should be equal to 2^p for some integer p. With 10 MHz spectrum bandwidth allocation for LTE, the number of data subcarriers N is defined as 600, also called occupied subcarriers. The nearest discrete FFT size $M > N$ is $M = 1024$. This corresponds to a sampling rate $fs = M \cdot \Delta f = 15.36$ MHz with $\Delta f = 15$ kHz as LTE subcarrier spacing.

Using the discrete FFT, the OFDM modulator can be implemented via Inverse Fast Fourier Transform (IFFT), as shown in Figure 11.14. After parallel-to-serial conversion the sampled time domain signal s_n is converted to the analogue signal by means of Digital-to-Analogue Convertor (DAC). Similar to OFDM modulation, the OFDM demodulator can be implemented using FFT processing with same size N of FFT window.

The high level architecture of the OFDM transceiver using FFT is depicted in Figure 11.15.

11.12 Cyclic Prefix

Mutual orthogonality of subcarriers permits demodulation of an ideal non-distorted OFDM signal without interference between subcarriers. In a wideband (time-dispersive) multipath channel, an inter-symbol interference may destroy orthogonality due to superposition of time-shifted multipath replicas of the subcarrier symbol. As a consequence, a multipath composed symbol may lose its original shape and spread outside its borders leading to inter-symbol interference between subcarriers. While effective symbol duration of the OFDM symbol is intended to be long enough compared with the delay spread of the channel, a small amount of inter-symbol interference may still result

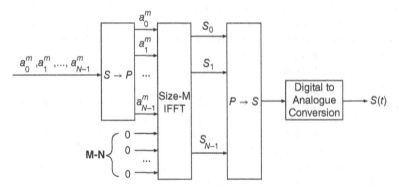

Figure 11.14 OFDM modulation by means of IFFT processing [11].

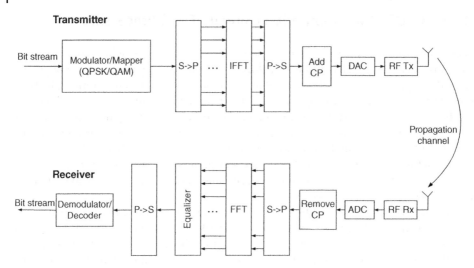

Figure 11.15 Architecture of an OFDM transceiver.

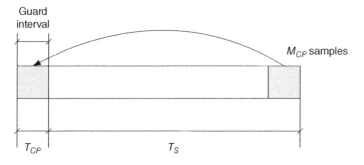

Figure 11.16 CP insertion.

in severe performance degradation. In the frequency domain, orthogonality between ideal subcarriers is ensured by means of the specific sinc-like frequency structure of the subcarriers. Relatively small inter-symbol interference in the time domain may result in modest distortion of the side lobe frequency structure of the subcarrier that, in turn, results in the loss of subcarrier orthogonality. In order to mitigate time dispersion, the LTE downlink composite signal include a time guard intervals T_{CP} between consecutive symbols filled up with the CP. As shown in Figure 11.16, the CP is a copy of the last M_{CP} samples of the IFFT output.

CP insertion thus increases the overall time period confining the OFDM symbol from T_s to $T_s + T_{CP}$, where T_{CP} is the length of the cyclic prefix. As a consequence, the effective OFDM symbol rate is reduced. On the other hand, the CP length should be not too short compared with the delay spread of the propagation channel. As long as the delay spread of the propagation channel is less than CP length, the subcarrier orthogonality can be achieved with relatively simple equalization.

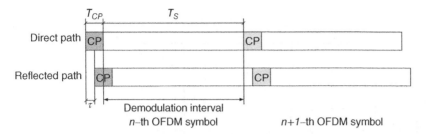

Figure 11.17 CP role in a multipath environment.

Ideally, the CO length T_{CP} should exceed the maximum multipath delay τ imposed by the propagation channel. This situation is schematically illustrated in Figure 11.17. The delay spread typically increases with cell size. On the other hand, increasing the length of the cyclic prefix, without a corresponding reduction in the subcarrier spacing Δf, implies an additional overhead and respective loss in throughput rate as well as in loss of power in demodulation of the received signal since only part $T_s/(T_s + T_{CP})$ is utilized by the demodulator. In practice, there is a trade-off in CP length between signal corruption due to the multipath and loss in demodulation power and effective throughput. The LTE system supports two different CP lengths in two different scenarios:

• Shorter (normal) CP in small-cell environments to minimize the CP overhead.
• Longer CP in environments with extreme time dispersion.

Another possible source of the OFDM performance degradation is a distortion of frequency synchronization between transmitter and receiver due to a channel Doppler spread. A multipath channel produces multiple replicas of the scattered with different angle of arrival and, respectively, different Doppler frequency shifts in a moving UE. As a result, random frequency errors may lead to random frequency shifts of subcarriers in frequency domain and respective degradation of orthogonality.

11.13 Channel Estimation and Reference Symbols

Decoding/demodulation in LTE receiver is based on coherent detection for both uplink and downlink signals. Coherent detection needs correct estimation of both amplitude and phase of transmitted signal. That, in turn, requires accurate estimation of the propagation channel.

A common approach to estimation of the channel response is to transmit known signals that carry no user information but specified sequence of the symbols. In LTE, Reference Signals (RS) also referred as Pilot signals are used for channel estimates. The RSs are inserted in known positions in the OFDM time-frequency two-dimensional downlink grid. The RSs usage in LTE is two-fold; on one hand, the RS supports channel estimate, on the other hand the RS position in frequency domain is used for cell identification. More details on RSs are provided in Section 11.25. Figure 11.18 illustrates the reference signal position in the downlink payload grid for the normal CP length.

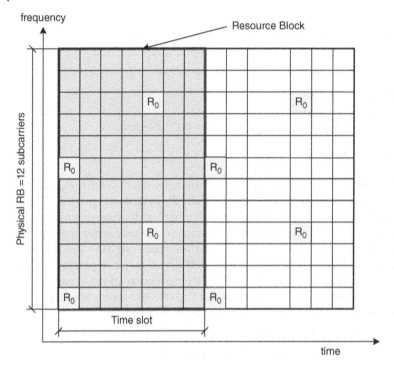

Figure 11.18 Cell-specific reference-symbol arrangement in the case of normal CP length for one antenna port.

The required spacing in time between the reference symbols can be obtained by considering the maximum Doppler spread (highest speed) to be supported, which for LTE corresponds to 500 km/h. This estimate follows approach in [12]. The Doppler shift is $f_d = (f_c \cdot v/c)$, where f_c is the carrier frequency, v is the UE speed in metres per second and c is the speed of light ($3 \cdot 10^8$ m/s). Considering $f_c = 2$ GHz and $v = 500$ km/h, then the Doppler shift is $f_d = 950$ Hz. According to Nyquist's sampling theorem, the minimum sampling frequency needed in order to reconstruct the channel is therefore given by $T_{sl} = 1/(2f_d) = 0.5$ ms under these assumptions. This implies that two reference symbols per slot (0.5 ms) are needed in the time domain order to estimate the channel correctly.

In the frequency axis of the time-frequency downlink lattice there is one reference symbol every six subcarriers on each OFDM symbol, as shown in Figure 11.18. However, the RSs are distributed within each Resource Block (RB) so that there is one reference symbol every three subcarriers. This spacing is related to the expected coherence bandwidth of the channel, which is in turn related to the channel delay spread. In particular, the studies performed during development phase of the LTE standard shown that with maximum RMS channel delay spread considered to be 991 ns, corresponding coherence bandwidth with 90% and 50% probability is estimated as $B_{c,90\%} = 20$ kHz and $B_{c,50\%} = 200$ kHz, respectively [12]. The RS placement every three subcarriers corresponds to frequency separation of 45 kHz, thus providing trade-off in resolution to frequency variation of channel response.

11.14 OFDM Subcarrier Spacing

Basically, two factors affect the selection of subcarrier spacing Δf:

- Subcarrier spacing Δf should be not too large in order to have the OFDM symbol length $T_s = 1/\Delta f$ long enough to minimize impact the relative cyclic-prefix overhead T_{CP}/T_s.
- Subcarrier spacing Δf should be large enough compared with Doppler shift spread of the channel.

The value accepted in LTE standard is $T_s = 66.67$ μs, the respective $\Delta f = 15$ kHz. The results of a study of subcarrier interference provided in [11] shown substantial performance with a ~20 dB Signal-to-Interference ratio given subcarrier spacing of $\Delta f = 15$ kHz and a Doppler spread of the channel $f_d \sim 1$ kHz.

11.15 Output RF Spectrum Emissions

The output UE transmitter spectrum consists of three components; the emission within the occupied bandwidth (channel bandwidth), the Out-Of-Band (OOB) emissions and the far out spurious emission domain, see Figure 11.19.

Channel bandwidth is defined as the bandwidth containing 99% of the total integrated mean power of the transmitted spectrum on the assigned channel. The OOB emissions are unwanted emissions immediately outside the assigned channel bandwidth resulting from the modulation process and non-linearity in the transmitter. This OOB emission limit is specified in terms of a spectrum emission mask and an Adjacent Channel Leakage power Ratio. A rectangular pulse shaping (Figure 11.11) of the OFDM signal produces a relatively large out-of-band emission, somewhere in order of 10% of channel bandwidth. Therefore, occupied bandwidth for all transmission bandwidth configurations (Resource Blocks) that is composed by a number of subcarriers shall be less than the channel bandwidth 1.4, 3, 5,…, 20 MHz, specified by spectrum allocation. This leads to the channel arrangements as shown in Table 11.4.

Taking an example of a 5 MHz channel bandwidth and subcarrier spacing of 15 kHz, 300 subcarriers result in 4.5 MHz of occupied bandwidth.

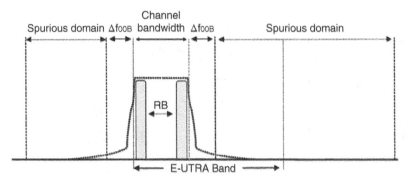

Figure 11.19 Transmitter RF spectrum [13].

Table 11.4 LTE channel arrangement.

Channel bandwidth, MHz	1.4	3	5	10	15	20
Number of occupied subcarriers	72	180	300	600	900	1200

11.16 LTE Multiple-Access Scheme, OFDMA

The channel bandwidth in LTE is shared by active users. In the downlink, the eNB schedules different subsets of available subcarriers to different users during each OFDM symbol interval, as shown in Figure 11.19. Similarly, in the uplink direction, each user transmits a different subset of subcarriers in each OFDM symbol interval, see Figure 11.20. Given the mutual orthogonality of subcarriers, such a method of user multiplexing is called Orthogonal Frequency Division Multiple Access or OFDMA.

As discussed in relation to CP insertion, perfect time alignment is necessary to ensure orthogonality of subcarriers. Apparently, time misalignment at the base station for uplink signals from different terminals should be less than the length of the cyclic prefix in order to manage the differences. The LTE uses approach similar to GSM; that is, eNB sends time alignment (time advance) commands to each terminal.

Another practical requirement for subcarrier orthogonality is a power alignment of received subsets of subcarriers from different users. Given the different path loss in radiolinks to different users, the received uplink signals from different users can be significantly different in signal strength. In theory, mutual orthogonality of subcarriers is preserved even in case of significantly different values of the signal strength in each subcarrier. The important nature of orthogonality in subcarriers is that it is ensured by the structure of side lobes to the same extent as the main lobe in frequency domain. In practice, however, the received signal is corrupted due to frequency errors and Doppler spread. If received strength of subcarriers is significantly different, then a stronger signal may produce significant interference caused by channel inflicted errors and vulnerability of the side lobes to errors.

In order to mitigate the channel impact, the LTE system needs to deploy some power control for uplink transmission. Normally, the system will command to reduce transmission power for terminals located near base station and equalize the received signal

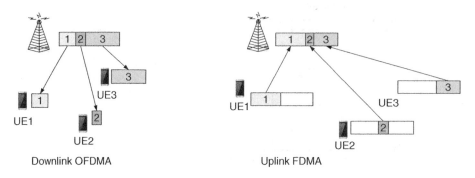

Downlink OFDMA   Uplink FDMA

Figure 11.20 OFDM as a user multiplexing/multiple-access scheme.

4G-Long Term Evolution (LTE) System

strength evenly across all users in the cell. This mechanism is similar to the solution to the 'near-far' problem in the CDMA system.

11.17   Single-Carrier FDMA (SC-FDMA)

The Single-Carrier FDA uses the Discrete Fourier Transform (DFT) as a precoding of the input data and then follows the normal OFDM procedure. This is the reason that SC-FDMA is also called DFT-spread OFDM. Figure 11.21 depicts signal generation in a SC-FDMA transmitter.

In SC-FDMA, a block of Q modulated symbols is first applied to a size-Q DFT. The output of the DFT represents a spectrum of the previously modulated data symbols. The output of the DFT is then mapped to the set of subcarriers in OFDM modulator where the OFDM modulator is implemented as a size-M inverse DFT (IDFT) with $M > Q$ and where the unused inputs of the IDFT are set to zero. The IDFT size M is selected as $M = 2^m$. for some integer m to allow computationally efficient IDFT implementation with Fast Fourier Transform algorithm; that is, IFFT. A cyclic prefix (CP) is then inserted for each transmitted block after serial-to-parallel conversion. The last stage in Figure 11.21 is performed with digital-to-analogue conversion.

The value Q is smaller than M and the remaining inputs to the IFFT are set to zero. The output of the IFFT is a signal with 'single-carrier' properties. It has low-power variations and a bandwidth that depends on parameter Q. An important feature of such signal generator is that instantaneous bandwidth of the generated signal can be varied dynamically by varying the block size Q of the DFT. Assuming a sampling rate f_s at the output of the IDFT, the nominal bandwidth of the IDFT output signal is then $BW = f_s \times Q/M$. A second feature particular to the IFFT process is that shifting the IDFT inputs to where the DFT outputs are mapped produces a shift of the transmitted output signal in the frequency domain.

The main benefit of SC-FDMA, compared to normal OFDM, is reduced variations in the instantaneous transmission power. Such variations are quantified using a PAPR value; that is, Peak-to-Average Power Ratio (PAPR) = peak signal power to average signal power level within one OFDM symbol period. High PAPR levels produce non-linear distortion in the transmitter amplifier and then in the output signal. In order to combat high PAPR impact, one needs to use sophisticated linearization schemes in an amplifier. The linearization is feasible technically and economically in the Base Station but not effective in the design of mobile handsets due to high cost and power consumption. The reduction of PAPR in SC-FDMA compared with OFDM can be estimated to be at 3–4 dB depending on the modulation type in the input data stream.

SC-FDMATransmitter

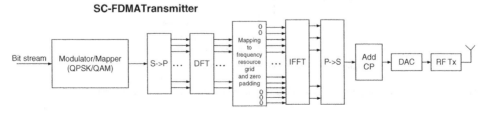

Figure 11.21 Block diagram of a SC-FDMA transmitter [14].

11.18   OFDMA versus SC-FDMA Operation

Figure 11.22 shows both the OFDMA and SC-FDMA simplified operation. The SC-FDMA is similar to the OFDM/OFDMA case. The main difference is that in the case of SC-FDMA there is additional processing before the IFFT: the modulated symbols (interpreted in this case as time signals) are fed to FFT processing. The outputs are the frequency components of the modulation symbols. Those frequency components are mapped to the allocated inputs of the IFFT and, from there, the normal OFDM processing continues.

The additional FFT processing block in SC-FDMA spreads the information of each bit over all subcarriers. In an SC-FDMA signal, each subcarrier used for transmission contains information about all transmitted modulation symbols since the input data stream has been spread by the FFT transform over the available subcarriers. In OFDMA, each subcarrier only carries information related to specific modulation symbols. The FFT output size is smaller than the IFFT input size. This is because the granted UL resources to one UE cannot exceed the total resources in the cell. Multiple UEs can be allocated in uplink, each one using different (groups of) subcarriers.

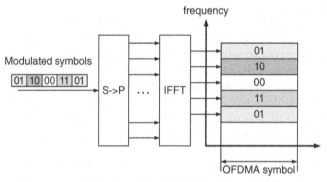

Downlink OFDMA symbol processing

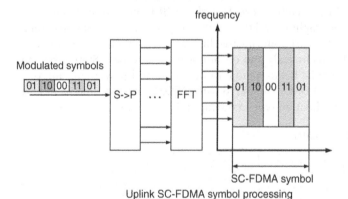

Uplink SC-FDMA symbol processing

Figure 11.22 Physical-layer processing: downlink link-OFDMA and uplink-SC-FDMA.

The SC-FDMA symbol duration is the same as the OFDMA symbol duration (66.7 µs) but whereas in OFDMA each modulated symbol lasts the whole OFDMA symbol duration, in SC-FDMA each modulated symbol lasts for '$1/n$' th of the SC-FDMA symbol (where n is the number of used sub- carriers, 1/5th in Figure 11.22 as an illustrative example)

In OFDMA there is one modulated symbol per subcarrier (see the top of Figure 11.22). In SC-FDMA each modulated symbol is spread across the used subcarriers. In the frequency domain, each OFDMA data symbol occupies 15 kHz and each SC-FDMA data symbol occupies $n \times 15$ kHz bandwidth (see the example at the bottom of Figure 11.22). It can be seen that, as the bandwidth increases, the modulated symbol duration decreases. Therefore, to double the data rate in the case of SC-FDMA, the total bandwidth needs to be doubled (as well as the number of FFT inputs in the transmitter) and the modulated symbol duration is halved.

11.19 SC-FDMA Receiver

The basic principle of demodulation of a SC-FDMA (DFTS-OFDM) signal is based on reverse operations involved in DFTS-OFDM signal generation. It means removal of CP after A/D conversion, size- M DFT (FFT) processing, removal of the $M - Q$ frequency samples that do not carry the useful signal and size- Q inverse DFT/FFT processing producing the block of Q modulated symbols, as shown in Figure 11.23. To combat signal distortion inflicted by radio channel in the time-frequency domain the LTE receiver uses frequency equalizer. The channel response W (Figure 11.23) is estimated from the CP portion of the signal and then used by frequency equalizer to correct the spectrum of the received signal at the input to iFFT for correct reconstruction of the block of transmitted symbols.

11.20 User Multiplexing with DFTS-OFDM

As illustrated in Figure 11.20, multiple access in ODFM is ensured by selecting different sets of subcarriers for different users. Flexible allocation of user bandwidth is performed by dynamically adjusting the transmitter DFT size Q and respective block of modulation symbols $a_0, a_1, a_2, \ldots, a_{Q-1}$. The mapping of the user resource block to a specific position within the channel bandwidth is performed by shifting the IDFT inputs to which the DFT outputs are mapped. The basic principle is illustrated in Figure 11.24.

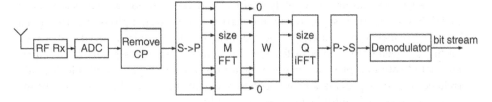

Figure 11.23 Basic principle of DFTS-OFDM demodulation [14].

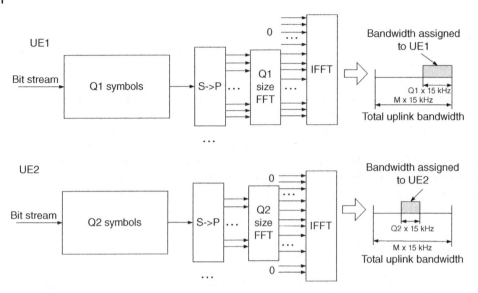

Figure 11.24 Uplink user multiplexing in the case of DFTS-OFDM.

11.21 MIMO Techniques

LTE is designed to utilize Multiple Input Multiple Output (MIMO) operation, including spatial multiplexing as well as precoding and transmit diversity. The basic principle of spatial multiplexing is sending signals from two or more different antennas with different data streams simultaneously. The receiver then uses a signal processing algorithm to separate and recover different data streams, hence increasing the peak data rates by a factor of 2 or more, depending on the number of transmit antennas. Apparently, spatial multiplexing can be achieved when the Signal-to-Interference-to-Noise Ratio (SINR) in receiving antennas is high enough to ensure quality reception.

With transmit diversity, the same signal is sent from different antennas and it can also be specifically coded with space time coding. At the receiver end, the signal is combined thus exploiting the independence of fading on radiolinks between multiple transmit/receive antennas. The transmit diversity is often used for communications at the cell edges in order to improve the SINR's exploitation of diversity gain.

The best results in MIMO operation are achieved when signals and channel responses in two receive antennas are completely uncorrelated. Base-station antennas are uncorrelated if they are spatially separated by about 10 or more wavelengths or use orthogonal polarization planes (cross-polarity). The spatial separation required for UE antennas is about half a wavelength. The most feasible solution is to use cross-polarized antennas. Uncorrelated antennas provide the potential for diversity and spatial multiplexing gain. Correlated antennas provide a robust coherence gain that is used for classical beamforming but no spatial multiplexing and/or diversity gain.

Under optimal conditions 2×2 SU-MIMO duplicates the peak user data rate. In a realistic environment 2×2 SU-MIMO results in a cell capacity enhancement of 10% for macro-cellular to 40% for micro-cellular deployment scenarios.

For spatial multiplexing, LTE introduces some specific terminology:

- A stream of data is associated with spatial layer.
- Number of transmitted layers (streams) is regarded as a rank of transmission.
- Independently encoded transport data block is called a codeword. The codeword is mapped to the transmission layer.

Figure 11.25 presents the typical MIMO configuration encompassing dual-codeword 2×2 DL Single-User MIMO Spatial Multiplexing (SU-MIMO). This MIMO scheme doubles the downlink peak user data rate by transmitting two separate data streams in parallel to a single UE. The two base station antennas transmit signals, two UE antennas receive two signals each and four channel responses h_{11}, h_{12}, h_{21}, h_{22}, form a channel matrix. The UE receiver performs joint demodulation of the two signals received by two antennas to recover the transmitted data streams. The two unknown transmit signals can demodulated using the estimated four channel responses and the possible transmit alphabet. In practice, joint demodulation is affected by noise and external interference.

This MIMO scheme uses assumption that the channel is time-invariant at least over the OFDM symbol period meaning that the channel is 'frozen' and can be described by a single channel matrix that does not change during that time. Also, the propagation channel is considered to be a narrowband frequency flat fading channel, which is quite a reasonable assumption with OFDM.

Using vector notation, the received signal can be presented as

$$r = \begin{pmatrix} r_1 \\ r_2 \end{pmatrix} = \begin{pmatrix} c_{11} & c_{12} \\ c_{21} & c_{22} \end{pmatrix} \begin{pmatrix} s_1 \\ s_2 \end{pmatrix} + \begin{pmatrix} n_1 \\ n_2 \end{pmatrix} = C \cdot s + n \tag{11.4}$$

Assuming no noise and that the channel matrix C is invertible, the vector $\vec{s}$, and thus both signals s_1 and s_2, can be perfectly recovered at the receiver, with no residual interference between the signals, by multiplying the received vector r with a matrix $W = C^{-1}$.

$$\begin{pmatrix} \hat{s}_1 \\ \hat{s}_2 \end{pmatrix} = W \cdot r = \begin{pmatrix} s_1 \\ s_2 \end{pmatrix} + C^{-1} \cdot n \tag{11.5}$$

Perfect restoration of transmitted signal as long as channel matrix is invertible is theoretically possible in the absence of noise. In a realistic scenario, equation (11.5) demonstrates that joint demodulation of the two data streams may increase the noise in receiver to the extent of the product the inverse channel matrix C^{-1} and noise n. The minimum effective noise increase is when channel matrix's rows and columns are linearly independent and matrix C is a normal non-singular matrix. In the case of spatial multiplexing or

Figure 11.25 MIMO principle with a 2×2 antenna configuration.

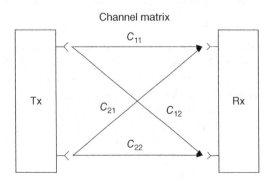

Channel matrix

C_{11}

Tx

C_{21} C_{12}

Rx

C_{22}

transmit diversity, this means that signals transmitted from different antennas (antenna ports) experience independent fading when passed through the propagation channel. The correlation of the fading in different antenna ports can be regarded as mutual interference in the received signal.

In the general case of N_T transmit antenna ports (layers), M_R receive antenna and S symbols (subcarriers) to be transmitted, equation (11.4) can be written as

$$R = C \cdot X + N, \tag{11.6}$$

where C is $N_T \times M_R$ channel matrix,

$$C = \begin{bmatrix} c_{11} & c_{12} & \cdots & c_{1M_R} \\ c_{21} & c_{22} & \cdots & c_{2M_R} \\ \cdots & \cdots & \cdots & \cdots \\ c_{N_T 1} & c_{N_T 2} & \cdots & c_{N_T M_R} \end{bmatrix} \tag{11.7}$$

X is a set of T data symbols transmitted in one block via N_T antennas,

$$X = \begin{bmatrix} x_{11} & x_{12} & \cdots & x_{1S} \\ x_{21} & x_{22} & \cdots & x_{2S} \\ \cdots & \cdots & \cdots & \cdots \\ x_{N_T 1} & x_{N_T 2} & \cdots & x_{N_T S} \end{bmatrix} \tag{11.8}$$

and $M_R \times S$ matrix N is additive Gaussian noise collected in M_R receive antennas.

11.21.1 Precoding

Precoding is used in LTE together with spatial multiplexing to make the signals from multiple antennas 'quasi–orthogonal', hence allowing for improved signal isolation at the receiver side, reducing mutual interference and improving SINR. The purpose of precoding is to minimize the errors in the receiver by exploiting some knowledge of the propagation channel. An optimal transmitter over the MIMO channel is based on a channel aware precoding that effectively includes beamforming and power allocation across a transmitted set of symbols and antennas. Ideally, instant channel characteristics need to be known to the transmitter side, otherwise the optimal transmitter can use some statistical information about the channel. The channel information is reported by the UE to the eNodeB in the uplink control information feedback messages.

Assuming that a block X of $N_T S$ data symbols is to be sent over the channel via N_T streams (or layers) of S symbols each. Stream i consists of symbols $[x_{i,1}, x_{i,2}, \ldots, x_{i,S}]$.

The precoded transmitted signal Y can now be written as

$$Y(X) = VPX \tag{11.9}$$

where V is an $N_T \times N_T$ transmit beamforming matrix and P is a $N_T \times N_T$ diagonal power-allocation matrix with $\sqrt{p_i}$ as its i-th diagonal element, where p_i is the power allocated to the i-th stream. The term VP in equation (11.9) is a precoder obtained by singular-value decomposition operation.

Using Singular-Value Decomposition (SVD), channel matrix C can be decomposed into parallel non-interfering subchannels (also Eigen-channels) by choosing a unitary beamforming matrix V (i.e. $V^H V$ is the identity matrix of size N_T) in such way that

$$C = U \sum V^H, \tag{11.10}$$

with U as a unitary matrix and $\sum \text{diag}(\lambda_i)$ is an $N_T \times N_T$ diagonal matrix with the λ_i eigenvalues of $C^H C$ as its diagonal elements. The symbol $\{\cdot\}^H$ denotes the Hermitian matrix operator. Thus, the i-th right singular vector of C, given by the i-th column of V, is used as a transmit beamforming vector for the i-th stream. The receiver combines the signals from its M_R antennas using weights $W = [w_1, w_2, \ldots, w_{M_R}]$. The received signal is then given by

$$Z = W \cdot R = WCX + WN \tag{11.11}$$

The optimal weight is a beamforming vector matched to the channel, the so-called Maximum Ratio Combining (MRC) solution $= C^H$. This can be seen as a spatial version of the well-known matched filter. Combining the SVD expression for channel matrix (11.10) and equation (11.11), one may obtain that the optimal beamformer for the ith stream is the ith left singular vector of C, obtained as the ith row of U^H; that is, u_i^H. The received ith stream is given by

$$u_i^H R = \lambda_i \sqrt{p_i}[x_{i,1}, x_{i,2}, \ldots, x_{i,S}] + u_i^H N \tag{11.12}$$

Since $\sum$ is a diagonal matrix, there is no interference between the spatially multiplexed signals at the receiver.

With a full-rank channel matrix C that equals $min(N_T, M_R)$, the multiplexed streams are uncorrelated and can be used for parallel data transmission thus increasing throughput and the spectral efficiency. Otherwise, some streams are correlated and spatial multiplexing gain shrinks to $rank(C)$. On the other hand, all parallel channels can be loaded with just a single data stream, which gives us a transmit diversity gain. The relation of rank to codeword-to-layer mapping is given in Table 11.5.

As the operation progresses, each UE is configured via RRC signalling to one of the number of transmission modes defined in the LTE system. In the first releases of LTE, 10 transmission modes have been defined [15]. The transmission mode defines what kind of downlink transmissions the UE should expect, such as transmit diversity or closed-loop spatial multiplexing and also modifies the channel-state feedback according to modes corresponding to the desired operation. We limit the current study to the four first transmission modes. The transmission modes 5–10 extend transmission up to eight antenna ports, please refer to [15] for full details:

- Transmission Mode 1: Transmission from a single eNodeB antenna port, no precoding.
- Transmission Mode 2: Transmit diversity with two or four antenna ports using Space Frequency Block Coding.

Table 11.5 Layer mapping.

	codeword 1	codeword 2
rank 1	layer 1	
rank 2	layer 1	layer 2
rank 3	layer 1	layer 2 & layer 3
rank 4	layer 1 & layer 2	layer 3 & layer 4

- Transmission Mode 3: Open-loop spatial multiplexing. This is an open-loop mode with the possibility to do rank adaptation based on the RI feedback. In the case of rank = 1 transmit diversity is applied similarly to transmission mode 2. With higher rank spatial multiplexing with up to four layers with large delay, Cyclic Delay Diversity (CDD) is used.
- Transmission Mode 4: Closed-loop spatial multiplexing. This is a spatial multiplexing mode with precoding feedback supporting dynamic rank adaptation. In the case of closed-loop precoding it is assumed that the network may select the precoder matrix based on feedback from the terminal. Based on measurements on the cell-specific reference signals, the terminal selects a suitable transmission rank and corresponding precoder matrix and reports to eNodeB in the Uplink Control Information message in the form of rank Indication (RI) and Precoder Matrix Indication (PMI). Network analysis information is provided by the UE and decides what precoding matrix to use in downlink transmission. In the case where eNodeB uses precoding based on the UE report, it just sends confirmation to the UE, otherwise it sends this to the mobile of the PMI of precoding to be used. The UE then needs to update its decoding procedure accordingly.

The LTE uses a limited set of predefined-by-precoding matrices for each transmission rank, they constitute a standard codebook. The precoding shall be done according to a codebook matrix selected by both UE and eNodeB in each transmission mode. In control signalling, between the base station and terminal only index (PMI) of the precoding matrix is sent over the radio interface.

11.21.2 Cyclic Delay Diversity (CDD)

In the case of open-loop spatial multiplexing, the feedback channel-state message from the UE only indicates the rank of the channel. In this mode, if the rank is greater than 1, LTE uses CDD, which involves transmitting the same set of OFDM symbols on the same set of OFDM subcarriers from multiple transmit antennas, with a different delay on each antenna applied before CP insertion, as shown in Figure 11.26. The delay is cyclic over the Fast Fourier Transform (FFT) size. Adding a delay before the CP insertion does not increase delay spread of the channel.

Adding a time delay is identical to applying a phase shift in the frequency domain. The same time delay is applied to all subcarriers, therefore phase shift will increase linearly across the subcarriers with increasing subcarrier frequency. When combined in the receiver antenna, each subcarrier from one antenna interferes constructively or

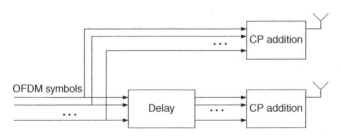

Figure 11.26 Cyclic delay diversity.

destructively with the delayed version from another antenna when combined in the receiver antenna.

The number of peaks and dips created by the CDD in the received signal spectrum across the subcarriers therefore depends on the length of the delay: as delay is increased, the number of peaks and troughs in the spectrum also increases. The time-delay/phase-shift equivalence means that the CDD operation can be implemented as a frequency domain precoder for the affected antenna(s), where the precoder phase changes on a per-subcarrier basis according to a fixed linear function. The advantage of frequency domain implementation is that it is not limited to delays corresponding to an integer number of samples.

In LTE, the frequency domain phase-shift values φ are π, $2\pi/3$ and $\pi/2$ for two-, three- and four-layer transmission, which equal $(T_s/2)$, $(T_s/3)$ and $(T_s/4)$ in the time domain, respectively. In FFT implementation of the size 2048, these values correspond to 1024, 682.7 and 512 sample delays, respectively. For the application of CDD in LTE, the eNodeB transmitter combines CDD delay-based phase shifts with additional precoding using fixed unitary DFT-based precoding matrices. For multilayer CDD operation, the mapping of the layers to antenna ports is carried out using precoding matrices selected from the spatial multiplexing codebooks. As the UE does not indicate a preferred precoding matrix in the open-loop spatial multiplexing transmission mode in which CDD is used, the particular spatial multiplexing matrices selected from the spatial multiplexing codebooks in this case are predetermined [16].

In the case of two transmit antenna ports, the predetermined spatial multiplexing precoding matrix W is always the same (the first entry in the two transmit antenna port codebook, which is the identity matrix). Thus, the transmitted signal can be expressed as follows:

$$Y = WDUX = \frac{1}{\sqrt{2}} \begin{bmatrix} 1 & 0 \\ 0 & 1 \end{bmatrix} \begin{bmatrix} 1 & 0 \\ 0 & e^{j\pi/2} \end{bmatrix} \frac{1}{\sqrt{2}} \begin{bmatrix} 1 & 0 \\ 0 & e^{-j\pi/2} \end{bmatrix} X, \tag{11.13}$$

where diagonal matrix D supports cyclic delay diversity.

In the case of four transmit antenna ports, v different precoding matrices can be used from the four transmit antenna port codebook where v is the transmission rank. These v precoding matrices can be applied in turn across groups of v subcarriers in order to provide additional decorrelation between the spatial streams.

11.22   Link Adaptation and Frequency Domain Packet Scheduling

LTE downlink uses the dynamic packet scheduling every transmission time interval (OFDM symbol period) by allocating some subset of subcarriers (physical resource) to the user and selecting transmission parameters including a modulation-and-coding scheme. The latter is referred to as link adaptation. Link adaptation of transmission parameters requires some knowledge of instantaneous channel conditions in the frequency domain related to subcarrier subset.

The Frequency Domain Packet Scheduling (FDPS) exploits frequency selective power variations on either the desired signal or interference, thus scheduling users with the best channel conditions on a selected set of subcarriers (resource block). Contrary to HSDPA where all users operate in the same bandwidth, in LTE, the user provided channel report indicates the distribution of channel states over both the cell area and frequency domain (subsets of subcarriers allocated). A condition for achieving highest cell throughput is therefore that the radio channel's effective coherence bandwidth is less than the system bandwidth, which is typically the case for cellular macro and micro cell deployments with system bandwidths equal to or larger than 5 MHz. On the other hand, FDPS implies that based on knowledge about the instantaneous channel conditions, different subcarriers can be used for transmission to or from different terminals.

Information on radio link channel conditions is required for both link adaptation and FDPS. The instantaneous downlink channel conditions can be estimated by terminal based on measurement of downlink pilot or reference signal (RS). The reference signal has a predetermined structure and is transmitted at constant power. Information about the instantaneous downlink conditions is then reported to the base station. Given some propagation delay, such a report may not reflect the condition at the time of transmission at base station. The more rapid the channel variation in the time domain, the less efficient the link adaptation. In addition to link adaptation based on measurement of the Reference Signal and possibly some prediction, the LTE system also deploys a Hybrid-ARQ mechanism with soft combining, which is described in Section 12.27.

11.23   Radio Protocol Architecture

The role of the LTE radio interface protocols is to setup, reconfigure and release the Radio Bearer that provides the means for transferring the EPS bearer over air-interface to/from the UE. There are two types of bearers in LTE:

1) Signalling Radio Bearer (SRB) that carries the RRC signalling messages and
2) User-Plane Radio Bearer that carries the user data.

The LTE radio interface protocol architecture is shown in Figure 11.27. The radio protocol stack includes the following entities: Radio Resource Control (RRC), Packet Data Convergence Protocol (PDCP), Radio Link Control (RLC), Medium Access Control (MAC) and Physical Layer (PHY). In reference to the OSI model, Layer 2 includes RLC, MAC and PDCP. Layer 3 consists of the Radio Resource Control (RRC) protocol, which is part of the control plane.

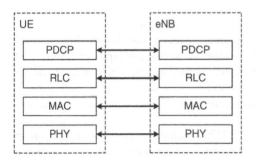

Figure 11.27 LTR radio interface user-plane protocol stack [1].

On the top of radio protocol stack layers there are protocols in the core network. Those protocols are established between the UE and the core network and generally referred to as Non-Access Stratum (NAS) signalling, they are transparent to radio layers.

11.23.1   User Plane

Figure 11.27 shows the downlink protocol stack over the radio interface for the user-plane, where PDCP, RLC and MAC sublayers perform respective functions for the user plane. The uplink protocol structure is similar to downlink. The functions of radio protocol stack for user plane are as follows:

- PDCP performs IP header compression and produces output PDCP-PDU.
- The RLC protocol is responsible for segmenting (and concatenation on uplink) of the PDCP-PDUs for radio interface transmission. It also performs error correction with the Automatic Repeat Request (ARQ) method.
- Medium Access Control (MAC): The MAC layer is responsible for scheduling the data according to priorities and multiplexing data to Layer 1 transport blocks. The MAC layer also provides error correction with Hybrid-ARQ.
- PHY performs coding, modulation, antenna and resource mapping.

11.23.2   Control Plane

The functions of radio protocol stack in control plane (Figure 11.28) are as follows:

- PDCP sublayer (terminated in eNB on the network side) performs ciphering and integrity protection;
- RLC and MAC sublayers (terminated in eNB on the network side) perform the same functions as for the user plane;
- RRC (terminated in eNB on the network side) performs broadcast, paging, RRC connection management, RB control, mobility functions, UE measurement reporting and control;

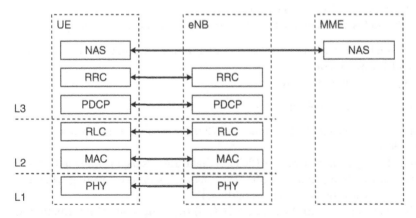

Figure 11.28 LTE radio interface control plane protocol stack [1].

- NAS control protocol (terminated in MME on the network side) performs among other things:
  - EPS bearer management;
  - Authentication;
  - ECM-IDLE mobility handling;
  - Paging origination in ECM-IDLE;
  - Security control.

11.23.3   Scheduler

The scheduler is a part of the eNodeB MAC layer, it controls the assignment of uplink and downlink resources in terms of resource-block pairs in time and the frequency domain. The eNodeB in each TTI = 1 ms interval takes a scheduling decision and sends scheduling information to the selected set of terminals. The resource allocation is based on resource blocks (RBs), each consisting of 12 subcarriers occupying 180 kHz.

Uplink and downlink scheduling are separated in LTE, and uplink and downlink scheduling decisions can be taken independently. The downlink scheduler dynamically controls to which UE to transmit the respective set of RBs. The uplink scheduling is supported by a scheduling grant request sent from the mobile to eNodeB and respective grant response to the mobile by eNodeB with transport format defined. Transport format selection (selection of transport-block size, modulation scheme and antenna mapping) and logical-channel multiplexing for downlink transmissions are controlled by the eNodeB. The UE just obeys the uplink scheduling grant. The uplink scheduler controls which terminals are to transmit on their respective uplink shared traffic channel and on which uplink time-frequency resources.

While the uplink transport format is dictated by eNodeB, the terminal still controls the logical channel multiplexing for uplink transmission.

11.23.4   Logical and Transport Channels

The MAC provides services to the RLC in the form of *logical channels*. A logical channel is defined by the type of information it carries.

The logical channel is associated with radio bearer and can be one of two types:

- a control channel that carries control and configuration information necessary for operating an LTE system, or
- a traffic channel carrying user data.

The set of LTE logical channels includes:

- The downlink Broadcast Control Channel (BCCH) for broadcasting the system control information.
- The downlink Paging Control Channel (PCCH) that transfers paging information and system information change notifications. This channel is used for paging when the network does not know the location cell of the UE.
- The downlink Common Control Channel (CCCH) transmits control information in conjunction with random access. This channel is used for UEs with no RRC connection with the network.

- The Dedicated Control Channel (DCCH) is a point-to-point bi-directional channel that transmits dedicated control information between a UE and the network. Used by UEs with an RRC connection.
- The Multicast Control Channel (MCCH) is a point-to-multipoint downlink channel used for transmitting MBMS control information from the network to the UE.
- The Dedicated Traffic Channel (DTCH) is a point-to-point channel used for transmission of user data to/from a terminal.
- The Multicast Traffic Channel (MTCH) is used for downlink transmission of MBMS services.

The physical layer offers information transfer services to MAC and higher layers in form of *transport channels*. A transport channel is defined by how and with what characteristics the information is transmitted over the radio interface. Data on a transport channel is organized into transport blocks. In each TTI, at most one transport block is transmitted over the radio interface to/from a terminal via a single RF/antenna path. In the case of spatial multiplexing (MIMO), there can be up to two transport blocks per TTI.

The parameters of transmission specifying how the transport block is to be transmitted over the radio interface are defined in Transport Format (TF). The transport format defines the transport-block size, the modulation-and-coding scheme and the antenna mapping. The transport format selection is a part of link adaptation to radio channel condition. By varying the transport format, the MAC layer can realize different data rates for transmission; that is, perform rate control.

The following transport channels are defined in LTE:

- The Broadcast Channel (BCH) has a fixed transport format. It is used for transmission of parts of the BCCH system information, Master Information Block (MIB).
- The Paging Channel (PCH) is used for transmission of paging information from the PCCH logical channel. The PCH supports discontinuous reception (DRX) to allow the terminal to save battery power by waking up to receive the PCH only at predefined time instants.
- The Downlink Shared Channel (DL-SCH) is the main transport channel used for transmission of downlink data in LTE. It supports key LTE features such as dynamic rate adaptation and channel-dependent scheduling in the time and frequency domains, hybrid-ARQ with soft combining and spatial multiplexing. It also supports DRX in reducing terminal power consumption while still providing an always-on experience. The DL-SCH is also used for transmission of the parts of the BCCH system information not mapped to the BCH. There can be multiple DL-SCHs in a cell, one per terminal scheduled in this TTI, and, in some subframes, one DL-SCH carrying system information.
- The downlink Multicast Channel (MCH) is used to support the MBMS.
- The Uplink Shared Channel (UL-SCH) is a major transport channel for uplink user data transmission. It supports the HARQ, dynamic link adaptation by varying the transmit power and potentially modulation and coding and both dynamic and semi-static resource allocation.
- The uplink Random Access Channel (RACH) is used in the uplink to respond to the paging message or initiate an access procedure to target the cell by sending some

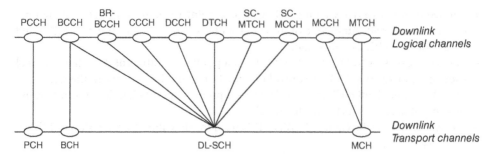

Figure 11.29 Downlink channel mapping [1].

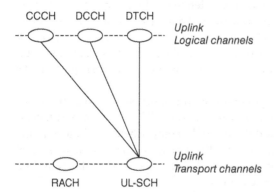

Figure 11.30 Uplink channel mapping [1].

preamble that indicates the attempt and can be used by eNodeB to estimate the timing advance.

MAC functionality includes multiplexing the different logical channels and mapping the logical channels to the appropriate transport channels. Figures 11.29 and 11.30 illustrate the mapping between downlink and uplink logical and transport channels, respectively.

11.23.5 Physical Layer

A physical channel corresponds to the set of time-frequency resource elements carrying the information from higher layers.

The physical channels defined in LTE include the following:

- The Physical Downlink Shared Channel (PDSCH) is the main physical channel that can carry both the DL-SCH and PCH transport channels.
- The downlink Physical Broadcast Channel (PBCH) carries system information required by the terminal in order to access the network. The coded BCH transport block is mapped to four subframes within a 40 ms interval. The 40 ms timing is blindly detected; that is, there is no explicit signalling indicating 40 ms timing. Each subframe is assumed to be self-decodable; that is, the BCH can be decoded from a single reception, assuming sufficiently good channel conditions.
- The downlink Physical Multicast Channel (PMCH) carries the MCH transport channel.

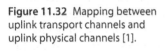

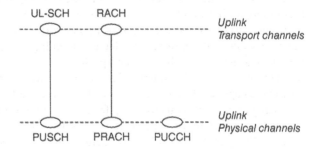

Figure 11.31 Mapping between downlink transport channels and downlink physical channels [1].

Figure 11.32 Mapping between uplink transport channels and uplink physical channels [1].

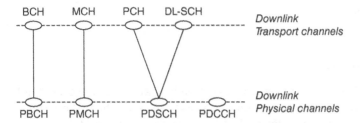

- The Physical Uplink Shared Channel (PUSCH) carries the UL-SCH transport channel. This is a main user-plane channel for uplink.

Figures 11.31 and 11.32 depict the mapping between transport and physical channels: LTE define some physical channels that have no corresponding transport channel. These channels carry control *necessary for proper reception and decoding of the downlink and uplink data transmission. Control information* channels include the following:

- The Physical Downlink Control Channel (PDCCH). The PDCCH informs the UE and (relay node (RN) in LTE-A) about the resource allocation of PCH and DL-SCH and Hybrid-ARQ information related to DL-SCH. The PDCCH carries the uplink scheduling grant
- The downlink Physical Hybrid-ARQ Indicator Channel (PHICH) carries Hybrid-ARQ ACK/NAKs in response to uplink transmissions
- The downlink Physical Control Format Indicator Channel (PCFICH). The PCFICH informs the UE (and the Relay Node) about the number of OFDM symbols used for the PDCCHs. This info is necessary to correctly decode PDCCH. The PCFICH transmitted in every downlink or special subframe
- The Physical Uplink Control Channel (PUCCH). The PUCCH carries:
  - Hybrid-ARQ ACK/NAKs in response to downlink transmission;
  - Scheduling Request (SR);
  - Channel-State Information (CSI) reports.

In addition to the physical channels, some physical signals are defined in LTE radio interface architecture. A downlink physical signal corresponds to a set of resource elements used by the physical layer but does not carry information originating from higher layers. The following downlink physical signals are defined:

- Reference signal

- Synchronization signal
- Discovery signal

11.23.6 RRC State Machine

The LTE RRC protocol defines two different states in terminal, RRC_CONNECTED and RRC_IDLE, depending on whether RRC connection has been established. Those states relate to two respective EPS mobility states. The left-hand side of Figure 9.34 (first introduced for 3G-WCDMA) illustrates the RRC state machine.

In the RRC_IDLE state, the terminal is registered in the network and allocated an ID that uniquely identifies the UR within the Tracking Area. There is no signalling connection that exists between the UE and the network. The UE monitors a paging channel during specific DRX intervals to detect incoming calls, acquires system information from broadcast control channels and performs neighbouring cell measurement and cell reselection. The DRX is configured by NAS and broadcasted on the BCCH. In fact, the terminal sleeps most of the time in order to reduce battery consumption and wakes up during certain intervals to check for the paging message. In RRC_IDLE, mobility is controlled by a UE that can perform cell reselection.

In the RRC_CONNECTED state, data transfer between UE and E-UTRAN is possible after the RRC context is established. The terminal is assigned then the Cell Radio Network Temporary Identifier (C-RNTI). During data transfer on the SCH, the UE monitors associate with the control channel for scheduling information and send back to the eNodeB channel-quality indicator. The mobile can also measure the neighbour cells' signal levels and report back to the eNodeB, monitor BCCH of the neighbour cells in RRC_CONNECTED state and acquire respective system information.

11.23.7 Time-Frequency Structure of the LTE FDD Physical Layer

The specification of physical-layer channels, modulation format and time frame structure is detailed in [16].

The size of any field in the time domain is expressed in a number of basic time units $T_{sample} = 1/f_s$ seconds, where f_s is a sampling rate and T_{sample} is a basic sampling interval. The sampling rate $f_s = \Delta f \times N_{FFT} = 15\,000 \cdot N_{FFT}$ is determined by maximum FFT size N_{FFT} used in LTE processing and carrier spacing $\Delta f = 15\,000$ Hz. Common implementation of FFT is done with a maximum size $N_{FFT} = 2048$. That corresponds to the sampling rate $f_s = 30.72$ MHz and respective basic time unit (basic sampling interval) $T_{sample} = 1/(15\,000 \times 2048) = 32.552$ ns. The OFDM symbol duration $T_s = 1/\Delta f$ and sampling interval related as $T_s = 2048 \cdot T_{sample}$ that results in $T_s = 66.7$ μs. It worth noting that the maximum LTE sampling rate is a multiple of the HSPA chip rate of 3.84 Mchip/s that ease the design of the multimode transceiver at the mobile terminal and base station.

In the time domain, downlink and uplink transmissions are organized into radio frames with $T_f = 307\,200 \times T_{sample} = 10$ ms duration. Each radio frame consists of 20 slots of length $T_{slot} = 15\,360 \cdot T_{sample} = 0.5$ ms, numbered from 0 to 19, as shown in Figure 11.33. For FDD LTE, two slots are organized into a subframe of 1 ms duration, 10 subframes are available for downlink transmission and 10 subframes are available for uplink transmissions in each 10 ms interval. Each time slot contains a number

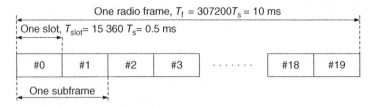

Figure 11.33 Frame structure of FDD LTE [16].

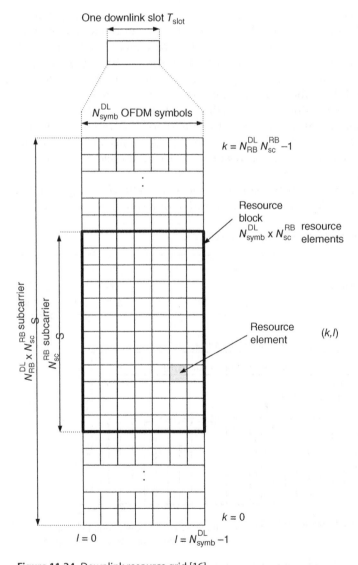

Figure 11.34 Downlink resource grid [16].

of OFDM symbols with a duration of T_s, seven symbols for the normal CP and six symbols for the long CP frame, respectively. Unless specified, we further consider only the normal CP frame.

The transmitted signal in each slot is described by one or several resource grids composed of the product of the number of resource block N_{RB} and number of OFDM symbols N_{symb}. $N_{symb} = 7$, in the case of a short frame. The resource grid is illustrated in Figure 11.34. The quantity N_{RB} depends on the transmission bandwidth configured in the cell and can be any value within the limits of $6 \leq N_{RB} \leq 110$ defined by LTE specifications.

A physical resource block (RB) consists of 12 subcarriers in the frequency domain and one timeslot in the time domain. The LTE specification introduces a resource element as a composition of a single subcarrier and OFDM symbol, see Figure 11.34. The resource block comprises $N_{symb} \times 12 = 7 \times 12 = 84$ resource elements in the case of s normal cyclic prefix, thus corresponding to one slot 0.5 ms in the time domain and 12×15 kHz $= 180$ kHz in the frequency domain The resource-block structure is defined above and over the timeslot interval. On the other hand, the TTI for scheduling is defined over one subframe that consists of two timeslots. The resource grid and resource-block structure as defined here applies to both the downlink and uplink frame structures. In the frequency domain, LTE carrier structure is composed of a number of resource blocks as shown in Table 11.6.

Figure 11.35 shows the relation between the channel bandwidth ($BW_{channel}$) and the transmission bandwidth configuration as a composition of resource blocks, N_{RB}. The

Table 11.6 Transmission bandwidth configuration N_{RB} in E-UTRA.

Channel bandwidth BW Channel [MHz]	1.4	3	5	10	15	20
Transmission bandwidth configuration N_{RB}	6	15	25	50	75	100
Number of occupied subcarriers	72	180	300	600	900	1200

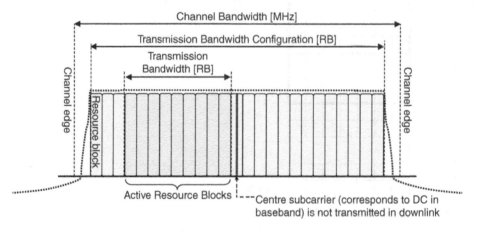

Figure 11.35 Definition of the channel bandwidth and transmission bandwidth configuration for one E-UTRA carrier [17].

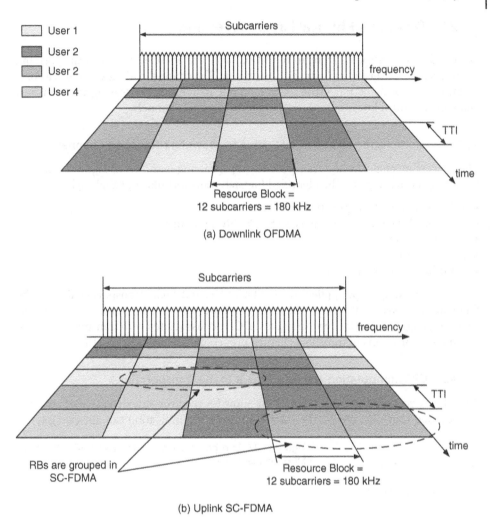

Figure 11.36 OFDMA transmission time-frequency grid.

channel edges are defined as the lowest and highest frequencies of the carrier f_c separated by the channel bandwidth; that is, at the $f_c \pm BW_{channel}$.

In the downlink configuration, there is an unused DC subcarrier that coincides with the carrier centre frequency. The DC subcarrier is not transmitted on the downlink. The uplink channel bandwidth configuration is similar to the downlink with a few differences. The uplink subcarriers are defined in such way that the centre frequency of uplink carrier is located between two subcarriers. Another difference is in a resource-block configuration for multiple access. As shown in Figure 11.36, the LTE downlink multiple-access scheme can allocate several non-consecutive resource blocks in one TTI for a single user. On the other hand, in the SC-FDMA uplink, the user is always allocated a solid set of consecutive resource blocks.

11.24 Downlink Physical Layer Processing

Data and control streams from/to the MAC layer are encoded to offer transport and control services over the radio transmission link. The channel coding scheme is a combination of error detection, error correcting, rate matching, interleaving and transport channel or control information) mapping onto physical channels.

11.24.1 Multiplexing and Channel Coding for Downlink Transport Channels

The processing chain for the transport block processing is shown in Figure 11.37. The following coding steps can be identified for each transport block of a DL cell:

- Add CRC to the transport block;
- Code block segmentation and code block CRC attachment;
- Channel coding;
- Rate matching;
- Code block concatenation.

The same coding steps applied to DL_SDH, PCH and MCH transport channels. The following sections briefly review the coding and multiplexing for transport channels as well as highlight coding for control signals (DCI) not associated with transport channels. The full specifications of transport channel processing are provided in [18].

11.24.2 CRC Computation and Attachment to the Transport Block

The CRC (Cyclic Redundancy Check) bits allow for receiver-side detection of errors in the decoded transport block. The corresponding error indication can, for example, be

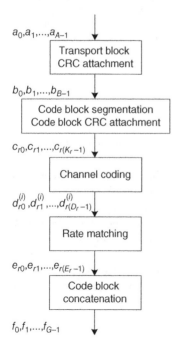

Figure 11.37 Transport block processing for DL-SCH, PCH and MCH channel coding [18].

used by the downlink hybrid-ARQ protocol as a trigger for requesting retransmissions. The CRC parity bits are generated by the cyclic generator polynomials. The details can be found in [18]. The length of CRC depends on channel configuration and can take the values 8, 16 or 24. For a DL-SCH, a CRC sequence of 24 bits is attached to each transport block.

11.24.3   Code Block Segmentation and Code Block CRC Attachment

If the input bit sequence to the code block segmentation is larger than the maximum code block size Z, segmentation of the input bit sequence is performed and an additional CRC sequence of 24 bits is attached to each code block. The maximum code block size is 6144 bits. The size of 6144 bits is defined to be optimal by the LTE turbo coder, which is always used for user data encoding. Therefore, code block segmentation ensures that the transport block is segmented into smaller blocks with the sizes supported by the turbo coder.

11.24.4   Channel Coding

Usage of coding scheme and coding rate for the different types of transport channels (TrCH) is shown in Table 11.7 [18].

The LTE turbo encoder implementation allows for parallel processing that substantially improves coding performance for high data rates. The scheme of turbo encoder actually includes the internal inter-leaver and two eight-state constituent encoders. The bits output from the turbo code internal inter-leaver are obtained by permutation of input bit stream. These output bits are to the input to the second eight-state constituent encoder.

11.24.5   Rate Matching for Turbo Coded Transport Channels

With rate matching, one needs to collect the encoded bit in the buffer and then select the exact number of bits to be extracted from the buffer to match the available physical resources for transmitting the transport block via the physical channel. Output of the turbo encoder with a third of the rate is tripled in length, since for one systematic

Table 11.7 Channel coding scheme and coding rate for Transport Channels (TrCHs).

TrCH	Coding Scheme	Coding Rate
UL-SCH	Turbo coding	1/3
DL-SCH		
PCH		
MCH		
SL-SCH		
SL-DCH		
BCH SL-BCH	Tail biting convolutional coding	1/3

(input) bit there are two parity bits. That constitutes three parallel bit streams used in the rate-matching scheme. These bit streams are separately interleaved and then collected in a virtual circular buffer: consecutive systematic bits first, followed by a sequence of alternating parity bits. The consecutive bits are then extracted from the circular buffer in a quantity to match the number of available resource elements in the Resource Blocks assigned for transmission. The details can be found in [18].

The rate-matching scheme operates on set of coded bits corresponding to one transport block and not on a single code block. This allows use of HARQ at the transport block level. With HARQ, the exact set of bits extracted for retransmission may start from a different point in the circular buffer.

11.24.6 Downlink Control Information Coding

Downlink Control Information may include downlink or uplink scheduling information, requests for aperiodic CQI reports and uplink power control commands. The coding steps for DCI are common steps for transport channel and consist of the following:

- Information element multiplexing;
- CRC attachment;
- Channel coding;
- Rate matching.

The control information coding schemes are given in Table 11.8 [18].

11.24.7 Physical Channel Processing

Once the data have been encoded, the code words are provided onwards for baseband processing of downlink physical channel, which is defined in terms of the following steps:

- scrambling of coded bits in each of the codewords to be transmitted on a physical channel,
- modulation of scrambled bits to generate complex-valued modulation symbols,

Table 11.8 Channel coding scheme and coding rate for control information.

Control Information	Coding Scheme	Coding Rate
DCI	Tail biting convolutional coding	1/3
CFI	Block code	1/16
HI	Repetition code	1/3
UCI	Block code	variable
	Tail biting convolutional coding	1/3
SCI	Tail biting convolutional coding	1/3

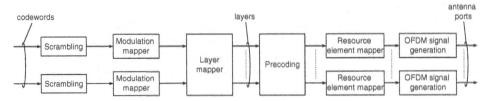

Figure 11.38 Physical-layer processing [16].

- mapping of the complex-valued modulation symbols onto one or several transmission layers,
- precoding of the complex-valued modulation symbols on each layer for transmission on the antenna ports,
- mapping of complex-valued modulation symbols for each antenna port to resource elements,
- generation of complex-valued time-domain OFDM signal for each antenna port.

The schematic diagram of the baseband processing is shown in Figure 11.38.

11.24.7.1 Bit-Level Scrambling

Downlink scrambling implies that the block of code word bits is multiplied (exclusive or operation) by a bit-level scrambling sequence. Different cells have assigned different scrambling sequences. The objective of scrambling is to prevent a situation where the channel decoder in UE may try to decode data from a neighbouring cell should the resource allocations happen to be identical between cells. By applying different scrambling sequences for neighbouring cells, the interfering signals behave like noise, thus ensuring processing gain with descrambling the channel code.

Downlink cell-specific scrambling is applied to all physical channels (except PMCH) and control signalling. The LTE downlink cell-specific scrambling sequence generator algorithm is initialized with the physical-layer cell identity. The PMCH channel is also scrambled but the generator is initialized with an MBSFN area identity, thus creating the scrambling sequence specific to a multicast area.

The scrambling sequence generator is reinitialized every subframe based on the identity of the cell, the subframe number (within a radio frame) and the UE identity.

11.24.7.2 Data Modulation

Following the scrambling stage, the data bits from each channel are mapped to complex-valued modulation symbols depending on the relevant modulation scheme, then mapped to layers and precoded mapped to REs and finally translated into a complex-valued OFDM signal by means of an Inverse Fast Fourier Transform (IFFT).

The downlink data modulation transforms the block of scrambled bits to a corresponding block of complex modulation symbols by means of a modulation mapper that relates a constellation diagram to the respective set of the input bits. The set of modulation schemes supported for the LTE downlink physical channels includes QPSK, 16QAM, 64QAM and 256QAM. Output of the modulation mapper is a block of complex symbols representing input codewords.

11.24.7.3   Layer Mapping

Layer mapping is closely associated with antenna mapping. In fact, antenna mapping is a two stage process comprised of layer mapping and specific precoding of the code-words for transmission via different antenna ports over the radio channel. With layer mapping the complex-valued modulation symbols $d^{(q)}(0), \ldots, d^{(q)}(M_{symb}^{(q)} - 1)$ for each of the code words q to be transmitted are mapped onto one or several layers $x(i) = [x^{(0)}(i) \ldots x^{(v-1)}(i)]^T$, $i = 0,1, \ldots, M_{symb}^{layer} - 1$, where v is the number of layers and M_{symb}^{layer} is the number of modulation symbols per layer.

The number of codewords is equal or less than the number of transmission layers. The transmission layer is associated with selected antenna port. The number of code words to be mapped is limited to two. For transmission on a single antenna port (one layer), one code word is directly mapped to that antenna port. The layer (antenna) mapping can be configured in different ways corresponding to different multiple antenna transmission schemes, including transmit diversity, beam forming and spatial multiplexing. LTE supports transmission using up to eight antenna ports depending on the exact multi-antenna transmission scheme.

The layer mapping for spatial multiplexing is done according to Table 11.9.

The layers in spatial multiplexing transmit multiple data streams in parallel as seen from the algorithm depicted in Table 11.9. The number of layers is referred to as the transmission rank, quantified by the Rank Indicator (RI). Performance of spatial multiplexing is measured by the UE by means of estimating the SNR at each transmission layer. The UE evaluates and feedbacks the measured RI to eNodeB for monitoring and adjustment to antenna mapping. For transmit diversity, the layer mapping is done according to Table 11.10. There is only one codeword and the number of layers v is equal to the number of antenna ports p used for transmission of the physical channel.

11.24.7.4   Precoding

Precoding uses a predefined 'codebook' to form the transmitted layers. The precoder takes as input a block of $x(i) = [x^{(0)}(i) \ldots x^{(v-1)}(i)]^T$, $i = 0,1, \ldots, M_{symb}^{layer} - 1$, from the layer mapping and generates a block of vectors $y(i) = [\ldots y^p(i) \ldots]^T$, $i = 0,1, \ldots, M_{symb}^{ap} - 1$, to be mapped onto resources on each of the antenna ports, where $y^p(i)$ represents the signal for antenna port. Here, M_{symb}^{ap} is a number of modulation symbols to transmit per antenna port for a physical channel, $M_{symb}^{ap} = M_{symb}^{layer}$.

Precoding for spatial multiplexing is defined by

$$\begin{bmatrix} y^{(0)}(i) \\ \vdots \\ y^{(P-1)}(i) \end{bmatrix} = W(i) \begin{bmatrix} x^{(0)}(i) \\ \vdots \\ x^{(v-1)}(i) \end{bmatrix} \tag{11.14}$$

where the precoding matrix $W(i)$ is of size $p \times v$ and $i = 0,1, \ldots, M_{symb}^{ap} - 1$, $M_{symb}^{ap} = M_{symb}^{layer}$.

In a closed-loop operation with CSI reporting, the values of $W(i)$ shall be selected among the precoder elements in the codebook configured in both the eNodeB and the UE. The matrices W are defined by codebook tables in [16]. The simple case for transmission mode 4 with two transmission layers mapped on two antenna ports is shown in Table 11.11.

Table 11.9 Codeword-to-layer mapping for spatial multiplexing [16].

Number of Layers	Number of Codewords	Codeword-to-Layer Mapping $i = 0,1, \dots, M_{symb}^{layer} - 1$	
1	1	$x^{(0)}(i) = d^{(0)}(i)$	$M_{symb}^{layer} = M_{symb}^{(0)}$
2	1	$x^{(0)}(i) = d^{(0)}(2i)$ $x^{(1)}(i) = d^{(0)}(2i + 1)$	$M_{symb}^{layer} = M_{symb}^{(0)}/2$
2	2	$x^{(0)}(i) = d^{(0)}(i)$ $x^{(1)}(i) = d^{(1)}(i)$	$M_{symb}^{layer} = M_{symb}^{(0)} = M_{symb}^{(1)}$
3	1	$x^{(0)}(i) = d^{(0)}(3i)$ $x^{(1)}(i) = d^{(0)}(3i + 1)$ $x^{(2)}(i) = d^{(0)}(3i + 2)$	$M_{symb}^{layer} = M_{symb}^{(0)}/3$
3	2	$x^{(0)}(i) = d^{(0)}(i)$ $x^{(1)}(i) = d^{(1)}(2i)$ $x^{(2)}(i) = d^{(1)}(2i + 1)$	$M_{symb}^{layer} = M_{symb}^{(0)} = M_{symb}^{(1)}/2$
4	1	$x^{(0)}(i) = d^{(0)}(4i)$ $x^{(1)}(i) = d^{(0)}(4i + 1)$ $x^{(2)}(i) = d^{(0)}(4i + 2)$ $x^{(3)}(i) = d^{(0)}(4i + 3)$	$M_{symb}^{layer} = M_{symb}^{(0)}/4$
4	2	$x^{(0)}(i) = d^{(0)}(2i)$ $x^{(1)}(i) = d^{(0)}(2i + 1)$ $x^{(2)}(i) = d^{(1)}(2i)$ $x^{(3)}(i) = d^{(1)}(2i + 1)$	$M_{symb}^{layer} = M_{symb}^{(0)}/2 = M_{symb}^{(1)}/2$
5	2	$x^{(0)}(i) = d^{(0)}(2i)$ $x^{(1)}(i) = d^{(0)}(2i + 1)$ $x^{(2)}(i) = d^{(1)}(3i)$ $x^{(3)}(i) = d^{(1)}(3i + 1)$ $x^{(4)}(i) = d^{(1)}(3i + 2)$	$M_{symb}^{layer} = M_{symb}^{(0)}/2 = M_{symb}^{(1)}/3$
6	2	$x^{(0)}(i) = d^{(0)}(3i)$ $x^{(1)}(i) = d^{(0)}(3i + 1)$ $x^{(2)}(i) = d^{(0)}(3i + 2)$ $x^{(3)}(i) = d^{(1)}(3i)$ $x^{(4)}(i) = d^{(1)}(3i + 1)$ $x^{(5)}(i) = d^{(1)}(3i + 2)$	$M_{symb}^{layer} = M_{symb}^{(0)}/3 = M_{symb}^{(1)}/3$
7	2	$x^{(0)}(i) = d^{(0)}(3i)$ $x^{(1)}(i) = d^{(0)}(3i + 1)$ $x^{(2)}(i) = d^{(0)}(3i + 2)$ $x^{(3)}(i) = d^{(1)}(4i)$ $x^{(4)}(i) = d^{(1)}(4i + 1)$ $x^{(5)}(i) = d^{(1)}(4i + 2)$ $x^{(6)}(i) = d^{(1)}(4i + 3)$	$M_{symb}^{layer} = M_{symb}^{(0)}/3 = M_{symb}^{(1)}/4$

(Continued)

Table 11.9 (Continued)

Number of Layers	Number of Codewords	Codeword-to-Layer Mapping $i = 0, 1, \ldots, M_{symb}^{layer} - 1$	
8	2	$x^{(0)}(i) = d^{(0)}(4i)$ $x^{(1)}(i) = d^{(0)}(4i+1)$ $x^{(2)}(i) = d^{(0)}(4i+2)$ $x^{(3)}(i) = d^{(0)}(4i+3)$ $x^{(4)}(i) = d^{(1)}(4i)$ $x^{(5)}(i) = d^{(1)}(4i+1)$ $x^{(6)}(i) = d^{(1)}(4i+2)$ $x^{(7)}(i) = d^{(1)}(4i+3)$	$M_{symb}^{layer} = M_{symb}^{(0)}/4 = M_{symb}^{(1)}/4$

Table 11.10 Codeword-to-layer mapping for transmit diversity.

Number of Layers	Number of Codewords	Codeword-to-Layer Mapping $i = 0, 1, \ldots, M_{symb}^{layer} - 1$	
2	1	$x^{(0)}(i) = d^{(0)}(2i)$ $x^{(1)}(i) = d^{(0)}(2i+1)$	$M_{symb}^{layer} = M_{symb}^{(0)}/2$
4	1	$x^{(0)}(i) = d^{(0)}(4i)$ $x^{(1)}(i) = d^{(0)}(4i+1)$ $x^{(2)}(i) = d^{(0)}(4i+2)$ $x^{(3)}(i) = d^{(0)}(4i+3)$	$M_{symb}^{layer} = \begin{cases} M_{symb}^{(0)}/4 & \text{if } M_{symb}^{(0)} \bmod 4 = 0 \\ (M_{symb}^{(0)} + 2)/4 & \text{if } M_{symb}^{(0)} \bmod 4 \neq 0 \end{cases}$ If $M_{symb}^{(0)} \bmod 4 \neq 0$ two null symbols shall be appended to $d^{(0)}(M_{symb}^{(0)} - 1)$

Table 11.11 Codebook for transmission on antenna ports {0,1} [16].

Codebook Index	Number of Layers v	
	1	2
0	$\frac{1}{\sqrt{2}} \begin{bmatrix} 1 \\ 1 \end{bmatrix}$	$\frac{1}{\sqrt{2}} \begin{bmatrix} 1 & 0 \\ 0 & 1 \end{bmatrix}$
1	$\frac{1}{\sqrt{2}} \begin{bmatrix} 1 \\ -1 \end{bmatrix}$	$\frac{1}{2} \begin{bmatrix} 1 & 1 \\ 1 & -1 \end{bmatrix}$
2	$\frac{1}{\sqrt{2}} \begin{bmatrix} 1 \\ j \end{bmatrix}$	$\frac{1}{2} \begin{bmatrix} 1 & 1 \\ j & -j \end{bmatrix}$
3	$\frac{1}{\sqrt{2}} \begin{bmatrix} 1 \\ -j \end{bmatrix}$	-

Table 11.12 Large-delay cyclic delay diversity [16].

Number of Layers v	U	$D(i)$
2	$\dfrac{1}{\sqrt{2}}\begin{bmatrix} 1 & 1 \\ 1 & e^{-j2\pi/2} \end{bmatrix}$	$\begin{bmatrix} 1 & 0 \\ 0 & e^{-j2\pi i/2} \end{bmatrix}$

The codebook index of the precoding matrix is reported by the UE as a 'PMI' in a CSI message. The eNodeB decides what codebook index to use and communicates it back to UE. In the case of open-loop transmission mode 3 with CDD, precoding for spatial multiplexing is defined by

$$\begin{bmatrix} y^{(0)}(i) \\ \vdots \\ y^{(P-1)}(i) \end{bmatrix} = W(i)D(i)U \begin{bmatrix} x^{(0)}(i) \\ \vdots \\ x^{(v-1)}(i) \end{bmatrix} \tag{11.15}$$

where the precoding matrix $W(i)$ is of size $p \times v$ and $i = 0,1, \ldots, M^{ap}_{symb} - 1$, $M^{ap}_{symb} = M^{layer}_{symb}$. The diagonal size $v \times v$ matrix $D(i)$ supporting cyclic delay diversity and the size $v \times v$ matrix U are both given by the respective tables in [16] for different numbers of layers v. Table 11.12 shows the matrices U and D for two-layer transmission.

In open-loop spatial multiplexing with CDD, the UE does not report preferred code book index PMI. The precoding matrix $W(i)$ is predetermined in this case. With two transmit antenna ports, matrix W is always an identity matrix given by

$$W(i) = \frac{1}{\sqrt{2}}\begin{bmatrix} 1 & 0 \\ 0 & 1 \end{bmatrix} \tag{11.16}$$

Precoding for transmit diversity is only used in combination with layer mapping for transmit diversity given in Table 11.10. The precoding operation for transmit diversity is defined for two and four antenna ports. For transmission on two antenna ports, $p = \{0, 1\}$, the output $(i) = [y^{(0)}(i)\, y^{(1)}(i)]^T$, $i = 0,1, \ldots, M^{ap}_{symb} - 1$, of the precoding operation is defined by

$$\begin{bmatrix} y^{(0)}(2i) \\ y^{(1)}(2i) \\ y^{(0)}(2i+1) \\ y^{(1)}(2i+1) \end{bmatrix} = \frac{1}{\sqrt{2}}\begin{bmatrix} 1 & 0 & j & 0 \\ 0 & -1 & 0 & j \\ 0 & 1 & 0 & j \\ 1 & 0 & -j & 0 \end{bmatrix}\begin{bmatrix} \mathrm{Re}(x^{(0)}(i)) \\ \mathrm{Re}(x^{(1)}(i)) \\ \mathrm{Im}(x^{(0)}(i)) \\ \mathrm{Im}(x^{(1)}(i)) \end{bmatrix}, \tag{11.17}$$

where in the input $i = 0,1, \ldots, M^{ap}_{symb} - 1, M^{ap}_{symb} = M^{layer}_{symb}$. The precoding and respective multi-antenna transmission techniques are applied in different ways depending on the type of physical signal or physical channel, according to Table 11.13 that provides a summary of diversity and spatial multiplexing techniques applied to LTE downlink signals and channels.

11.24.7.5 Mapping to Resource Elements

For each of the antenna ports used for transmission of the physical channel, the block of complex-valued symbols has to be grouped into sets of resource blocks assigned for transmission by MAC scheduler and mapped in sequence to the time-frequency

Table 11.13 Multi-antenna technique for physical channels.

Physical Channel/Signal	Transmit Diversity	Spatial Multiplexing	CDD
Reference signal	NO	NO	NO
Primary synchronization signal	NO	NO	NO
Secondary synchronization signal	NO	NO	NO
Physical broadcast channel	YES	NO	NO
Physical downlink control channel	YES	NO	NO
Physical hybrid-ARQ indicator channel	YES	NO	NO
Physical control format indicator channel	YES	NO	NO
Physical multicast channel	YES	YES	NO
Physical downlink shared channel	YES	YES	YES

resource grid. Each resource block consists of 84 resource elements (12 subcarriers during seven OFDM symbols). However, some of the resource elements within a resource block may not be available for the transport channel transmission as they are reserved in the resource grid for broadcast of system information or associated control signalling. Such resource elements include:

- downlink reference signals,
- downlink control information signals carried by PDCCH,
- synchronization signal or PBCH.

11.24.7.6 Downlink Reference Signals
The current LTE standard defines six types of downlink reference signals:

- Cell-specific Reference Signal (CRS)
- MBSFN reference signal
- UE-specific Reference Signal (UE-RS) associated with PDSCH
- DeModulation Reference Signal (DM-RS) associated with E-PDCCH or MPDCCH
- Positioning Reference Signal (PRS)
- CSI Reference Signal (CSI-RS)

There is one reference signal transmitted per downlink antenna port. Downlink reference signals are predefined signals occupying specific resource elements within the downlink time-frequency grid. Only one type, namely the *cell-specific reference signal*, is considered here.

Structure of a Single Reference Signal A single cell-specific reference signal consists of *reference symbols* of predefined values inserted within the first and fourth (third last) OFDM symbol of each slot. Reference symbols are separated over six subcarriers in the frequency domain. Within each *resource-block pair*, consisting of 12 subcarriers during one 1 ms subframe (two timeslots), there are *eight reference symbols*. Figure 11.39 shows an example of mapping the cell-specific reference signals on the downlink resource grid.

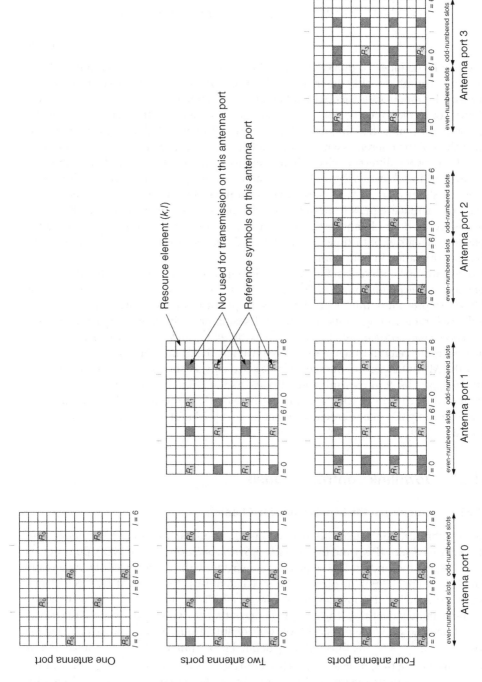

Figure 11.39 Mapping of downlink reference signals to the resource grid (normal cyclic prefix) [16].

All the reference signals (cell-specific, UE-specific or MBSFN-specific) are QPSK modulated. The signal can be written as

$$r_{l,n}(m) = \frac{1}{\sqrt{2}}(1 - 2 \cdot c(2m)) + j\frac{1}{\sqrt{2}}(1 - 2 \cdot c(2m + 1)), \; m = 0,1,\ldots,2N_{RB}^{max,DL} - 1$$

(11.18)

where m is the index of the RS, n is the slot number within the radio frame and 'l' is the symbol number within the time slot. The pseudorandom sequence $c(i)$ is comprised of a pseudorandom sequence, with different initialization values depending on the type of RSs. For the cell-specific RSs, the sequence is reinitialized at the start of each OFDM symbol, with a value that uniquely defined by the cell identity, N_{ID}^{cell} and slot number [16]. The values of the reference symbols vary between different reference-symbol positions and also between different cells. When reference signal sequence (11.18) is mapped to resource grid, as shown in Figure 11.39, it shifts in the frequency domain to a cell-specific reference shift $v_{shift} = N_{ID}^{cell} \bmod 6$. Such a pattern allows avoidance of time-frequency collision of reference signals from different cells for up to six neighbour cells. Reference signals are used for channel estimation (signal quality and strength). Distributing the reference signals in both time and frequency domains allows the UE to complete the channel estimation in both domains.

There are 504 different reference-signal sequences defined for LTE, where each sequence corresponds to one of 504 different physical-layer cell identities. During the cell-search procedure the terminal detects the physical-layer identity of the cell as well as the cell frame timing. Thus, from the cell-search procedure, the terminal knows the *reference-signal sequence of the cell* (given by the *physical-layer cell identity*) as well as the start of the reference-signal sequence (given by the frame timing).

In the MIMO case, the number of reference signals should match the number of antenna ports used in transmission. This assists the UE in estimating the propagation channel for each transmitting antenna. In these cases, the resource elements allocated to reference signals in one antenna cannot be occupied in the other antennas as shown in Figure 11.39.

11.25 Downlink Control Channels

11.25.1 Structure of the Synchronization Channel

Two synchronization signals are defined on each downlink carrier, the *Primary Synchronization Signal* (PSS) and the *Secondary Synchronization Signal* (SSS). Both signals are generated by random like sequences associated with physical Cell ID. As already stated, there are 504 unique physical-layer cell identities. The physical-layer cell identities are grouped into 168 unique physical-layer cell-identity groups, each group containing three unique identities. The grouping is such that each physical-layer cell identity is part of one and only one physical-layer cell-identity group. A physical-layer cell identity $N_{ID}^{cell} = 3N_{ID}^{(1)} + N_{ID}^{(2)}$ is thus uniquely defined by a number $N_{ID}^{(1)}$ in the range of 0 to 167, representing the physical-layer cell-identity group, and a number $N_{ID}^{(2)}$ in the range of 0 to 2, representing the physical-layer identity within the physical-layer cell-identity group.

Table 11.14 Root indices for the primary synchronization signal [16].

$N_{ID}^{(2)}$	Rootlindex u
0	25
1	29
2	34

The sequence $d(n)$ used to generate three types of the PSS from a frequency domain Zadoff–Chu sequence according to

$$d_u(n) = \begin{cases} e^{-j\frac{\pi u n(n+1)}{63}} & n = 0,1,\ldots,30 \\ e^{-j\frac{\pi u(n+1)(n+2)}{63}} & n = 31,32,\ldots,61 \end{cases} \tag{11.19}$$

where the Zadoff–Chu root sequence index u is given by Table 11.14.

11.25.2 Time-Domain Position of Synchronization Signals

The time-domain positions of the synchronization signals within the frame depends on the frame type structure that, in turn, depends on operation mode FDD or TDD. In the case of FDD (Frame structure Type 1) the PSS is transmitted within the last symbol of the first slot of subframes 0 and 5, while the SSS is transmitted within the second last symbol of the same slot; that is, just prior to the PSS, see Figure 11.40.

11.25.3 Frequency Domain Structure of Synchronization Signals

11.25.3.1 PSS Structure
The PSS occupies a 1.08 MHz bandwidth composed of six resource blocks; that is, 72 subcarriers centred around a DC carrier that is not transmitted, as shown in Figure 11.41. The OFDM modulator for PSS is composed of 62 subcarriers with a Zadoff–Chu binary sequence extended with five zeros at the edges.

The central position around the DC carrier of the PSS allows for a bandwidth-independent cell and frame synchronization as well as initial cell access.

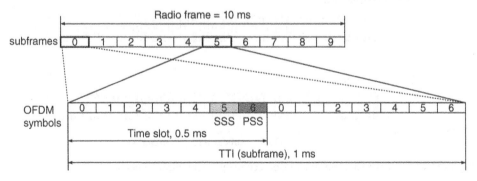

Figure 11.40 Time-domain position of PSS and SSS for FDD.

11.25.3.2   SSS Structure

Each SSS sequence is constructed in the frequency domain by interleaved concatenation of two length of 31 bits binary sequences d_1 and d_2 that produce a secondary synchronization code. Both sequences d_1 and d_2 are created by a cyclic shift of the same length; a 31 binary sequence. The cyclic shift indices of the binary sequences d_1 and d_2 are derived from a function of the physical-layer cell-identity group (table 6.11.2.1–1 in TS 36.211, Ref. [16]). In the frequency domain, the sequences d_1 and d_2 are interleaved as shown in Figure 11.42 and alternated between transmission in the subframe 0 (slot 0) and subframe 5 (slot 10) (Figure 11.41 and Figure 11.42).

The concatenated sequence is scrambled with a scrambling sequence that depends on the primary synchronization signal and unique physical Cell ID [16]. Contrary to PSS, the SSS sequence is BPSK modulated, the signal is transmitted with 1.08 MHz bandwidth (6 RB = 72 subcarriers of 15 kHz) and placed in the end of the first and eleventh slots (slots 0 and 10) of the 10 ms frame or, in other words, in subframes 0 and 5, see Figure 11.40. Similar to PSS, the SSS occupies 72 subcarriers that are centred on but not including the DC subcarrier in subframes 0 and 5 (FDD, Frame structure type 1).

The alternated structure of SSS code in subframes 0 and 5 enables the UE to determine the 10 ms radio frame timing from a single reading of a secondary synchronization signal. The position of synchronization signals in resource grid is shown in Figure 11.43.

11.25.4   PBCH

The PBCH carries some of the system information content of BCH transport channel related to initial access to the cell. This information is contained in the Master Information Block (MIB), which is transmitted logically in an interval of 40 ms. One

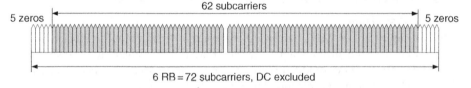

Figure 11.41 Definition and structure of PSS.

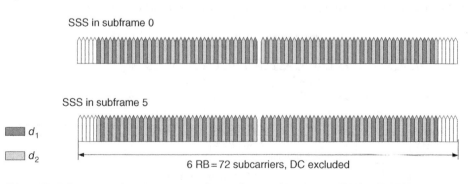

Figure 11.42 Frequency domain structure of secondary synchronization signal in FDD LTE.

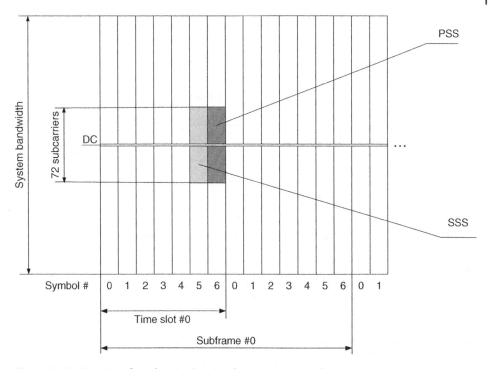

Figure 11.43 Mapping of synchronization signals to a resource grid.

MIB comprises 14 information bits and 10 spare bits. The 16 CRC bits are attached to the MIB. After channel coding with a third of the rate and repetition to match total 1920 bits (with normal cyclic prefix) for a 40 ms transmission interval for PBCH, the coded sequence is scrambled and QPSK modulated. Then, the overall block is segmented in to four equal size self-decodable units that are mapped in the frequency domain on the minimum transmission bandwidth of six resource blocks centred over DC carrier; that is, 72 subcarriers of 1.08 MHz bandwidth. Each unit is transmitted in the PBCH resources in every radio frame (see Figure 1.57), thus a MIB is decodable every 10 ms. Transmission of MIB is thus effectively spread over a 40 ms period exploiting time diversity in the radio propagation channel. The PBCH is mapped to the first subframe of each radio frame, it follows immediately after the PSS and SSS, as shown in Figure 11.44.

As a result, the effective coding rate for PBCH is $40/1920 = 1/48$ in each 40-ms TTI. Given that only 14 information bits of MIB are transmitted over a 40 ms interval, the effective data rate is very low; 350 bps. Apparently, the total number of resource elements used to transmit MIB with 14 information bits is very large implying a very large effective processing gain caused by repetition and channel coding. On the other hand, the UE in the neighbouring cell should be able to decode the BCH in the case of a week PBCH signal; that is, a low signal-to-interference ratio. If the SIR is high the UE can decode the MIB from one subframe, otherwise the UE can receive the other MIBs and perform soft combining.

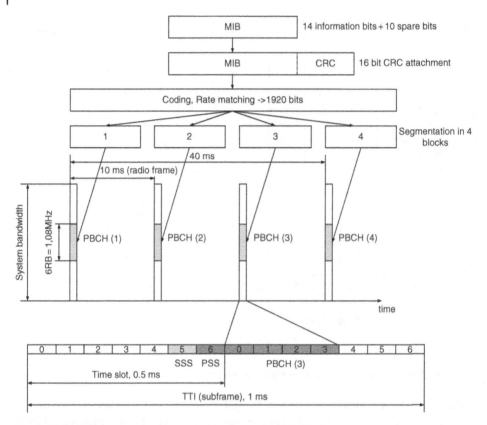

Figure 11.44 PBCH structure and mapping to resource grid [12].

11.25.5 Physical Control Format Indicator Channel: PCFICH

The PCFICH carries information about the number of OFDM symbols used for transmission of PDCCHs in a subframe; that is, the size of control region. The UE must correctly decode the PCFICH in order to process the control channel information dedicated to that user. The PCFICH consists of two bits of information representing the number of OFDM symbols (up to four) used for PDCCH [16]. This information is encoded with a 32-bit pattern. The coded bits are scrambled with a cell and subframe number specific scrambling sequence, QPSK modulated and then mapped to 16 subcarriers. The mapping to resource elements is defined in terms of quadruplets of complex-valued symbols. The PCFICH then constitutes four non-adjacent resource-element groups (REGs) in the frequency domain. The separation of REGs in the frequency enables frequency diversity with some gain to correct decoding of PDCCH. The concept of splitting the information into several REGs is also used for other downlink control channels.

The specific position of PCFICH depends on the physical Cell ID. A scrambling sequence generator for PCFICH bits is initialized with a root index that depends on the physical Cell ID and frame number. This allows avoidance of collision of messages from neighbour cells and reduces inter-cell interference, as shown in Figure 11.45.

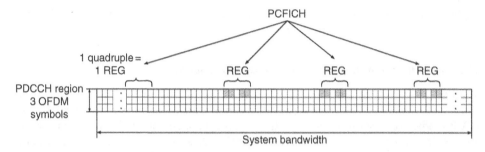

Figure 11.45 Example of mapping of PCFICH to a resource grid.

11.25.6 PDCCH

The PDCCH carries scheduling assignments and other control information in the form of Downlink Control Information (DCI) messages dedicated to a single UEs or group of UEs. The DCI may contain associated control signalling information, such as: resource-block assignments, assignments type, Modulation-and-Coding Scheme (MCS), HARQ feedback information, power control commands for uplink transmission and so on. The length of PDCCH is indicated in PCFICH channel. The PDCCH supports multiple formats [16].

A physical control channel is transmitted on an aggregation of one or several consecutive control channel elements (CCEs) where a control channel element corresponds to nine resource-element groups. The number of resource-element groups not assigned to PCFICH or PHICH is N_{REG}. The CCEs available in the system are numbered from 0 to $N_{CCE} - 1$, where $N_{CCE} = \lfloor N_{REG}/9 \rfloor$. The PDCCH supports multiple formats as listed in Table 11.15. Multiple PDCCHs can be transmitted in a subframe.

The number of CCEs for a specific PDCCH is determined by the eNodeB according to the channel conditions of the dedicated user. If the PDCCH is intended for a UE with a good downlink channel (e.g. close to the eNodeB), then one CCE could be sufficient. However, for a UE located near the cell edge, several CCEs may be required for correct decoding.

The PDCCH information block is appended with 16 bits CRC that are scrambled with the 'UE identity code'. This allows the intended UE to use a blind decoding to search for the designated DCI. Further processing for PDCCH includes convolutional channel coding, rate matching and scrambling with a cell-specific scrambling sequence [18]. Scrambled bits are QPSK modulated and mapped to the set of REGs. The PDCCHs are transmitted on the same set of antenna ports as the PBCH, and transmit diversity is

Table 11.15 Supported PDCCH formats [16].

PDCCH Format	Number of CCEs	Number of Resource-Element Groups	Number of PDCCH Bits
0	1	9	72
1	2	18	144
2	4	36	288
3	8	72	576

applied if more than one antenna port is used. An example of mapping of PDCCH on a resource grid is given in Figure 11.46.

11.25.7   PHICH, Physical Hybrid-ARQ Indicator Channel

The PHICH carries the hybrid-ARQ feedback information to the UEs. The feedback contains the ACK or NACK of a previous uplink transmission by particular UE. The HARQ indicator is set to 0 for a positive ACKnowledgement (ACK) and 1 for a Negative ACKnowledgement (NACK). The HARQ indicator is repeated three times and each triple is orthogonally Walsh spread to four complex symbols. The Walsh sequence is BPSK modulated and mapped on to respective REGs. The processing is shown in Figure 11.47 [12].

Multiple PHICHs mapped to the same set of resource elements constitute a PHICH group, where PHICHs within the same PHICH group are separated through different orthogonal Walsh sequences.

11.26   Mapping the Control Channels to Downlink Transmission Resources

Figure 11.46 illustrates mapping of the various downlink channels on to a time-frequency resource grid. The PCFICH is transmitted in the first symbol every subframe to inform the UE about the number of symbols used for the PDCCH channel.

The PCFICH is mapped to four resource-element groups, followed by allocating the resource-element groups required for the PHICH. The resource-element groups left after the PCFICH and PHICH are used for the different PDCCHs in the system. The unused resource-element groups can be used for additional PDCCHs.

11.27   Uplink Control Signalling

There are two types of uplink control signalling:

1) Associated control signalling required to process the current data packet. This control information is always transmitted together with uplink data, examples include transport format indicators, 'New-Data' Indicators (NDIs) and MIMO parameters.
2) Control Information signalling which is transmitted independently of uplink user data packet. Examples include HARQ acknowledgements (ACK/NACK) of received downlink data packets, Channel-State Information to assist downlink transmission and Scheduling Requests (SRs) for uplink transmissions.

The HARQ-ACK/NACK bits are sent in response to downlink data packets on the Physical Downlink Shared Channel (PDSCH), one ACK/NACK bit per downlink codeword is sent in the feedback.

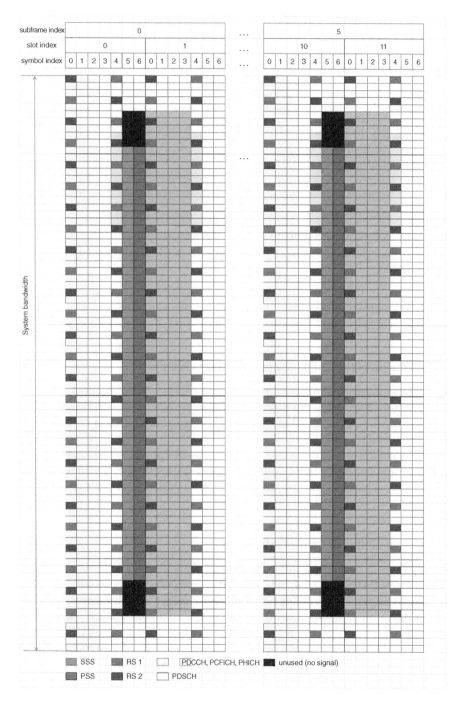

Figure 11.46 Illustration of a LTE downlink resource grid.

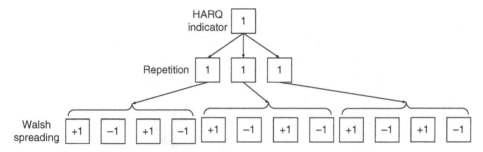

Figure 11.47 PHICH signal construction [12].

Twenty bits per subframe can be allocated for Channel-State Information (CSI), which includes:

- Channel-Quality Indicators (CQIs) to support link adaptation,
- MIMO feedback information such as Rank Indicators (RIs) and Precoding Matrix Indicators (PMIs).

The Uplink Control Information (UCI) signalling can be sent either on a standalone Physical Uplink Control Channel (PUCCH) or mapped on a shared Physical Uplink Traffic Channel (PUSCH).

11.27.1 Processing of the Uplink Shared Transport Channel

Figure 11.48 shows the processing structure for the UL-SCH transport channel. The coding steps for user data are similar to downlink coding, however, uplink processing may include coding and multiplexing of Uplink Control Information (UCI) for transmission on PUSCH together with data. The dashed blocks in Figure 11.48 shows that option.

The control signalling of UCI is realized on PUSCH by allocating the dedicated control resource that is valid during the uplink subframe scheduled for the UE. The multiplexing of user data and UCI requires an application of channel interleaver in order to correctly map the UCI on the resource grid.

11.27.2 Channel Coding of Control Information

Control data arrives at the coding unit in the form of channel-quality information (CQI and/or PMI), HARQ-ACK and rank indication (RI). Different coding rates for the control information are achieved by allocating different number of coded symbols for its transmission. The number of channel-quality bits (CQI) is variable and depends on the transmission format/type report to eNodeB. With rank indication bits, the UE provides a feedback to eNodeB on performance of the MIMO based on the measurement of SNR in each receive antenna. The meaning of RI value is discussed in Section 11.21.

11.27.3 Multiplexing and Channel Interleaving

The control and data are multiplexed into an uplink symbol block on UL_SDH after separate coding and modulation are applied to different control fields (ACK/NAK, CQI,

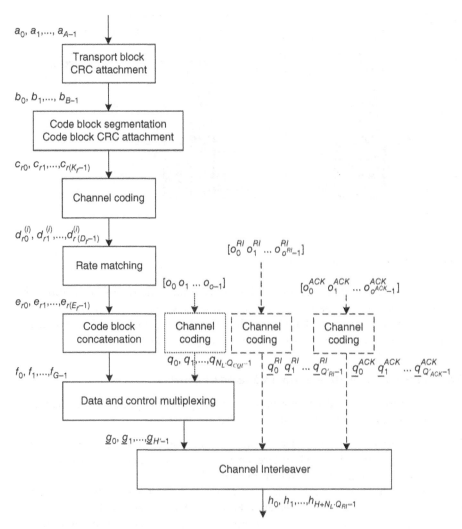

Figure 11.48 Transport-block processing for UL-SCH [18].

PMI, RI). The control and data multiplexing is performed such that HARQ-ACK information is present on both slots and is mapped to resources around the demodulation reference signals. In addition, the multiplexing ensures that control and data information are mapped to different modulation symbols.

The channel interleaver implements a mapping of modulation symbols onto the time-frequency resource grid while ensuring that the HARQ-ACK and RI information are present on both slots in the subframe. HARQ-ACK information is mapped to resources around the uplink demodulation reference signals while RI information is mapped to resources around those used by HARQ-ACK. The RI is placed on symbols next to ACK/NACK. CQI/PMI symbols are placed at the beginning of the SC-FDMA symbols and they are spread over all the available SC-FDMA symbols. The example of mapping is presented in Figure 11.49.

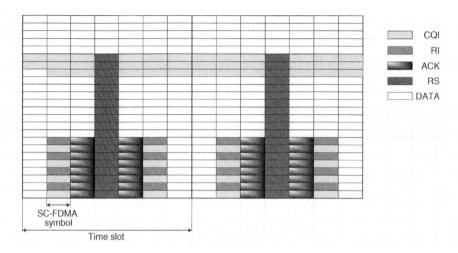

Figure 11.49 Mapping of UCI to PUSCH [19].

As the control information has specific locations around the reference symbols, the physical-layer control information is separately coded and placed in a predefined set of modulation symbols. Thus the channel interleaver in Figure 11.49 can be regarded as a mapper of user and control information to resource grid in uplink subframe.

11.27.4 Processing for Physical Uplink Shared Channel

The processing for baseband signal representing the physical uplink shared channel is performed in the following steps as shown in Figure 11.50.

- *Bit-Level Scrambling:* The objective of uplink scrambling is the same as for the downlink; that is, to randomize the interference and thus ensure that the processing gain provided by the channel code can be fully utilized.
- *Modulation:* A block of scrambled bits is mapped to a block of complex-valued symbols. The modulation schemes used are QPSK, 16QAM and 64QAM.
- *Layer Mapping*: Layer mapping of the complex-valued modulation symbols onto one or several transmission layers. This operation is similar to downlink. The number of transmission layers in uplink is limited to four.
- *Transform Precoding*: Transform precoding is also called DFT precoding. The block of M modulated complex symbols is fed through a size-M DFT coder, where M corresponds to the number of subcarriers assigned for the transmission.
- *MIMO Precoding*: Precoding for MIMO transmission. This operation is similar to downlink precoding. The DFT processed data stream from a DFT precoder can be

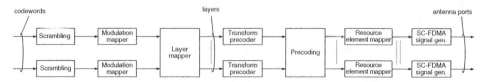

Figure 11.50 Uplink physical channel processing [16].

divided in multiple streams depending on MIMO scheme applied and then precoded with channel matrix according to respective uplink coding book [16].

- *Mapping*: This is the mapping of precoded symbols to resource elements in each antenna port. For each antenna port used for transmission of the PUSCH in a subframe the block of complex-valued symbols is multiplied with the amplitude scaling factor in order to conform to the transmit power controlled by a power control mechanism and mapped to physical resource blocks on respective antenna port assigned for transmission. The scheduler assigns a set of subframes with a number of subcarriers to be used for the uplink transmission of PUSCH. Mapping of the DFTS-OFDM symbol to resource grid is illustrated in Figure 11.22. Two of 14 OFDM symbols in the subframe are used for uplink demodulation reference signals and not available for PUSCH transmission. One additional symbol may be reserved for the transmission of the sounding reference signal.

Generation of the complex-valued time-domain SC-FDMA signal for each antenna port is done by OFDM subcarrier shaping (IFFT block in Figure 11.21) of the block of DFT symbols that come from the mapper to resource grid in the previous step.

11.27.5   Physical Uplink Control Channel, PUCCH

An alternative way to transmit uplink control information (UCI) is to use a standalone physical uplink control channel, PUCCH. In case if the terminal has not been assigned an uplink resource for UL-SCH transmission, the control information (channel-state reports, hybrid-ARQ acknowledgements and scheduling requests) is transmitted on uplink resources (resource blocks) specifically assigned for uplink L1/L2 control on PUCCH. In such an arrangement, the uplink user traffic channel, PUSCH, caries user data only. Simultaneous transmission of PUCCH and PUSCH from the same UE is supported if enabled by higher layers. Data arrives to the coding unit in the form of indicators for measurement indication, scheduling request and HARQ acknowledgement.

Different forms of UCI channel coding are used depending on allocated format for PUCCH. The PUCCH supports seven different formats (with an eighth added in Release 10), depending on the information to be signalled. The details of mapping between PUCCH formats and UCI can be found in [16] and [18].

PUCCH consists of a frequency resource of one resource block (12 subcarriers) and a time resource of one subframe. The PUCCH is mapped at the edges of the uplink system bandwidth, as shown in Figure 11.51 in order to support contiguous resource grid for user traffic. Additionally, PUCCH is protected with a kind of frequency diversity realized by means of symmetrical frequency hopping of each slot containing PUCCH, see Figure 11.51.

11.27.6   Multiplexing of UEs Within a PUCCH

The PUCCH can be shared by different users. Different UEs can be separated on PUCCH by means of Frequency Division Multiplexing (FDM) or Code Division Multiplexing (CDM). With FDM, additional PUCCH resource blocks are normally assigned next to each other while still located at the edges of system bandwidth. The CDM is used inside

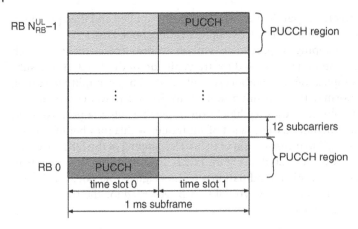

Figure 11.51 Mapping to physical resource blocks for PUCCH format 2/2a/2b [16].

a single PUCCH resource block. The CDM can be realized using different zero autocorrelation codes for PUCCH assigned to different terminals [10].

Another approach to user multiplexing is by using cyclic time shifts of a sequence with suitable orthogonal properties. It can be done by assigning different Zadoff–Chu sequences to different UEs. The same Zadoff–Chu sequences compose the waveform of reference signals, they present orthogonal phase rotations (or cyclic time shift) of a cell-specific length-12 frequency domain sequence. As discussed in [10], PUCCH structure can be arranged in a way to combine both different cyclic time shifts of a base sequence and time-domain code multiplexing based on different orthogonal block spreading codes to support multiple UEs simultaneously.

11.27.7 Physical Random Access Channel (PRACH)

Only one PRACH resource can be configured in the FDD uplink subframe. The periodicity of PRACH resources can be scaled according to the expected RACH load. Frequency of PRACH resources allocation can vary from PRACH occurrence every subframe to one PRACH per 20 ms. PRACH transmits random access preamble that consists of a preamble sequence and a preceding cyclic prefix as shown in Figure 11.52.

Due to the wide range of operational environment, the LTE defines multiple preamble formats (see table 5.7.1–1, Random access preamble parameters in [16] for details).

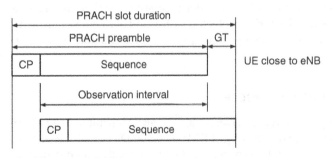

Figure 11.52 Random access with PRACH [16].

Those formats differ in the duration of cyclic prefix and preamble sequence that can span 800 μs for small and medium cells to 1600 μs for large cells up to ~100 km. The design constraints for preamble parameters are discussed in detail in [12]. The length of preamble is driven by a trade-off between sequence length and requirement to fit into single subframe and accommodate a guard period after the preamble that is necessary due to time delay between different terminals within the cell. The length of the cyclic prefix is also adjusted to a radio propagation environment, the longer CP assists in mitigating a long delay spread in larger cells.

In each cell, there are 64 preamble sequences available. The set of parameters defining the generation of specific preamble sequence from the Zadoff–Chu root sequence are signalled as a part of system information for PRACH configuration. When performing an unsynchronized random access attempt, the UE randomly selects one set and respective preamble sequence. Since the preamble sequence is constructed from the Zadoff–Chu sequence it retains its native correlation properties. The 'quasi-orthogonality' of Zadoff–Chu sequence minimizes interference from overlapping PRACH attempts by different users in the cell in case of collision. PRACH occupies six RB with 1.08 MHz bandwidth that provides an acceptable resolution ~±0.5 μs for the timing estimate in uplink transmission of PUCCH and PUSCH.

11.28 Uplink Reference Signals

As in the downlink, the LTE Single-Carrier Frequency Division Multiple-Access (SC-FDMA) uplink incorporates Reference Signals (RSs) for data demodulation and channel sounding. The roles of the uplink RSs include enabling channel estimation to aid coherent demodulation, channel-quality estimation for uplink scheduling, power control, timing estimation and direction-of-arrival estimation to support downlink beamforming. Two types of RS are supported on the uplink:

1) *DeModulation RS* (DM-RS), associated with transmissions of uplink data on the Physical Uplink Shared CHannel (PUSCH) and/or control signalling on the Physical Uplink Control CHannel (PUCCH). These RSs are used for channel estimation for coherent demodulation.
2) *Sounding RS* (SRS), not associated with uplink data and/or control transmissions, and primarily used for channel-quality estimation to enable frequency selective scheduling on the uplink.

The uplink Demodulation RSs are time-multiplexed with the data symbols. The DMRSs of a given UE (User Equipment) occupy the same bandwidth (i.e. the same Resource Blocks, RBs) as its PUSCH/PUCCH data transmission. Thus, the allocation of orthogonal (in frequency) sets of RBs to different UEs for data transmission automatically ensures that their DM RSs are also orthogonal to each other.

The SRSs, if configured by higher-layer signalling, are transmitted on the last SC-FDMA symbol in a subframe; SRS can occupy a bandwidth different from that used for data transmission. UEs transmitting SRS in the same subframe can be multiplexed via either Frequency or Code Division Multiplexing (FDM or CDM, respectively).

Key requirements for the uplink Reference Signal are:

- Constant amplitude in the frequency domain thus facilitating efficient channel estimation;
- Low amplitude variation in the time domain;
- Good autocorrelation properties for accurate channel estimation.

Zero cross-correlation properties between different RSs to reduce interference from RSs transmitted on the same resources in other (or, in some cases, the same) cells.

The reference signal sequence is defined by a cyclic shift of a base sequence $\bar{r}_{u,v}(n)$, $0 \leq n < M_{sc}^{RS}$, $M_{sc}^{RS} = mN_{sc}^{RB}$ is a length of base sequence expressed in a number of subcarriers. The same set of base sequences is used for demodulation and sounding reference signals. The definition of the base sequence $\bar{r}_{u,v}(0), \ldots, \bar{r}_{u,v}(M_{sc}^{RS} - 1)$ depends on the sequence length M_{sc}^{RS}. Long base sequences with length $M_{sc}^{RS} \geq 3 \cdot RB$ are generated by means of Zadoff –Chu sequence, shorter sequences are generated by orthogonal sequence of complex numbers of unit amplitude. [16]. Both basic sequences satisfy to requirements stated previously.

Basic RS sequence has a periodic autocorrelation function that is zero except for the zero shift value. The cyclic, or circular, shifts of a sequence are orthogonal to each other. Thus, a cyclic shift applied to a basic RS sequence provides a means to derive multiple orthogonal sequences from a single RS sequence and to multiplex different UEs in common available system bandwidth.

11.28.1 Mapping of Reference Signals to the Uplink Frame Structure

LTE defines DM-RS reference signal in each uplink time slot. In case of normal CP PUSCH the DM-RS is placed in the middle (4th symbol) of the time slot while in case of extended CP it occupies the third symbol. For standalone control channel, PUCCH, the exact position of DM-RS depends on the type of uplink UCI to be transmitted.

The Sounding Reference Signal is placed in the last SC-FDMA symbol in a specifically configured subframe, as shown in Figure 11.53. The subframes in which SRS can be configured are indicated by cell-specific broadcast signalling.

The eNodeB in LTE may either request an individual SRS transmission from a UE or configure a UE to transmit SRS periodically until terminated.

In order to support frequency selective scheduling between multiple UEs, it is necessary for eNodeB to obtain instantaneous information on channel quality across the range of uplink system bandwidth at the moment used by active UEs in different location within the cell. As a means to achieve that, LTE introduces the SRS with different sounding bandwidth in a way that SRSs from different users can overlap. This method is named Interleaved FDMA with a RePetition Factor (RPF) of two that is applied in the frequency domain such that the SRS signal occupies every second subcarrier within the allocated sounding bandwidth, see Figure 11.54.

A frequency domain decimation factor of two provides the spacing between occupied subcarriers of an SRS signal with a comb-like spectrum. The UE receives configurable

Data	Data	Data	DMRS	Data	Data	SRS
Symbol # 0	1	2	3	4	5	6

Figure 11.53 Uplink subframe configuration with an SRS symbol.

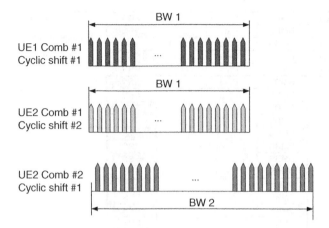

Figure 11.54 Multiplexing of SRS transmissions from different terminals [11].

SRS parameters indicating the 'transmissionComb' index (0 or 1) nn to transmit to the SRS, the sounding bandwidth and SRS sequence length. The SRS bandwidth (in RBs) must be an even number, due to the RPF of two, with a minimum sounding bandwidth of 24 subcarriers and minimum SRS sequence length of 12. Simultaneous SRS can be transmitted from multiple UEs using the same RBs and the same offset of the comb, using different cyclic time shifts of the same basic RS sequence to achieve orthogonal separation. An example of multiplexing the different SRSs from different UEs is shown in Figure 11.54. Figure 11.55 shows a time slot structure with SRSs from multiple users.

The SRSs overlapping different user bandwidths provide a means to estimate channel quality in different fractions of system bandwidth corresponding to different locations of terminals. This information can be used for dynamic changes in the frequency selective uplink channel scheduling for different UEs.

11.29 Physical-Layer Procedures

Major physical-layer procedures in LTE include cell search, synchronization, power control, link adaptation, HARQ, timing alignment and random access.

11.29.1 Cell Search

The first step in access to a radio cell is a cell-search procedure by which a UE acquires time and frequency synchronization with a cell and detects the physical-layer Cell ID of that cell. There are three levels of synchronization:

1) symbol timing acquisition,
2) carrier frequency synchronization that eliminates frequency mismatch between base and UE oscillators as well as Doppler frequency shift and
3) sampling clock synchronization that is the finest level necessary for signal demodulation.

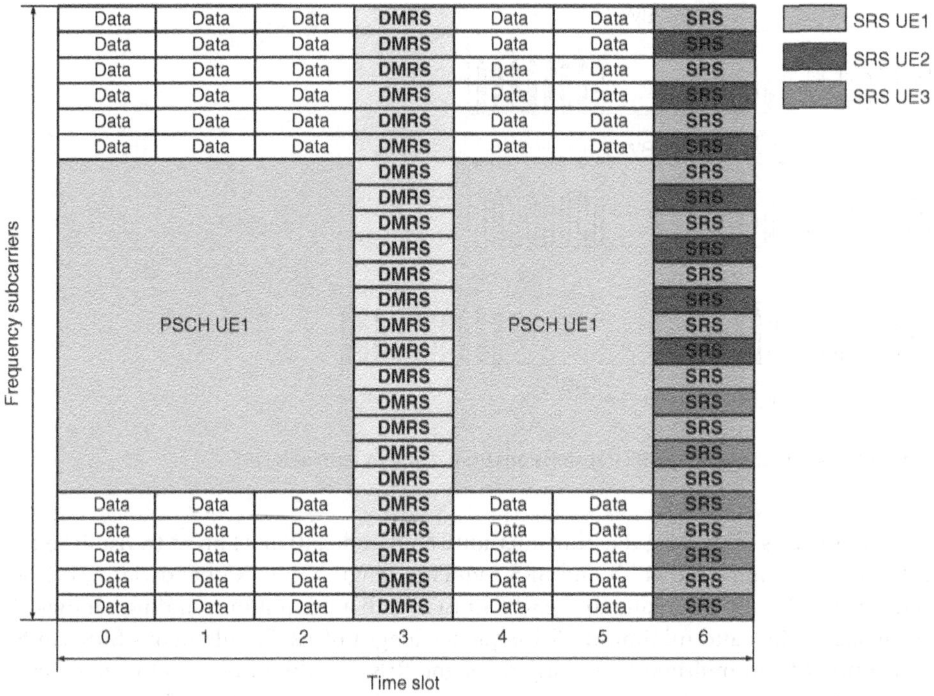

Data	Data	Data	DMRS	Data	Data	SRS
Data	Data	Data	DMRS	Data	Data	SRS
Data	Data	Data	DMRS	Data	Data	SRS
Data	Data	Data	DMRS	Data	Data	SRS
Data	Data	Data	DMRS	Data	Data	SRS
Data	Data	Data	DMRS	Data	Data	SRS
PSCH UE1			DMRS	PSCH UE1		SRS
			DMRS			SRS
			DMRS			SRS
			DMRS			SRS
			DMRS			SRS
			DMRS			SRS
			DMRS			SRS
			DMRS			SRS
			DMRS			SRS
			DMRS			SRS
			DMRS			SRS
			DMRS			SRS
			DMRS			SRS
Data	Data	Data	DMRS	Data	Data	SRS
Data	Data	Data	DMRS	Data	Data	SRS
Data	Data	Data	DMRS	Data	Data	SRS
Data	Data	Data	DMRS	Data	Data	SRS
Data	Data	Data	DMRS	Data	Data	SRS
0	1	2	3	4	5	6

Frequency subcarriers (vertical axis) — Time slot (horizontal axis)

Legend: SRS UE1, SRS UE2, SRS UE3

Figure 11.55 Sounding RS symbol structure with RPF = 2 [12].

The cell-search procedure is performed by the UE with assistance of the Primary Synchronization Signal (PSS) and the Secondary Synchronization Signal (SSS) transmitted on the downlink. Location of synchronization signals on resource grid differs depending on system operation mode FDD or TDD while the structure of the signals is the same for both duplex modes. PSS and SSS provide radio frame and slot synchronization, as well as the physical-layer group and Cell ID.

After synchronization, the UE can decode the PBCH deriving the system bandwidth, PHICH configuration and the current System Frame Number (SFN). The PBCH contains four bits of information about the system bandwidth allocated to a downlink, which results in 16 different bandwidth configurations measured in resource blocks (RBs). The rest of system information is contained in System Information Blocks (SIBs) that are scheduled on regular shared resources, DL-SCH. The presence of system information on DL-SCH in a subframe is indicated by the transmission of a corresponding PDCCH marked with specific scrambling identifier. The main information derived from SIBs is related to uplink configuration and RACH configuration.

After retrieval of this data, the UE is ready to perform the random access procedure. At this stage the only downlink is adjusted, therefore the random access process is contention based.

11.29.2 Random Access Procedure

Two different random access procedures have been specified for LTE, contention-based and contention-free random access. The contention-based procedure is that one described next, it also called a non-synchronized access procedure [20]. The contention-based random access procedure is initiated by the mobile, when the UEs transmit randomly selected preambles on a common resource to establish a network connection or request resources for uplink transmission in four steps described next.

Contention-free random access is initiated by the network. It is used for a time alignment during handover and when RRC connected UE needs to be synchronized for downlink data reception. The network sends and the UE receives an index of dedicated preamble sequence to transmit then on PRACH. This message is received either in the handover command or in PDCCH. In addition, eNodeB transmits some information on frequency and time resources for PRACH. Since eNodeB knows the identity (C-RNTI) of the UE since the UE is already in RRC CONNECTED state, the contention resolution (with Steps 3 and 4) is not needed in that case.

At the physical-layer, the random access procedure involves transmission of random access preamble and random access response. The main objective is an acquisition of uplink timing. The procedure is performed in the following steps:

Step 1

The UE selects a random access preamble and transmits it in the time and frequency resources of the PRACH with initial PRACH power settings derived from SIBs. The base station may estimate time delay in transmission between eNodeB and user terminal.

Step 2

The UE monitors the PDCCH for a downlink response from base station within a time window. The start and end of monitoring window are configured by the eNodeB and broadcasted as part of the cell-specific system information.

On receiving the access preamble, the eNodeB estimates uplink timing and responds with timing advance command contained in the Random Access Response (RAR) message. The RAR is scheduled with a special identifier called a Random Access Radio Network Temporary Identifier (RA-RNTI).

The identity RA-RNTI is constructed using with frequency and time resource of the preamble selected in PRACH and should match those parameters on reception by the UE. The RAR to different users are bundled together. The eNodeB response indicates the sequence numbers of the observed preambles and, for all acknowledged PRACH, it provides:

- A timing advance command
- An uplink grant for transmission on PUSCH including information on channel configuration, such as frequency hopping, power control command for uplink transmission and so on.
- A temporary allocation for identity called Cell Radio Network Temporary Identifier (C-RNTI), which is used for addressing PUSCH grants and DL-SCH assignments in further steps of the procedure. The TC-RNTI identity is can be granted permanently to the UE after contention resolution in following steps of random access procedure.

If multiple UEs have collided by selecting the same signature sequence in the same preamble and time-frequency resource, they each receive the same RAR. If the UE does not receive a RAR within the configured time window, it retransmits the preamble. The RAR responses in the contention-based and contention-free based procedures are identical, however, the provision of C-RNTI is not mandatory in contention-free access, since the UE already has a valid C-RNTI identity being in RRC CONNECTED state.

Step 3

On reception of the RAR message the UE switches directly to UL-SCH and schedules the first uplink message on PUSCH. The message includes RRC connection request, temporary C-RNTI provided in RAR message in Step 2 and unique UE identity (48 bits). The uplink transmission in Step 3 makes use of HARQ.

If collision has happened in Step 1, collided UE would have received the same temporary C-RNTI in the RAR. The uplink messages from these terminals will also collide occupying the same time-frequency resource. The eNodeB may not be able to decode any colliding message from contending UEs. In that case, the UEs will restart random access procedure. In case one UE's message is decoded, the eNodeB selects the UE unique identity ignoring the messages from other collided UEs with the same temporary C-RNTI. This is to be resolved in Step 4 with the help of a contention-resolution message.

Step 4

The contention-resolution message is sent on the DL-SCH. The message is addressed to all UE(s) with the same temporary C-RNTI. It contains the UE identity of the decoded uplink message. This transmission is supported with the HARQ. The collided UEs attempt to decode the contention-resolution message. The outcome may be as follows:

- The UE correctly decodes the message and detects its own unique identity. The UE transmits to eNodeB a positive ACKnowledgement, 'ACK'.
- The UE correctly decodes the message and discovers mismatch between its own and received UE identity. The UE does not transmit anything in HARQ, it needs to restart random access procedure from the Step 1.
- The UE fails to decode the message, send nothing back to eNodeB, needs to restart random access procedure from Step 1.

11.29.3 Link Adaptation

Link adaptation operates in combination with the dynamic packet scheduler, as discussed in Section 11.23. The link adaptation supplies information on selected modulation-and-coding scheme (MCS) to the scheduler to be applied to the resource blocks dedicated to a given user. The decision for MCS is based on CSI feedback messages from participating users in the cell. Also, link adaptation decisions may be based on HARQ acknowledgements.

Three types of link adaptation are performed according to the channel conditions, the UE capability such as the maximum transmission power and maximum transmission bandwidth and so on, and the required QoS such as the data rate, latency, packet error rate and so on. Three link adaptation methods are as follows.

- Adaptive transmission bandwidth;
- Transmission power control;
- Adaptive modulation and channel coding rate.

Adaptive Modulation and Coding (AMC) allows application of various modulation schemes and channel coding rates based on the channel conditions. For LTE, the 3GPP standard has defined several Modulation Coding Schemes (MCS) that provide, besides differences in the modulation patterns (QPSK, 16QAM, 64QAM) also differences in the coding rate (i.e. ratio of coding portion to the total radio packet length). The more complex symbols are supported (4 for QPSK, 64 for 64 QAM etc.) and the smaller the ratio of the coding portion, the higher the throughput per symbol but the lower the robustness of the MCS against bad SINR conditions. Generally speaking, the closer to the serving base station, the higher MCS may be used. Conversely, when the UE moves further away from the base station, the SINR decreases due to:

- increased signal attenuation and
- increased interference coming from neighbouring cells.

11.29.4   Power Control

LTE power control is applied to uplink transmission only. There is no downlink power control. The objective of power control in LTE is chiefly to reduce terminal power consumption and minimize the power dynamic range in the eNodeB receiver.

The LTE uplink power control aims to keep the Power Spectral Density (PSD) (watt/Hertz) constant for a particular UE instead of controlling absolute power level. When MCS and data rate changes, transmitted bandwidth also changes leading to a respective change in absolute power level of the UE. The power control ensures that absolute power level of the UE is adjusted in a way to keep PSD (W/Hz) unchanged, see Figure 11.56. Compared with 3G, the power control mechanism in LTE is relatively slow with a loop delay of 5 ms and also differs in irregular steps in power adjustment.

With power control operation, the eNodeB uses path loss estimation from open-loop mechanism to set a coarse operating point for the transmission PSD that estimated as sufficient in current cell propagation environment. This estimate includes input from the UE's measurement of downlink path loss based on the power level of the downlink Reference Signal.

The overall power control mechanism in LTE includes rescheduling uplink transmission resources to different users based on an instant transmit power used by the UE across the cell. In order to assist in resource redistribution, the UE can report its available power headroom to the eNodeB. The eNodeB uses the power headroom report to predict what additional bandwidth can be used by reporting terminal. In a case where the uplink resource grant requires more transmission power than the UE has available, the UE signals to eNodeB a negative headroom report. The eNodeB will reduce the size in RB of the next uplink grant and reallocates free resources to the other UEs in the cell.

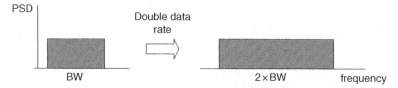

Figure 11.56 UL power control with variable data rate [19].

11.29.5 Paging

A terminal should be able to monitor paging messages in PDSCH during predetermined time intervals and paging cycles. Any information on paging interval and a specific subframe within that interval where the paging message could be sent is provided in PDCCH. A group of UEs is allocated a Paging Group Identity (P-RNTI). If the terminal detects a group identity used for paging (the P-RNTI), it will process the corresponding downlink paging message transmitted on the PCH. The paging message includes the identity of the terminal(s) being paged.

11.29.6 HARQ

Hybrid-ARQ (HARQ) is a combination of forwards error-correction coding and ARQ. The basic principle is to correct a part of packets with help of error-correction coding and identify uncorrected table errors with the help of CRC code. The receiver then requests still-corrupt packets to be retransmitted. The UE provides feedback to the eNodeB on PUCCH with an ACK/NACK message. In the case where NACK is received from the UE, the eNodeB will schedule retransmission of the affected packet via PDSCH. The same procedure applies to the HARQ for uplink. Downlink ACK/NAKs in response to uplink transmissions are sent on PHICH.

In LTE the Hybrid-ARQ can be used together with soft combining. Soft combining means that corrupted packet is not discarded; instead that packet is stored in the buffer memory and then combined with retransmitted packet. With soft combining, the retransmitted packet is an exact copy of initial packet with same coding and matching rate. Such combining is also called Chase combining [20]. Assuming that errors in both packets are uncorrelated, a combined packet has a higher probability to be decoded without errors. In case of a successful CRC check with erroneous decoding, the UE will send positive acknowledgement (ACK) to the eNode. The eNodeB then proceeds with the next data packet applying a new HARQ process. The ARQ operation thus involves a SAW (Stop-And-Wait) process due to analysis and retransmission of corrupted packets.

The HARQ operation in LTE can also be based on principle of incremental redundancy [21]. For incremental redundancy, each retransmission applies different parameters in a rate-matching operation, therefore different subsets of information bits can be transmitted in each retransmission thus defining a different redundancy version (RV) of the initial data packet. To enable continuous data flow, eight HARQ SAW processes spanning eight subframes are available for FDD LTE in both uplink and downlink. The timing diagram illustrating the HARQ SAW process is provided in Figures 11.57 and 11.58.

Each HARQ process is identified with a unique HARQ process IDentifier (HARQ ID) and requires a separate soft buffer allocation in the receiver for combining the retransmissions. As defined in [1], the downlink uses asynchronous adaptive HARQ. Every downlink transmission is then accompanied by explicit signalling of control information, which includes HARQ ID, redundancy version (RV) and the indicator of a new packet.

The uplink HARQ can be either synchronous non-adaptive and asynchronous adaptive. The uplink non-adaptive HARQ operation is based on a predefined RV sequence 0, 2, 3, 1, 0, 2, 3, 1,… for successive transmissions of a packet, explicit control signalling

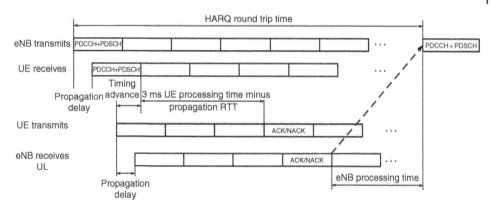

Figure 11.57 Timing diagram of the downlink HARQ [12].

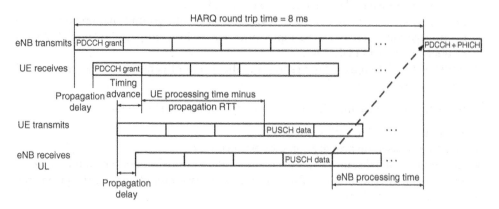

Figure 11.58 Timing diagram of the uplink HARQ [12].

is not required in that mode. For the asynchronous adaptive uplink HARQ operation, accompanying control signalling is necessary to indicate the redundancy version RV.

11.30 LTE Radio Dimensioning

The radio planning process in the general case consists of the following steps:

- *Dimensioning*: Computation of the number of sites to serve a certain area to fulfil customer requirements.
- *Nominal planning*: Creation of a nominal coverage plan with a planning tool.
- *Detailed planning*: Capacity analysis with a planning tool. Site validation. BTS Parameter planning (i.e. frequency, paging groups, site data built with default parameters).
- *Pre-launch optimization*: Cluster acceptance. Drive test measurements, analysis and changes implementation. Data build assessment/consistency.

LTE air-interface dimensioning aims to provide the first estimation of the sites volumes to be accounted when introducing the LTE radio network. Dimensioning should also provide an estimate of cell ranges and cell capacity. The results are then used in the radio network planning process. Some network planning parameters have to be decided first and then used as input parameters to dimensioning tool. The input parameters to the air-interface dimension are as follows:

- *General parameters*: define possible operating bands, system (channel) bandwidth, UE power class.
- *Equipment parameters*: provide equipment specifications, antenna configurations, gains and losses, System Overhead Parameters and so on.
- *Radio propagation parameters*: consist of parameters which describe wave propagation in a specific radio environment; for example, channel models, mobility and so on.
- *Capacity requirement parameters*: some parameters of this group are necessary in E-UTRAN for a correct definition of power and resource sharing among users.
- *Interference parameters*: represents the group of parameters related to interference calculation; for example, cell loads, G-factor and so on.
- *Radio propagation prediction parameters*: represents the group of parameters related to propagation models; for example, intercept points, propagation model slopes, eNodeB antenna and UE heights.
- *Radio Network Configuration*: represents the group of parameters that specify cell layout and sectorization.

The output of the air-interface dimensioning could be the following:

- Maximum Allowable Path Loss, MAPL
- Cell ranges based on propagation models' formulas
- Site-to-Site (ISD) distance and site areas for different site layouts, which finally yields an estimation of the number of sites and site density required to cover the given area
- Cell and site throughputs

11.30.1   LTE Coverage Dimensioning: Link Budget

The link budget is based on a minimum throughput requirement at the cell edge. By defining the service throughput value prior to the link budget calculation, it is possible to roughly estimate bandwidth and power allocation per one user; this allows for an appropriate estimation of the scheduling behaviour in a real system. The general parameters used for link budget calculation are described next.

Operating Band and Channel Bandwidth
For link budget calculation only the carrier centre frequencies are considered. The system bandwidth configuration impacts factors such as overhead ratio and total cell throughput. The best network performance (regarding maximum peak data rates and cell throughputs) is achieved by the deployment of maximum 20 MHz bandwidth. With narrow bandwidth scheduling, gain is decreased leading to relative performance degradation.

Tx/Rx Path Parameters
The eNodeB transmission power has to be set with respect to known hardware restrictions. The typical eNodeB output power/sector can be set from 8 to 100 W depending on the vendor of eNodeB equipment. Transmission power per eNodeB antenna should comply with the deployment specific requirements and law regulations in the

given country (power spectral density limitations). As a general rule, low Tx power is to be set for small bandwidth and high Tx power for a large system bandwidth.

The UE transmit power depends on the UE power class. UE Power Class 3 with 23 dBm output power can be assumed for dimensioning. The eNodeB allocates constant power per subcarrier in downlink transmission, which is configurable by the operator as an O&M parameter. The total eNodeB power is shared among all subcarriers, no matter how many of them are used for data allocation. The lower the number of subcarriers assigned to the user, the less power is received at the UE. If cell coverage is limited by the eNodeB Tx power, the TX power should be increased or channel bandwidth decreased.

UE output power is shared between only subcarriers assigned to transmission. When UE operates with smaller amount of subcarriers, it may increase the Tx power per subcarrier thus improving the uplink coverage. It should be noted that uplink power control is not taken into account in the link budget calculation since the focus is on estimating maximum possible coverage maximum output power for UE at the cell edge. The improvement in uplink coverage does not necessarily lead to better UL performance. Smaller amount of scheduled resources may require higher modulation-and-coding scheme (MCS) with better SINR (Signal-to-Interference-and-Noise Ratio) requirements. Additionally, channel interleaving over smaller amount of resources degrades frequency diversity that may impact UL performance.

Antenna Gain and Feeder Loss

The antenna gain value depends on the antenna type and is usually indicated in the technical data sheets of the antenna manufacturer. Typical values for antenna gain is 18 dBi for 2×2 MIMO DL and 0 dBi for UL, 3 dBi gain can be accepted for PCIMCIA. Feeder loss is normally omitted for feederless sites; only jumper connection losses of 0.4 dB can be accounted for.

Other Losses

Other losses depend on scenario specific assumptions. For indoor coverage calculation it is required to assume a certain penetration loss depending on the environment (i.e. glass surface, concrete walls, room dividers etc.). Body loss should be considered only when dimensioning is done for a handset (voice service at the cell-edge). On the other hand, body losses are irrelevant when LTE deployment scenario is about providing the broadband internet service to home CPE (Customer Premises Equipment) with the rooftop/outside-mounted antenna. Dimensioning for two-sector sites located along highways/tracks should be done with additional in-car/in-train loss. Default values for body loss and in-car loss are 3 dB and 6 dB, respectively.

Noise Figure

A typical noise figure for a base station receiver is 2–3 dB. The UE noise figure can be assumed to be 7 dB.

11.30.1.1 Physical-Layer Overhead Factors

The correct amount of the resources allocated to the given user can be estimated only if the whole amount of system overhead and BLER (Block Erasure Rate) target are taken into account. In order to determine the amount of radio resources and power that are available for user data transmission, it is necessary to subtract the system overhead portion that is consumed by signalling messages and control fields from the overall amount of physical resources provided by the channel bandwidth configuration.

Detailed estimation of system overheads depends on numerous parameters of specific configuration and transmission mode and can be obtained from sophisticated SW tool normally used by RF planners.

Typical values for DL overhead estimates can vary from 34% for 1.4 MHz bandwidth to 29% for 20 MHz bandwidth. Similar overhead estimates for the uplink vary from 39 to 23% for 1.4 MHz and 20 MHz channel bandwidth, respectively.

MCS Selection in the Link Budget MCS selection for the link budget is simplified by avoiding procedures of the Link Adaptation mechanism controlled by RRM. Since coverage dimensioning focuses on estimating the maximum possible cell range, the MCS scheme should be chosen as providing the best coverage and fulfilling a certain cell-edge throughput requirement.

As a general rule, the 3GPP standard has specified that the same MCS has to be applied to all groups of resource blocks belonging to the same L2 PDU scheduled to one user within one TTI by a single antenna stream. The same rule applies to multiple streams transmission. The MCS index is stated in the Transport Format together with information about the Transport-Block Size, resource allocation and transmission rank (i.e. one or two code words). The Transport Format is provided by the MAC layer.

A set of 29 different MCS indexes is defined for LTE [19]. They are listed below in Table 11.16 for downlink and Table 11.17 for uplink. As observed, every MCS is

Table 11.16 Modulation and TBS index table for PDSCH [19].

MCS Index I_{MCS}	Modulation Order Q_m	TBS Index I_{TBS}
0	2	0
1	2	1
2	2	2
3	2	3
4	2	4
5	2	5
6	2	6
7	2	7
...		
22	6	20
23	6	21
24	6	22
25	6	23
26	6	24
27	6	25
28	6	26/26A
29	2	reserved
30	4	
31	6	

Table 11.17 Modulation, TBS index and redundancy version table for PUSCH [20].

MCS Index I_{MCS}	Modulation Order Q'_m	TBS Index I_{TBS}
0	2	0
1	2	1
2	2	2
3	2	3
4	2	4
5	2	5
6	2	6
7	2	7
	...	
22	6	20
23	6	21
24	6	22
25	6	23
26	6	24
27	6	25
28	6	26
29	reserved	
30		
31		

unambiguously mapped to a certain modulation order (number of bits per symbol) related to modulation scheme: BPSK, QPSK and QAM.

Every MCS index is assigned with a TBS index that relates to the transport-block size. The Transport-Block Size (TBS) reflects the amount of user data bits sent during one TTI (1 ms) and depends on the number of scheduled PRBs. Table 11.18 shows some of the available configurations. A complete set of tables can be found in 3GPP TS 36.213 [19]. The maximum number of scheduled PRBs in Table 11.3 is 110 and this is more commonly available in the 20 MHz bandwidth.

The number of bits per transport block (per TTI) in Table 11.18 determines the user throughput TB_{rate} that impacts link budget calculation. The targeted user throughput Thr_{user} is a minimum acceptable net throughput for single user at the cell edge. The net throughput is defined by both transport-block size (which, in turn, depends on MCS) and Block Error Rate (BLER).

The BLER target represents a QoS requirement for the air-interface and is related to a retransmission ratio. The TB_{rate} required to achieve net user throughput U_{thr} can be obtained by taking retransmission into account (assuming this is the BLER after the 1st transmission):

$$TB_{rate} = {Thr_{user}}/{(1-BLER)} \tag{11.20}$$

Table 11.18 Transport-block size table in bits versus number of the scheduled PRB and TBS index [20].

TBS Index	Number of Scheduled PRBs									
	1	2	3	4	5	6	...	100	...	110
0	16	32	56	88	120	152	...	2792	...	3112
1	24	56	88	144	176	208	...	3624	...	4008
2	32	72	144	176	208	256	...	4584	...	4968
3	40	104	176	208	256	328	...	5736	...	6456
4	56	120	208	256	328	408	...	7224	...	7992
5	72	144	224	328	424	504	...	8760	...	9528
...	...	...	...	...	...	...	...	...	...	...
26	712	1480	2216	2984	3752	4392	...	75 376	...	75 376

Considering example of the cell edge target for user throughput of 256 kbps and 10% BLER, the chosen transmission scheme should be capable of delivering

$$TB_{rate} = [256 \text{ kbps} / (1 - 0.1)] = [256 \text{ kbps} / 0.9] = 284.5 \text{ kbps}$$

The 284.5 kbps effective throughput dictates a Transport-Block Size of at least 285 bits. There could be several choices for a TBS index and respective MCS selection satisfying requirement for a transport-block capacity, please refer to full tables in [TS 36.213]. As an example, with four scheduled PRBs the system would establish a 328 bits Transport Block and use QPSK modulation (TBS index = 5 leads to MCS index = 5, which implies QPSK modulation). This choice might not be an optimal one, given requirements for RF coverage related to SINR target value. Optimal solution for MCS and number or scheduled RBs can be traded off with required SINR targets. The lower the SINR target value, the better link performance and RF coverage. When selecting the TBS index for downlink, the less possible MCS index usually provides an optimal choice for MCS and number of PRBs combination. In the uplink, the trade-off between SINR and MCS might be achieved by adjusting power per PRB.

Another impact to SINR comes from the observed effect of turbo coding degradation for shorter coded blocks. The degradation is more evident for the number of PRB $\leq$ 10 and low MCS indexes and can lead to increase in required SINR target in the order of ~2 dB.

11.30.1.2 Multi-Antenna Systems

For the purpose of dimensioning, the usual configuration is one Tx antenna at the UE and two Rx antennas at eNodeB; that is, there is only receive diversity on uplink, no MIMO or transit diversity assumed for the downlink.

If the TBS is already known, one can easily obtain the maximum throughput achievable in the total channel assuming that all users are homogenous (meaning all users are using the same TBS configuration) and the cell works with 100% resource utilization. This leads to the following formula:

$$MaxMCSThr = TBS \cdot \frac{N_{RB}^{total}}{N_{RB}^{used}} [kbps] \tag{11.21}$$

MaxMCSThr – overall channel throughput that can be achieved if all resources are used;

TBS – Transport-Block Size;

N_{RB}^{total} – total number of PRBs within the available bandwidth;

N_{RB}^{used} – number of PRBs used by a single user.

The reciprocal value of the expression $N_{RB}^{total}/N_{RB}^{used}$ indicates the percentage of resources R_u used by a single user:

$$R_u = \frac{N_{RB}^{used}}{N_{RB}^{total}} \tag{11.22}$$

11.30.1.3 Required SINR

SINR is the power ratio of useful signal to total interference coming from neighbouring cells plus thermal noise. From the point of link budget evaluation, SINR values are considered as the targets for which transmission using a certain MCS can still be accomplished with a predefined quality (BLER). This target SINR value is also called 'required SINR'.

11.30.1.4 Link Budget Margins

The MAPL calculation is based on a particular SINR target that defines receiver sensitivity level and linked to throughput requirements. Several additional margins also need to be taken into account to reflect SINR degradations due to propagation impact and interference from other cells. As with the other systems, one of the additional margins is caused by lognormal fading, the values of which can be taken from Table 5.1 in Chapter 5, given accepted location probability.

The Rayleigh fading margin is normally omitted from calculations assuming that fast scheduling and link adaptation with TTI = 1 ms is sufficient to mitigate the fast fading effect.

11.30.1.5 Interference Margin

The LTE is deployed with frequency reuse 1, the same as 3G WCDMA. While the causes are similar, the interference estimation calculus is different.

Intra-Cell Interference

Intra-cell interference refers to any interference caused by the active terminals within the same cell. The 3G systems suffer from the cell breathing effect because of working with one frequency (frequency reuse = 1) and multiplexing the users based on the orthogonal codes assignment. In OFDMA there is no interference between users in the same cell, occupied subcarriers cannot interfere with each other because they are orthogonal. A higher number of active users causes a decrease of the available resources but it does not influence the G-factor (wideband C/I).

Inter-Cell Interference

Downlink and uplink interference between cells is to be taken into account. Unless the cells are synchronized in time and frequency, the orthogonality between subcarriers in different cells does not exist.

Calculation of Downlink Interference Margin

Given constant power per subcarrier, the power received by a single mobile is given by the following expression:

$$S = \frac{P_{RB} \cdot N_{RB}^{used}}{L} \tag{11.23}$$

S –   power per user/per UE;

P_{RB} –   power per resource block;

N_{RB}^{used} –   number of resource blocks assigned to user;

L –   signal attenuation factor ($\sim R^n$, R = range, n = propagation exponent)

In general, interference is comprised of two components intra-cell and inter-cell interference: $I_{own} + I_{oth}$. Since there is no need to consider intra-cell interference, I_{own} can be set to 0.

Interference power coming from another cell I_{oth} is given as follows (the result should be aggregated over all neighbour cells assuming corresponding L figures):

$$I_{oth} = \frac{P_{RB} \cdot N_{RB}^{used}}{L} \cdot \eta \cdot \frac{1}{G}, \tag{11.24}$$

where

η – cell load (the average amount of occupied frequency resources during the time of interest);

G – G-factor, corresponds to the C/I ratio (the offset between the target signal level C and any interference affecting the transmission). In this situation, the only part of channel bandwidth to be considered is the one on which the user receives data. The power coming from other cells needs to be multiplied by the factor η in order to take into account interfering cells' load.

The interference margin IM can be defined as a relative increase of the total noise level in presence of interference compared to a thermal noise only:

$$IM = \frac{I + N}{N} = \frac{I_{oth} + N}{N} \tag{11.25}$$

Introducing SINR by the formula

$$SINR = \frac{S}{I + N} = \frac{S}{I_{oth} + N}, \tag{11.26}$$

and redefining the expression for noise level

$$N = \frac{S}{SINR} - I_{oth} \tag{11.27}$$

One can obtain the following equations for downlink interference margin

$$IM = \frac{I_{oth} + \dfrac{S}{SINR} - I_{oth}}{\dfrac{S}{SINR} - I_{oth}} = \frac{\dfrac{S}{SINR}}{\dfrac{S}{SINR} - I_{oth}} = \frac{\dfrac{S}{SINR}}{\dfrac{S}{SINR} - \dfrac{P_{RB} \cdot N_{RB}^{used}}{L} \cdot \eta \cdot \dfrac{1}{G}} - I$$

$$M = \frac{\dfrac{1}{SINR}}{\dfrac{1}{SINR} - \eta \cdot \dfrac{1}{G}},$$ (11.28)

or, in logarithmic form

$$IM = -10 \cdot \log\left(1 - SINR \cdot \eta \cdot \frac{1}{G}\right)$$ (11.29)

Uplink Interference Margin

There are no analytical estimates for uplink interference margin. The values obtained from the system-level simulations are used instead in link budget calculations. The margin depends on system bandwidth and cell load, typical values are 2 dB for 50% load and 3 dB for 85% cell load with 10 MHz bandwidth.

11.30.1.6 Maximum Allowable Path Loss (MAPL)

The MAPL formula expresses the maximum allowable attenuation of the radio wave traversing the air interface. Together with the propagation model it is used for cell range estimation. MAPL is usually presented in a logarithmic scale. The formula comprises all the gains and losses that can be experienced in the system. The formula is the same for DL and UL and is independent from the chosen operating band.

$$MAPL^{DL} = P_{Tx}^{eNB} + G_{antenna}^{eNB} - L_{feeder}^{eNB} - L_{ins}^{TMA} - S_{Rx}^{UE} - M_{LNF} - IM^{DL}$$
$$+ G_{antenna}^{UE} - M_{BPL},$$ (11.30)
$$MAPL^{UL} = P_{Tx}^{UE} + G_{antenna}^{UE} - S_{Rx}^{eNB} - M_{LNF} - IM^{UL} + G_{antenna}^{eNB} - L_{feeder}^{eNB} - M_{BPL},$$

The parameters in (11.30) are as follows:

$MAPL$	– maximum allowable path loss;
S_{Rx}	– receiver sensitivity;
M_{LNF}	– lognormal fading margin;
IM	– interference margin;
$G_{antenna}$	– antenna gain;
M_{BPL}	– building penetration loss.
L_{feeder}	– feeder loss;
P_{Tx}	– total transmission power;
L_{ins}^{TMA}	– mast head amplifier insertion loss (when TMA is installed).

Receiver Sensitivity The general formula for receiver sensitivity level S_{Rx} is an applied addition to the required SNR on top of the receiver noise floor. The noise floor is determined by power of thermal noise and receiver noise figure, as discussed in Chapter 5. In the LTE case the SINR is used as SNR value, and formula for receiver sensitivity level S_{Rx} takes the form:

$$S_{Rx} = N_T + SINR + F_{noise},$$ (11.31)

where

$SINR$ – signal-to-interference-and noise ratio, dB;

N_T – thermal noise power, dBm;

F_{noise} – receiver noise figure, dB.

The thermal noise power is given by the following equation:

$$N_T = N_0 B, \tag{11.32}$$

Where:

N_0 is a spectral power density of the thermal noise, $N_0 = -174$ dBm/Hz,
B is a receiver bandwidth.

There is important difference in definition of receiver bandwidth for uplink and down-link in the link budget equations (11.30) and (11.31). The UE receives the total downlink OFDMA system bandwidth for further processing, whereas the bandwidth of the eNB receiver selected for a given user is limited to a number of resource blocks assigned to a given user for the uplink SC-FDMA transmission. It is convenient to define the equation (11.31) for receiver sensitivity in terms of power resource blocks assigned for transmission:

$$S_{Rx} = N_T^{RB} + SINR + F_{noise} + 10\log(N_{RB}) \tag{11.33}$$

N_T^{RB} is the noise power per resource block,

N_{RB} is the number of resource blocks assigned to transmission.

Given the size of the resource block in the frequency domain, 15kHz · 12 subcarriers = 180 kHz, the thermal noise per resource block is

$$N_T^{RB} = N_0 + 30 + 10\log(180) = -121.5 \text{ dBm}$$

11.30.1.7 Required SINR

From the point of link budget evaluation, SINR values are considered the targets for which the transmission using a certain MCS can still be accomplished with a prede-fined quality (BLER). The SINR requirements can be obtained from detailed simulation reports generated by specialized SW dimensioning tools. In a link budget estimation, the low index MCS with QPSK modulation is used in common calculations, which lead to SINR values in order of up to 7 dB for 1Tx–1Rx in the downlink and ~0.6 to 6 dB for 1Tx–2Rx in the uplink, depending on number of assigned power resource blocks and coding scheme.

11.30.2 Cell Range and Cell Capacity

The evaluation of the cell range R_{cell} in LTE is similar to other technology and based on combining the MAPL values and radio propagation model for the path loss $L(R)$, where R is a distance between transmitter and receiver and solving equation $MAPL = L(R_{cell})$

for cell range R_{cell}. The cell range can be used then in estimation of the cell capacity. Calculation of an average cell throughput is based on the method that utilizes the spectrum can be obtained from the dynamic system-level simulator. The simulation commonly takes into effect scheduling, MIMO configuration, power and admission control, link adaptation mechanism, the UE mobility, channel fading model and so on. The other input parameters significantly affecting results are Inter-Site Distance, defined as ISD = $2 \times R_{cell}$ and cell load. The results can presented in terms of spectral efficiency bps/Hz that can be scaled with different system bandwidth in operation.

11.31 Summary

The main features of LTE are as follows:

- LTE uses OFDM technology with up to 2048 of 15 kHz wide subcarriers with a sinc-square shaped subcarrier spectrum. Resultant OFDM symbol duration 66.7 µs is significantly longer than the channel delay spread. Operation of OFDM relies on orthogonality between OFDM subcarriers that could only be achieved with perfect time alignment of the OFDM symbols and precise synchronization of transmitter and receiver frequencies.
- Time alignment of transmitted subcarriers in a time dispersive channel is achieved by cyclic-prefix insertion that implies that the last part of the OFDM symbol is copied and inserted at the beginning of the OFDM symbol.
- The frequency alignment between eNodeB transmitter and UE receiver is achieved by means of acquiring the Reference Signal (RS) inserted in the downlink time-frequency resource grid. The reference signal is a predefined cell-specific signal in the cell-specific position on the resource grid that became known to UE after acquisition of the synchronization channel. The RS is used for channel estimation and coherent demodulation of the received signal.
- In OFDM the user data is transmitted in parallel across multiple orthogonal narrowband subcarriers. Each subcarrier only transports a part of the whole transmission.
- In the time domain, LTE transmissions are organized into (radio) frames of length 10 ms, each of which is divided into ten equally sized subframes of length 1 ms. Each subframe consists of two equally sized slots of length Tslot = 0.5 ms, with each slot consisting of a number of OFDM symbols including the cyclic prefix.
- A resource element, consisting of one subcarrier during one OFDM symbol, is the smallest physical resource in LTE. Resource elements are grouped into resource blocks, where each resource block consists of 12 consecutive subcarriers = $15 \times 12 = 180$ kHz in frequency domain and one 0.5 ms slot in the time domain. Each resource block thus consists of 7 (symbols) × 12 (subcarriers) = 84 resource elements in the case of a normal cyclic prefix and $6 \times 12 = 72$ resource elements in the case of an extended cyclic prefix.
- Although resource blocks are defined over one slot, the basic time-domain unit for dynamic scheduling in LTE is one subframe, consisting of two consecutive slots. The reason for defining the resource blocks over one slot is that distributed downlink transmission and uplink frequency hopping) are defined on a slot or resource-block basis. The minimum scheduling unit consists of two time-consecutive resource blocks within one subframe.

- The LTE uplink uses Single-Carrier FDMA, SC-FDMA. The main difference from OFDMA is that in the case of SC-FDMA there is additional processing before the IFFT: the modulated symbols (interpreted in this case as time signals) are fed to the FFT processing. The outputs are the frequency components of the modulation symbols; that is, each symbol spread over allocated bandwidth. Those frequency components are mapped to the allocated inputs of the IFFT and from there, the normal OFDM processing continues. The SC-FDMA ensures smaller peak-to-average power ratio (PAPR) of the signal variations compared with downlink OFDMA and lower complexity frequency acquisition.
- As in the downlink, the LTE uplink incorporates Reference Signals (RSs) for data demodulation and channel sounding. The roles of the uplink RSs include enabling channel estimation to aid coherent demodulation, channel-quality estimation for uplink scheduling, power control and timing estimation.
- The LTE incorporates frequency selective (channel-dependent) scheduling in both downlink and uplink. The downlink channel status information provided to eNodeB in UE periodic reports via uplink control channel (PUCCH). The uplink channel estimate made available to eNodeB from detection of uplink sounding reference signals.
- In contrary to WCDMA, the LTE deploys timing advance procedure to ensure time alignment of the received uplink time slots. This is necessary for orthogonal multiplexing the received OFDM symbols and, consequently, to eliminate intra-cell interference between users' signals on the uplink.
- There is no intra-cell interference in LTE, neither in downlink nor in uplink, however, intra-cell interference is present and its impact should be taken into account in RF planning and dimensioning.
- LTE RF planning need to consider a physical-layer cell-identity planning, CELL ID. The isolation between cells which are assigned the same physical-layer cell identity has to be maximized in order to ensure that UE never simultaneously receive the same identity from more than a single cell.
- The LTE network architecture is flat including only a packet-service nodes, MME and SGW/PGW. The distinctive feature of LTE is direct control plane interface between eNodeBs, X2, that supports handover related signalling between participating eNodeBs. The voice can be supported in LTE using SIP agent in UE and IP voice server or IMS in Core network. Alternative implementation is a 'fall back' option to 3G/GSM, whatever NW is deployed.

References

1 Evolved Universal Terrestrial Radio Access (E-UTRA) and Evolved Universal Terrestrial Radio Access Network (E-UTRAN); Overall description; 3GPP TS 36.300 V14.0.0 (2016–09).
2 General Packet Radio Service (GPRS) enhancements for Evolved Universal Terrestrial Radio Access Network (E-UTRAN) access (Release 14), 3GPP TS 23.401 V14.3.0 (2017–03).
3 Architecture enhancements for non-3GPP accesses (Release 14), 3GPP TS 23.402 V14.3.0 (2017–03).

4 Policy and charging control architecture (Release 14), 3GPP TS 23.203 V14.3.0 (2017–03).

5 3G Security; Security architecture (Release 14), 3GPP TS 33.102 V14.1.0 (2017–03).

6 3G security; Network Domain Security (NDS); IP network layer security (Release 14), 3GPP TS 33.210 V14.0.0 (2016–12).

7 Network Domain Security (NDS); Authentication Framework (AF) (Release 14), 3GPP TS 33.310 V14.0.0 (2016–12).

8 Rationale and track of security decisions in Long Term Evolved (LTE) RAN/3GPP System Architecture Evolution (SAE) (Release 9), 3GPP TR 33.821 V9.0.0 (2009–06).

9 Evolved Universal Terrestrial Radio Access (E-UTRA); User Equipment (UE) procedures in idle mode (Release 14), 3GPP TS 36.304 V14.0.0 (2016–09).

10 Holma, H. and Toskala, A. (eds). *LTE for UMTS – OFDMA and SC-FDMA Based Radio Access*, John Wiley & Sons, Ltd, 2009.

11 Dahlman, E., Parkvall, S. and Sköld, J. *4G LTE/LTE-Advanced for Mobile Broadband*, Elsevier Ltd, 2011.

12 Sesia, S., Baker M. and Toufik, I. *LTE – The UMTS Long Term Evolution: From Theory to Practice*, John Wiley & Sons, Ltd, 2009.

13 Evolved Universal Terrestrial Radio Access (E-UTRA); User Equipment (UE) radio transmission and reception (Release 14), 3GPP TS 36.101 V14.3.0 (2017–03).

14 Kreher, R. and Gaenger, K. *LTE Signalling, Troubleshooting, and Optimization*, John Wiley & Sons, Ltd, 2011.

15 Evolved Universal Terrestrial Radio Access (E-UTRA); Physical layer procedures (Release 14), 3GPP TS 36.213 V14.2.0 (2017–03).

16 Evolved Universal Terrestrial Radio Access (E-UTRA); Physical channels and modulation (Release 14), 3GPP TS 36.211 V14.2.0 (2017–03).

17 Evolved Universal Terrestrial Radio Access (E-UTRA); Base Station (BS) radio transmission and reception (Release 14), 3GPP TS 36.104 V14.1.0 (2016–09)].

18 Evolved Universal Terrestrial Radio Access (E-UTRA); Multiplexing and channel coding (Release 14), 3GPP TS 36.212 V14.2.0 (2017–03).

19 Evolved Universal Terrestrial Radio Access (E-UTRA); Physical layer procedures (Release 14), 3GPP TS 36.213 V14.2.0 (2017–03).

20 Chase, D. Code combining – a maximum-likelihood decoding approach for combining and arbitrary number of noisy packets, *IEEE T. Communications*, 33 (May 1985): 385–393.

21 Pursley, M.B. and Sandberg, S.D. Incremental-redundancy transmission for meteor-burst communications, *IEEE T. Communications*, 39 (May 1991): 689–702.

12

LTE-A

A new concept for a mobile system with capabilities beyond LTE Release 8 has been developed in ITU working groups. This concept, as applied to the entire network system, is called International Mobile Telecommunications-Advanced (IMT-Advanced). 3GPP has started to develop the standard for radio networks under the name LTE-Advanced (LTE-A) with the first release in 3GPP Release 10.

The requirements for radio networks from IMT-Advanced include:

- peak data rates up to 1 Gbps for nomadic (low mobility case) and 100 Mbps for the high mobility case;
- support for up to 100 MHz bandwidth;
- improved spectral efficiency in different environments.

An additional 3GPP requirement is for backward compatibility with 3GPP Release 8 LTE including support for inter-RAT mobility between LTE-A and LTE, GSM/EDGE, HSPA and cdma2000.

The LTE-A also has to support flexibility in spectrum allocation. Given differences in the available spectrum in different countries, LTE-A should support asymmetric bandwidth allocation for FDD downlink/uplink as well as non-contiguous spectrum allocation. The E-UTRA operating bands for LTE/LTE-A are listed in Table 12.1.

System performance targets for 3GPP Release 10 (LTE-A) are as follows:

- cost efficient data rates 3 Gbps for downlink and 1.5 Gbps for uplink;
- spectral efficiency up to 30 bps/Hz;
- higher capacity with increased number of simultaneously active subscribers;
- improved performance at cell edges.

Main new functionalities introduced in LTE-Advanced are:

- Carrier Aggregation (CA),
- enhanced MIMO,
- support for Relay Nodes (RN) and
- Coordinated multipoint transmission and reception (CoMP).

Introduction to Mobile Network Engineering: GSM, 3G-WCDMA, LTE and the Road to 5G,
First Edition. Alexander Kukushkin.
© 2018 John Wiley & Sons Ltd. Published 2018 by John Wiley & Sons Ltd.

Table 12.1 E-UTRA operating bands [1].

E-UTRA Operating Band	Uplink (UL) Operating Band BS Receive UE Transmit $F_{UL\_low}-F_{UL\_high}$	Downlink (DL) Operating Band BS Transmit UE Receive $F_{DL\_low}-F_{DL\_high}$	Duplex Mode
1	1920–1980 MHz	2110–2170 MHz	FDD
2	1850–1910 MHz	1930–1990 MHz	FDD
3	1710–1785 MHz	1805–1880 MHz	FDD
4	1710–1755 MHz	2110–2155 MHz	FDD
5	824–849 MHz	869–894 MHz	FDD
6	830–840 MHz	875–885 MHz	FDD
7	2500–2570 MHz	2620–2690 MHz	FDD
8	880–915 MHz	925–960 MHz	FDD
9	1749.9–1784.9 MHz	1844.9–1879.9 MHz	FDD
10	1710–1770 MHz	2110–2170 MHz	FDD
11	1427.9–1447.9 MHz	1475.9–1495.9 MHz	FDD
12	699–716 MHz	729–746 MHz	FDD
13	777–787 MHz	746–756 MHz	FDD
14	788–798 MHz	758–768 MHz	FDD
15	Reserved	Reserved	FDD
16	Reserved	Reserved	FDD
17	704–716 MHz	734–746 MHz	FDD
18	815–830 MHz	860–875 MHz	FDD
19	830–845 MHz	875–890 MHz	FDD
20	832–862 MHz	791–821 MHz	FDD
21	1447.9–1462.9 MHz	1495.9–1510.9 MHz	FDD
22	3410–3490 MHz	3510–3590 MHz	FDD
231	200–2020 MHz	2180–2200 MHz	FDD
24	1626.5–1660.5 MHz	1525–1559 MHz	FDD
25	1850–1915 MHz	1930–1995 MHz	FDD
26	814–849 MHz	859–894 MHz	FDD
27	807–824 MHz	852–869 MHz	FDD
28	703–748 MHz	758–803 MHz	FDD
29	N/A	717–728 MHz	FDD[b]
30	2305–2315 MHz	2350–2360 MHz	FDD
31	452.5–457.5 MHz	462.5–467.5 MHz	FDD
32	N/A	1452–1496 MHz	FDD[b]
33	1900–1920 MHz	1900–1920 MHz	TDD
34	2010–2025 MHz	2010–2025 MHz	TDD
35	1850–1910 MHz	1850–1910 MHz	TDD
36	1930–1990 MHz	1930–1990 MHz	TDD

Table 12.1 (Continued)

E-UTRA Operating Band	Uplink (UL) Operating Band BS Receive UE Transmit $F_{UL\_low}-F_{UL\_high}$	Downlink (DL) Operating Band BS Transmit UE Receive $F_{DL\_low}-F_{DL\_high}$	Duplex Mode
37	1910–1930 MHz	1910–1930 MHz	TDD
38	2570–2620 MHz	2570–2620 MHz	TDD
39	1880–1920 MHz	1880–1920 MHz	TDD
40	2300–2400 MHz	2300–2400 MHz	TDD
41	2496–2690 MHz	2496–2690 MHz	TDD
42	3400–3600 MHz	3400–3600 MHz	TDD
43	3600–3800 MHz	3600–3800 MHz	TDD
44	703–803 MHz	703–803 MHz	TDD
45	1447–1467 MHz	1447–1467 MHz	TDD
46	5150–5925 MHz	5150–5925 MHz	TDD[h), i)]
47	5855–5925 MHz	5855–5925 MHz	TDD
48	3550–3700 MHz	3550–3700 MHz	TDD
64		Reserved	
65	1920–2010 MHz	2110–2200 MHz	FDD
66	1710–1780 MHz	2110–2200 MHz	FDD[d)]
67	N/A	738–758 MHz	FDD[b)]
68	698–728 MHz	753–783 MHz	FDD
69	NA	2570–2620 MHz	FDD[b)]
70	1695–1710 MHz	1995–2020 MHz	FDD10[j)]

Notes

a) Band 6, 23 is not applicable
b) Restricted to E-UTRA operation when carrier aggregation is configured. The downlink operating band is paired with the uplink operating band (external) of the carrier aggregation configuration that is supporting the configured Pcell.
c) A UE that complies with the E-UTRA Band 65 minimum requirements in this specification shall also comply with the E-UTRA Band 1 minimum requirements.
d) The range 2180–2200 MHz of the DL operating band is restricted to E-UTRA operation when carrier aggregation is configured.
e) A UE that supports E-UTRA Band 66 shall receive in the entire DL operating band.
f) A UE that supports E-UTRA Band 66 and CA operation in any CA band shall also comply with the minimum requirements specified for the DL CA configurations CA_66B, CA_66C and CA_66A-66A.
g) A UE that complies with the E-UTRA Band 66 minimum requirements in this specification shall also comply with the E-UTRA Band 4 minimum requirements.
h) This band is an unlicensed band restricted to licensed-assisted operation using Frame Structure Type 3.
i) In this version of the specification, restricted to E-UTRA DL operation when carrier aggregation is configured.
j) The range 2010–2020 MHz of the DL operating band is restricted to E-UTRA operation when carrier aggregation is configured and TX-RX separation is 300 MHz. The range 2005–2020 MHz of the DL operating band is restricted to E-UTRA operation when carrier aggregation is configured and TX-RX separation is 295 MHz.

12.1 Carrier Aggregation

A straightforward way to achieve higher data rates compared with LTE Release 8 is to use more bandwidth. Given availability of spectrum and backward compatibility with previous releases (Release 8 and Release 9) any increase in bandwidth in LTE-Advanced has to be provided by means of aggregation of the LTE carriers with a choice of bandwidth of 1.4, 3, 5, 10, 15 or 20 MHz.

Each aggregated carrier is referred to as a component carrier (CC) in LTE-A. A maximum five CCs can be combined to form a maximum bandwidth of 100 MHz. The configuration of carriers can be asymmetric in DL versus UL with the number of uplink CCs always less compared with DL. With carrier aggregation, a number of serving cell is fixed as one per component carrier.

Table 12.2 illustrates different options of carrier aggregation in LTE-A. Fragmentation of the available spectrum makes contiguous carrier aggregation challenging in practice. LTE-A introduces various options for carrier aggregation: contiguous and non-contiguous carrier aggregation, either intra-band or inter-band aggregation. Figure 12.1 illustrates these possibilities.

Some details of the spectrum allocation for both intra- and inter-band CA can be found in series of technical reports: 3GPP TR36.714, 36.715, 36.833, 36.853, 36.854, 36.857, 36.836, 36.864, 36.865, 36.979 and 36.899.

Given the scarcity of spectrum and the UE capabilities, different carrier aggregation configurations are supposed to be introduced gradually. In Release 11 there are two component carriers DL and one or two component carriers in the UL when carrier aggregation is used. Figure 12.2 illustrates uplink parallel transmission of two DFT-spread OFDM component carriers with respective PUCCH and PUSCH channel arrangements.

The RF requirements for carrier aggregation in different configuration are still to be specified in 3GPP Working groups. With both contiguous and non-contiguous carrier aggregation, each CC limited to a maximum of 110 Resource Blocks in the frequency domain.

Table 12.2 Possible carrier aggregation options.

Transmitter Bandwidth	No and Type of Component Carriers	Band	Duplex scheme
UL: 40 MHz; DL:80 MHz	UL: Contiguous 2 × 20 MHz CCs DL: Contiguous 4 × 20 MHz CCs	3.5 GHz	FDD
UL/DL 100 MHz	Contiguous 5 × 20 MHz	2.3 GHz	TDD
UL: 40 MHz; DL:80 MHz	UL: Non-contiguous 20 MHz + 20 MHz CCs DL: Non-contiguous 2 × 20 MHz + 2 × 20 MHz CCs	3.5 GHz	FDD
UL/DL 100 MHz	Contiguous 5 × 20 MHz	3.5 GHz	TDD
UL: 10 MHz; DL:10 MHz	UL: Non-contiguous 5 MHz + 5 MHz CCs DL: Non-contiguous 5 MHz + 5 MHz CCs	900 MHz	FDD

Intra-band contiguous

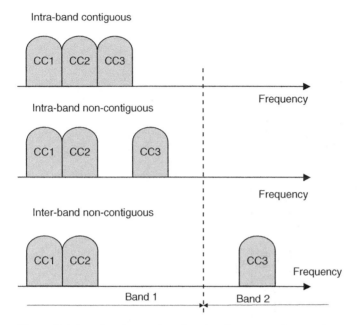

Figure 12.1 Intra-band versus inter-band carrier aggregation: contiguous and non-contiguous.

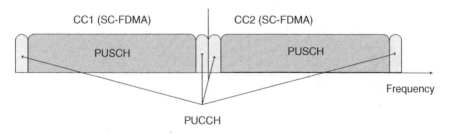

Figure 12.2 Uplink carrier aggregation.

In Release 13 LTE CA is planned to expand up to 32 CCs (potentially 640 MHz of bandwidth) and hence provide a major leap in the achievable data rates for LTE as well as in the flexibility to aggregate large numbers of carriers in different bands. This enhancement can be practically useful in conjunction with provision of access for LTE operation in unlicensed spectrum where large blocks of spectrum are available.

Carrier aggregation involves parallel processing and transmission of the number of LTE type transport blocks each mapped to respective DL-SCH for each CC, as shown in Figure 12.3 [2]. The processing of each DL-SCH in the physical layer is performed independently for each component carrier. The independent physical processing for CC includes control signalling, coding, HARQ, scheduling, modulation, antenna mapping and so on. Carrier aggregation is confined to the MAC layer only where logical channels are multiplexed in one (or two for spatial multiplexing) transport block(s) per component carrier. Figure 12.4 shows the structure of the uplink configured with carrier aggregation.

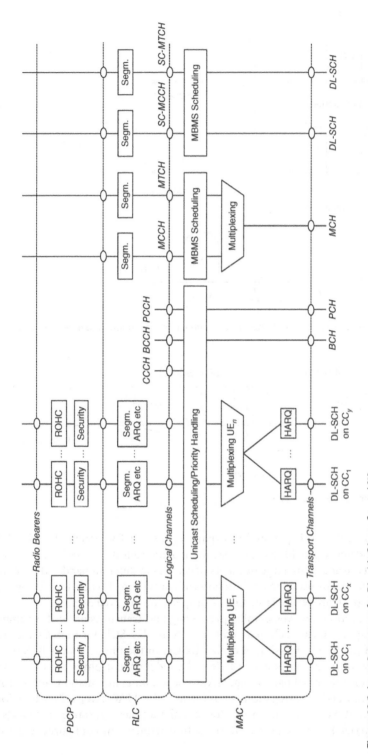

Figure 12.3 Layer 2 structure for DL with CA configured [2].

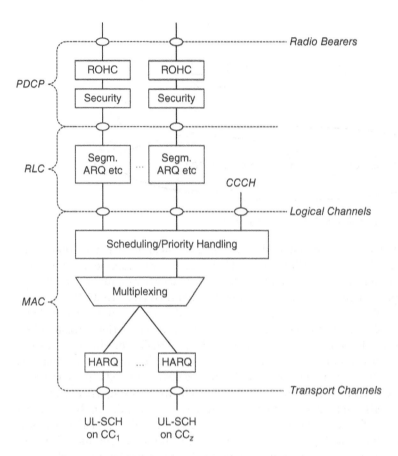

Figure 12.4 Layer 2 structure for UL with CA configured [2].

In both uplink and downlink, there is one independent hybrid-ARQ entity per CC and one transport block is generated per TTI per CC in the absence of spatial multiplexing. Each transport block and its potential HARQ retransmissions are mapped to a single serving cell associated with the component carrier.

UE may simultaneously receive or transmit on one or multiple CCs depending on its capabilities:

- UE with single timing advance capability for CA can simultaneously receive and/or transmit on multiple CCs corresponding to multiple serving cells sharing the same timing advance (multiple serving cells grouped in one Timing Advance Group: TAG);
- UE with multiple timing advance capability for CA can simultaneously receive and/or transmit on multiple CCs corresponding to multiple serving cells with different timing advances (multiple serving cells grouped in multiple TAGs). E-UTRAN ensures that each TAG contains at least one serving cell;
- A non-CA capable UE can receive on a single CC and transmit on a single CC corresponding to one serving cell only (one serving cell in one TAG).

When CA is configured, the UE only has one RRC connection with the network. The serving cell that provides RRC connection is referred as the Primary Cell (PCell). In the

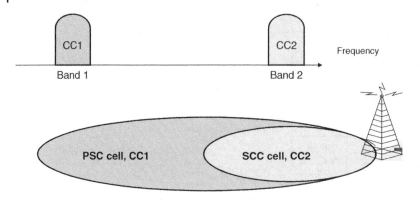

Figure 12.5 RF coverage with inter-band carrier aggregation.

downlink, the carrier corresponding to the PCell is the Downlink Primary Component Carrier (DL PCC) while in the uplink it is the Uplink Primary Component Carrier (UL PCC). The PCell provides the NAS mobility information, security input (key and RACH procedure), supports handover and receives the PUCCH (Primary PUCCH).

Depending on UE capabilities, Secondary Cells (SCells) can be configured to form together with the PCell a set of serving cells. In the downlink, the carrier corresponding to a SCell is a Downlink Secondary Component Carrier (DL SCC) while in the uplink it is an Uplink Secondary Component Carrier (UL SCC). The configured set of serving cells for a UE therefore always consists of one PCell and one or more SCells. The additional PUCCHs can be configured on SCell, the PUCCH SCell. The reconfiguration, addition and removal of SCells can be performed by RRC. At intra-LTE handover, RRC can also add, remove or reconfigure SCells for usage with the target PCell. When adding a new SCell, dedicated RRC signalling is used for sending all required system information of the SCell; that is, while in connected mode, UEs need not acquire broadcasted system information directly from the SCells. When a PUCCH SCell is configured, RRC configures the mapping of each serving cell to the Primary PUCCH group or Secondary PUCCH group; that is, for each SCell whether the PCell or the PUCCH SCell is used for the transmission of ACK/NAKs and CSI reports.

One may notice that different serving cells (CCs) may have different RF coverage areas. Figure 12.5 illustrates inter-band carrier aggregation. The PCell in Figure 12.5 is supported within larger RF coverage area compared with secondary cells at with component carrier in higher frequency range. The PCell provides both RC connection and user data communication while the secondary serving cell SCell is responsible for user data only. In this example, the inter-band carrier aggregation is only possible in area of overlapping coverage areas for different component carriers, namely, in SCell coverage area.

12.2 Enhanced MIMO

A major enhancement in LTE-A multiple antenna technique is the introduction of 8×8 MIMO in the downlink and 4×4 in uplink. The number of transmission layers is selected by rank adaptation to channel conditions. The additional multi-antenna transmission schemes are supported with a number of additional transmission modes defined in LTE-A. The eNodeB informs the UE via RRC signalling about transmission mode in use that identifies the MIMO scheme. The new MIMO schemes can be supported only with newly introduced terminals. Three new UE categories have been introduced in Release 10, namely Category 6, 7 and 8. The latter (Cat 8) supports 8×8 spatial multiplexing and carrier aggregation with five component carriers. As a means of mitigating the propagation channel distortions to the signal the LTE-A adds the Cell-specific Reference Signals (CRS) to the resource grid after precoding, one CRS per antenna port, this is the same procedure as in LTE. The CRS inserted into antenna domain (after precoding) allows to estimate propagation channel over the entire transmission band. The UE uses a propagation channel response estimate derived from the CRS together with codebook knowledge on precoding to decode/demodulate the received data.

In Release 10, a new reference signal, the Demodulation Reference Signal (DM-RS) is added to the different data streams before precoding. The DM-RS is a predefined signal that is precoded, as shown in Figure 12.6. This makes possible to apply non-codebook precoding. The user-specific DM-RS aims to estimate effective channel, the channel together with the precoder and, consequently, is inserted in the layer domain before the precoder. As an additional advantage, the user-specific DM-RS can be used for enhanced multiuser beamforming.

The DM-RS reference signals targeting PDSCH demodulation have the following properties [4]:

- UE-specific; that is, the PDSCH and the demodulation reference signals intended for a specific UE are subject to the same precoding operation.
- Present only in resource blocks and layers scheduled by the eNodeB for transmission.
- DM-RS reference signals transmitted on different transmission layers are mutually orthogonal at the eNodeB.

The DM-RS are generated using pseudorandom Gold sequence with a length of 31 initialized with a radio network temporary identifier (RTNI) assigned to user as well as other parameters such as cell ID and scrambling index [3]. Mapping of DM-RS to resource grid is performed using a combination of Code Division Multiplexing (CDM) and Frequency Division Multiplexing (FDM). The example of mapping DM-RS for rank 3 and 4 is illustrated in Figure 12.7 [4]. The DM-RS for the first layer and that for the second layer are multiplexed by means of CDM by using Orthogonal Cover Code over

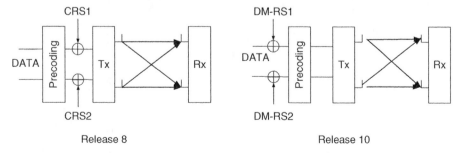

Figure 12.6 MIMO DL with precoding and reference signal for demodulation. DM-RS is UE and data stream specific signal.

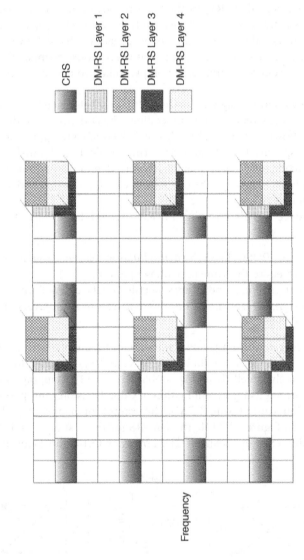

Figure 12.7 DM-RS pattern for rank 3 and 4 [4].

two consecutive resource elements in the time domain. The DM-RSs for the third layer and that for the fourth layer are shifted in frequency domain relative to ones of layer 1 and 2 and multiplexed in time domain by means of CDM using OCC over two consecutive resource elements. The DM-RS for first and second transmission layers and that for third and fourth layers are multiplexed by means of frequency division multiplexing (FDM).

Additional reference signals have been introduced in LTE-A from 3GPP Release 10. These reference signals, CSI-RS, carry Channel-State Information. The CSI-RS has a significantly lower time/frequency density being multiplexed over longer period over several subframes, thus implying less overheads compared to the cell-specific reference signals, CRS. The CSI-RS serves to estimate the physical channel over the entire transmission band to estimate variation of amplitude and phase in the channel and hence are inserted in the antenna domain; that is, after the precoding, similar to CRS in Figure 12.6.

The drive for CSI-RS introduction is two-fold. On one hand, the LTE-A introduced enhancement in MIMO with up to eight antenna ports in the first Release 10 with the aim of possible expansion in future. This alone invokes a need for change, since LTE in Release 8 has only four CRSs per cell to be mapped to four antenna ports for downlink MIMO. On the other hand, CRS design constraints were aimed to support channel estimation for coherent demodulation in extreme channel conditions with very fast channel variations in both the time and frequency domains. This resulted in relatively high overhead ~5% of time-frequency resources used for CRS. The enhanced MIMO with eight antenna ports (Transmission modes 9 and 10) is hardly used in extreme channel conditions, implying that less informative, but less demanding for resources, Channel-State Information may be sufficient, if used together with DM-RS, for effective demodulation of the PDSCH.

In uplink MIMO spatial multiplexing of up to four layers is supported by LTE-Advanced. With a single-user spatial multiplexing transmission (SU-MIMO) scheme, up to two transport blocks can be transmitted from a scheduled UE in a subframe per uplink component carrier. Each transport block has its own MCS level. Depending on the number of transmission layers, the modulation symbols associated with each of the transport blocks are mapped onto one or two layers according to the same principle as in Release 8 LTE E-UTRA downlink spatial multiplexing. The transmission rank can be adapted dynamically. The SU-MIMO is applied with closed-loop control and code book-based precoding similar to downlink transmission. Based on measurement of uplink Sounding RS the ENB estimates quality of uplink channel and select the rank and precoding index from the codebook. The eNB notifies the UE of selected precoding and provides information on resource allocation on PDCCH.

12.3 Coordinated Multi-Point Operation (CoMP)

Coordinated Multi-Point (CoMP) operation is a kind of spatial diversity technique for improving network performance at cell edges thus increasing throughput and mitigating inter-cell interference. The CoMP transmission and reception techniques utilize multiple transmit and receive antennas from multiple antenna locations, which may belong either to the same or different physical cells. Several deployment scenarios are considered for CoMP, either for homogeneous or heterogeneous networks.

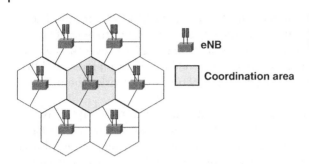

Figure 12.8 Scenario 1: Homogeneous network with intra-site CoMP [5].

eNB

Coordination area

Homogeneous network is based on the cells of one type, predominantly-macrocells where base stations belong to same type and power class. Since downlink MIMO transmission is controlled in each base station independently from its neighbours, the inter-cell interference coordination (ICIC) is performed by specific message exchange over X2 interface. The latency introduced by a backhaul carried X2 messages restricts the ICIC to relatively slow, semi-static operation. The performance targets for CoMP imply a fast dynamic coordination that can be supported only with optical fibre backhaul. Two scenarios for CoMP are considered in a homogeneous network:

1) Intra-site CoMP, Figure 12.8. Coordination between the cells (sectors) controlled by the same macro base station, no backhaul connection is needed. This scenario has some similarities with a soft handover in the 3G network.
2) Coordination between cells belonging to different radio sites from a macro network connected via fibre front-haul. This is typically distributed architecture (see Annex 1) with remote radio heads (RRH) or radio modules connected to centrally located system module/baseband unit (BBU) via CPRI or OBSAI optical fibre PR3 interface, as shown in Figure 12.9.

A heterogeneous network solution is based on deployment several types of cells within the macrocell as an umbrella cell with high-power TRX providing network coverage combined with low-power nodes (pico-, femtocell) as the enablers of increased throughput in high demand traffic areas. The small cell may experience severe interference from high-power macrocell node due to close proximity and different power classes. Two CoMP scenarios are considered for the heterogeneous network [5]:

1) Coordination between a macrocell and low-power RRHs (nodes) within macro coverage, each low-power node does control its own cell with Cell ID different from macrocell, see Figure 12.10.
2) Network with low-power RRHs within the macrocell coverage (same as 1) where the transmission/reception points created by the RRHs have the same cell IDs as the macro cell, see Figure 12.10. This is the case of distributed antenna architecture.

12.3.1 CoMP Categories

Each CoMP techniques can be classified into one of the following categories.

1) *Joint Processing* (JP): Data for a UE is available at more than one point in the CoMP cooperating set of transmission points for a time-frequency resource.

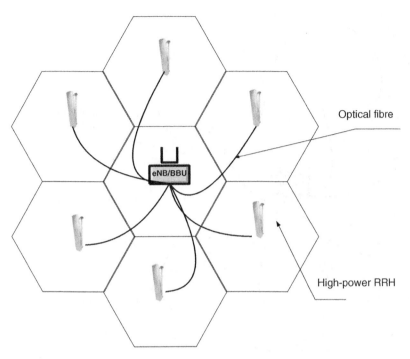

Optical fibre

eNB/BBU

High-power RRH

Figure 12.9 Scenario 2: Homogeneous network with high Tx power RRHs.

2) *Joint Transmission* (JT): Simultaneous data transmission from multiple geographically separated points to a single UE or multiple UEs in a time-frequency resource. Data to the UE is simultaneously transmitted from *multiple* points; for example, to (coherently or non-coherently) improve the received signal quality and/or data throughput.

3) *Dynamic Point Selection* (DPS)/muting: Data transmission from one point (within the cooperating g set of transmission points) in a time-frequency resource. The transmitting/muting point may change from one subframe to another including varying over the RB pairs within a subframe. Data is available simultaneously at multiple points. DPS enables UE to be dynamically scheduled by the most appropriate transmission point chosen on the basis of best propagation channel conditions. DPS may be combined with JT in which case multiple points can be selected for data transmission in the time-frequency resource.

4) *Coordinated Scheduling/Beamforming* (CS/CB): Data for a UE is only available at and transmitted from one point in the CoMP cooperating transmission point set (DL data transmission is done from that point) for a time-frequency resource. The user scheduling/beamforming decisions are made with coordination among points corresponding to the CoMP cooperating set. The transmitting points are chosen semi-statically.

A hybrid category of JP and CS/CB may be possible. Data for a UE may be available only in a subset of points in the CoMP cooperating set for a time-frequency resources, but user scheduling/beamforming decisions are made with coordination among points corresponding to the CoMP cooperating set. For example, some points in the

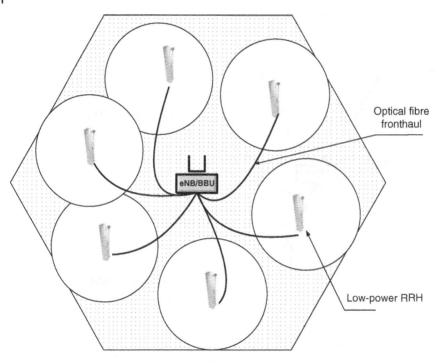

Figure 12.10 Scenario 3/4: Network with low-power RRHs within the macrocell coverage.

cooperating set may transmit data to the target UE according to JP while other points in the cooperating set may perform CS/CB.

12.3.2 Downlink CoMP

In CoMP downlink transmission, the downlink physical shared channel (PDSCH) is transmitted from multiple cells with precoding using DM-RS among coordinated cells. When two or more cells transmit on the same frequency in the same subframe this is called a Joint Transmission, as illustrated in Figure 12.11. The UE receiver performs then a coherent combining of both signals. Non-coherent JT study is described in [6]. This form of CoMP places a high demand onto the backhaul network because the data to be transmitted to the UE needs to be sent to each eNB that will be transmitting it to the UE. This may easily double or triple the amount of data in the network dependent upon

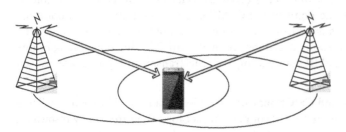

Figure 12.11 Joint transmission CoMP.

how many eNBs will be sending the data. In addition to this, joint processing data needs to be sent between all eNBs involved in the CoMP area.

The special case of the intra-site CoMP (Scenario 1 in a homogeneous network) is similar to softer handover in WCDMA and, as such, sets no requirements for the additional load on transport.

In centralized RAN architecture (C-RAN) where a single system module/BBU controls multiple RRHs, it can manage efficient implementation of joint scheduler for JT using dynamic adaptation of BBU resource pool to channel condition and traffic load. The UE supports JT with a report of Precoding Matrix Indicator and corresponding Channel-Quality Indicator (CQI) to a centralized scheduler controlling the participating joint transmission points. Another form of downlink CoMP is a Dynamic Point Selection (DPS) when PDSCH is transmitted from one cell dynamically selected. The scheduling is accompanied by antenna beamforming and coordinated among coordinated cells. In Dynamic Point Selection CoMP, the additional load on backhaul network is considerably reduced since UE data is transmitted to one eNB at the time. The only scheduling decisions and beamforming details needs to be coordinated between multiple eNBs.

The DPS is assisted by the UE that reports the index of its preferred transmission point based on channel measurements (the TP with the highest received SINR) and respective CSI. The selected TP may dynamically change from one subframe to the next, with 1 ms TTI. The muting associated with DPS may silence the other transmitting point from CoMP set, those TP may not transmit any data thus increasing the SINR in received signal.

Coordinated Scheduling/Coordinated beamforming is a high complexity technique that may require division of whole network in set of smaller clusters with centralized scheduling per cluster. The objective of that scheduling is to decide which transmission point(s) within the cluster should transmit to which UE in each subframe. A coordinated beam pattern is created by coordinating the precoders (beamforming matrices) in the cooperating TPs in a predefined manner in order to reduce interference variation and enable efficient link adaptation. The inter-TP interference is suppressed by coordination of Precoding Matrix indicators known to both the UE and base station. Relying on a TP-specific beam pattern in the time and/or frequency domain, the central controller cycles the fixed set of beams of participating TPs. Based on reports of the best resources and associated channel quality in each cycling period, the UE can be scheduled in the subframes or subbands where the coordinated beams provide the best channel conditions.

12.3.3 Uplink CoMP

CoMP reception in uplink can be implemented in two ways: Joint Reception and Coordinated Scheduling.

Joint Reception: The basic concept of Joint Reception and is to utilize antennas at different sites. By coordinating between the different eNBs it is possible to form a virtual antenna array. With Joint Reception, the PUSCH is received in multiple cells and then combined and processed to produce the final output signal, as illustrated in Figure 12.12. This technique allows for signals that are very low in strength, or masked by interference in some areas to be receiving with few errors.

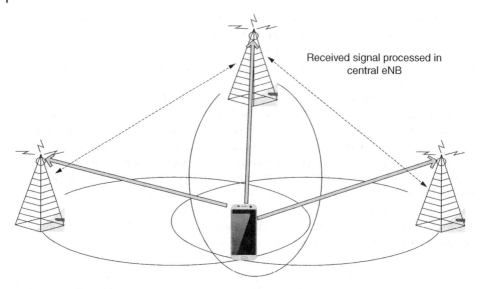

Received signal processed in central eNB

Figure 12.12 Uplink CoMP reception.

In case of decentralized conventional homogeneous network, large amounts of data need to be transferred between the TPs.

Coordinated scheduling: This scheme operates by coordinating the scheduling decisions among the RX points/cells for reception of data and reference signals from UEs. If inter-eNodeB coordination is supported, information needs to be signalled between the eNodeBs. As in the case of the downlink, this format provides a much reduced load in the backhaul network since only the scheduling decisions messages needs to be transferred between the different eNBs that are coordinating with each other.

The additional processing required for multiple site reception and transmission as well as communication between the different sites could add significantly to network latency. To overcome this, it could be possible if the different sites may be connected together in a form of Centralized RAN, or C-RAN. The UL CoMP may require some enhancements in radio interface; in particular, in power control and possibly control channel overhead with additional SRS resources [5].

Inter/Intra-site Backhauling Support For Downlink CoMP: In all scenarios described in this section transmission points may be viewed either as belonging to the same eNB or different eNBs. Those scenarios encompass different deployment architectures, depending on backhaul quality between points. Two cases are being considered:

Point-to-point fibre (zero latency and infinite capacity backhaul) applicable to scenarios 2, 3, 4.

- Higher latency and limited capacity backhaul applicable to scenarios 2 and 3. Backhaul links between macro eNodeBs may be used. Backhauling links may include inband relays, out-of-band relays or a combination.
- Depending on backhaul technology, latency and capacity may be asymmetric in the two directions connecting two points.

For scenarios 2 and 3 described in this section, points may also belong to different eNBs. In this case, backhaul information exchanges may require some standardization support. Note that the case of a higher latency and limited capacity backhaul is most relevant for this deployment architecture.

The backhaul capacity required by CoMP depends on the system bandwidth, the number of UEs served in a CoMP way, the size of the cluster and the network topology in a backhaul (e.g. ring or star). The technology used in backhaul (fibre or MW) affects the latency. In case of leased transport the additional security protocol stack adds additional delays. This means that implementation of inter-site CoMP with coordination over the X2 interface is constrained by availability of low-latency backhaul transport. The preferable option for CoMP deployment is a C-RAN architecture.

Another requirement for X2 based inter-site CoMP is a perfect phase synchronization, which is easily achievable with C-RAN architecture but may require additional means, like GPS receivers deployed in TPs connected via backhaul.

12.4   Relay Nodes

Relaying in LTE is another way of enhancing system performance at the cell edge. LTE relaying is supported with Relay Node (RN), which is usually a low-power base station that provides enhanced coverage and capacity at cell edges. LTE relaying is different to the use of a repeater that re-broadcasts the signal. A relay node in LTE receives, demodulates and decodes the data, applies any error correction and then retransmits a new signal. In Relaying, the UEs communicate with the Relay Node, which in turn communicates with a Donor eNB (DeNB).

Besides edge coverage enhancement there are several scenarios where LTE relaying brings considerable benefits.

RF Coverage Extension: LTE relays can be used instead of repeater for providing RF coverage in otherwise difficult to cover areas. With no need to install a complete base station with respective backhaul, the RN can be quickly installed so that it fills in the coverage blackspot.

Network Capacity Enhancement: When capacity is to be increased by deployment of the small cells in a high density network, the LTE relay nodes can substitute the eNBs without separate backhaul. Physically, typical Relay Node is much smaller than eNB (due to relatively low output power and absence transport block and associated processing) and therefore can allow flexible installation. The RN is wirelessly connected to the Donor eNB (DeNB) via a modified version of the E-UTRA radio interface-Un interface. The DeNB is in turn connected to the core network, see Figure 12.13.

The Relay Node supports the eNB functionality meaning it terminates the radio protocols of the E-UTRA radio interface, and the S1 and X2 interfaces. In addition to the eNB functionality, the Relay Node also supports a subset of the UE functionality, for example physical layer, layer-2, RRC and NAS functionality, in order to wirelessly connect to the DeNB replicating the mobile terminal.

12.4.1   Relay Radio Access

The relaying technology in LTE-A is based on the Layer 3 relay. The RF downlink signal from DeNB is decoded and demodulated in Relay Node and then undergoes full Layer 1

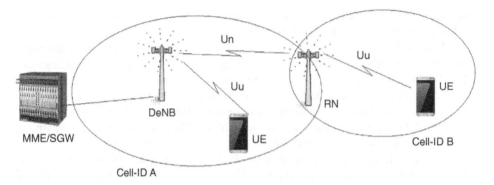

Figure 12.13 Relay configuration.

to Layer 3 processing as per the Uu radio interface requirements for retransmitting the data to the mobile station. With that processing, both the intercell interference and noise at the backhaul link are eliminated and a clear signal is transmitted on the downlink from RN to the UE. The UE sees the RN as a normal eNB that terminates the Uu radio interface protocol. However, the RN has a cell ID different from DeNB, as shown in Figure 12.9. This way the UE can distinguish the relay cell from macrocell covered by DeNB.With respect to the relay node's usage of spectrum, its operation can be classified into:

- Type 1. Inband relaying, in which case the DeNB-RN wireless backhaul link (Un) shares the same carrier frequency with RN-UE radio access links (Uu). The only way to operate access and bakhaul radio frames in a Time-Division Multiplexing scheme, which implies that, on the downlink carrier frequency at a given time, the RN either transmits on the access link or receives on the backhaul link. Likewise, at a given time on the uplink carrier frequency the RN either receives on the Uu access link or transmits on the Un backhaul link.
- Type 1a. Outband relaying, in which case the DeNB-RN link (Uu) operates on different carrier frequency from RN-UE radio access link, Uu. This transmission scheme does not invoke and changes to radio interface lower level protocol stack but leads to introduction of additional carrier. If deployed, the frequency domain isolation between backhaul and access links benefits in terms of reduced intercell interference.

For both inband and outband relaying, it is possible to operate the eNB-to-relay link on the same carrier frequency as eNB-to-UE links. At least 'Type 1' and 'Type 1a' RNs are supported by LTE-Advanced at current stage of standardization.

The Relay Node of either Type 1 or 1a is characterized by the following:

- It controls cells, each of which appears to a UE as a separate cell distinct from the donor cell.
- The cells shall have their own Physical Cell ID (as defined in LTE Release 8) and transmit their own synchronization channels, reference symbols and so on.
- In the context of single-cell operation, the UE receives scheduling information and HARQ feedback directly from the RN and sends its control channels (SR/CQI/ACK) to the RN.

12.4.2 Relay Architecture

The architecture for supporting RNs is shown in Figure 12.14 [2]. The RN terminates the S1, X2 and Un interfaces. The DeNB provides S1 and X2 proxy functionality between the RN and other network nodes (other eNBs, MMEs and S-GWs). The S1 and X2 proxy functionality includes passing UE-dedicated S1 and X2 signalling messages as well as GTP data packets between the S1 and X2 interfaces associated with the RN and the S1 and X2 interfaces associated with other network nodes. Due to the proxy functionality, the DeNB appears as an MME (for S1-MME), an eNB (for X2) and an S-GW (for S1-U) to the Relay Node.

Figure 12.15 details functional entities in relaying architecture. As shown, the S1-MME interfaces terminates in DeNB due to its proxy role for UE. The Relay GW in Figure 12.9A hosts a 'home eNB GW' type of functionality, which is transparent to the relay, the core network of the UE and other eNBs. The relay node accommodates two logical entities: eNB and UE with respective protocol states

The user-plane protocol stack and packet delivery process in relay mode are shown in Figure 12.16 and 12.17, respectively. With proxy GW architecture packet delivery process is performed as follows:

- The downlink UE packet is mapped to UE bearer at the PGW serving the UE and the packet is sent in the corresponding UE bearer GTP tunnel to the donor eNB, one GTP tunnel is created per UE bearer.
- The donor eNB classifies the incoming packets into RN radio bearers based on the QCI of the UE bearer and switches the UE bearer GTP tunnel from the SGW/PGW to another UE bearer GTP tunnel towards the RN (one-to-one mapping). As pointed out in Figure 12.9C, the EPS bearers of different UEs connected to the RN with similar QoS are mapped in one radio bearer over the Un interface.

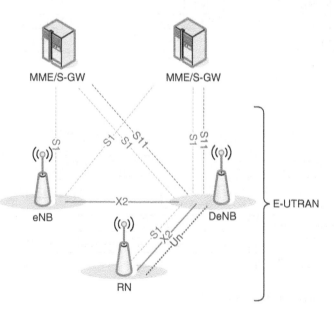

Figure 12.14 Overall E-UTRAN architecture supporting relay node [2].

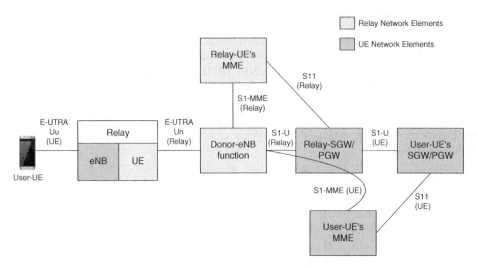

Figure 12.15 Functional entities in relaying architecture [7].

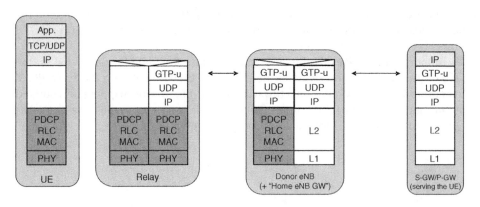

Figure 12.16 User-plane protocol stack [7].

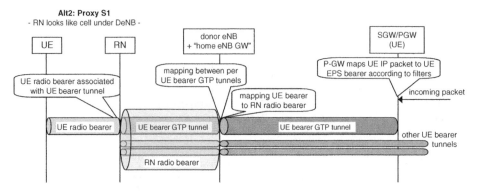

Figure 12.17 Packet delivery with relay architecture [7].

- The RN associates the received packet with the corresponding UE radio bearer based on the per UE bearer GTP tunnel.

In the uplink, the RN performs the UE bearer to RN bearer mapping, which can be done based on the QCIs of UE bearers.

The control plane protocol architecture and signalling connections is shown in Figure 12.18.

During attach and bearer setup procedures the S1-AP messages are sent between the MME and the DeNB, and between the DeNB and the RN. Upon the DeNB receiving the S1-AP messages, it translates the UE IDs between the two interfaces by means of modifying the S1-AP UE IDs in the message but leaving other parts of the message unchanged. This operation corresponds to an S1-AP proxy mechanism and would be similar to the HeNB GW function. The S1-AP proxy operation is transparent for the MME and the RN. That is, as seen from the MME it looks like as if the UE would be connected to the DeNB, while from the RN's perspective it would look like as if the RN would be talking to the MME directly. The S1-AP messages encapsulated by SCTP/IP are transferred over an EPS data bearer of the RN where the PGW functionality for the RN's EPS bearers is incorporated into the DeNB.

The S1 interface relations and signalling connections are shown in Figure 12.19. One S1 interface relation is established between the RN and the DeNB while another S1 interface is between the DeNB and the MME (serving the UE). Note that the RN has to maintain only one S1 interface (to the DeNB), while the DeNB maintains one S1 interface to each MME in the respective MME pool (one S1 connection to each MME serving the UE).

As shown in Figure 12.19, there is S1 interface relation and S1 signalling connection corresponding to the RN as a UE, going from the DeNB to the MME serving the RN. A similar logical structure applies to X2 interface between DeNB and eNB, see Figure 12.14. As can be seen from Figures 12.16 and 12.18, on the Uu interface between UE and RN, all control plane (RRC) and user plane (PDCP, RLC and MAC) protocols are terminated in RN. On the Un interface between RN and eNB, the user plane is based on standardized protocols (PDCP, RLC, MAC). The control plane on Un uses RRC (for the RN in its role as UE) protocol.

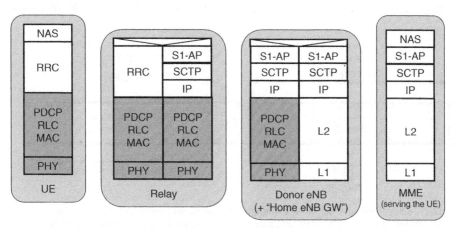

Figure 12.18 Control-plane protocol stack in relay architecture [7].

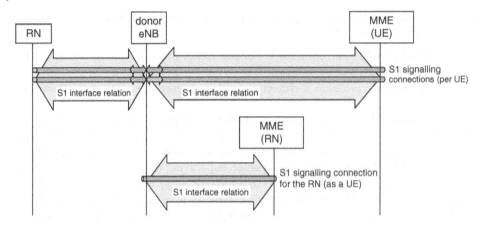

Figure 12.19 Signalling connection in relay architecture.

12.4.3   Resource Assignment for DeNB-RN Link in a Type 1 Relay

In order to allow inband relaying, some resources in the time-frequency space are set aside for the backhaul link (Un) and cannot be used for the access link (Uu).

The Relay Node is not transmitting to terminals when it is supposed to receive data from the donor eNodeB; that is, to create 'gaps' in the relay-to-UE transmission. These 'gaps' during which terminals (including Release 8 terminals) are not supposed to expect any relay transmission can be created by configuring MBSFN subframes as shown in Figure 12.20. Relay-to-DeNB transmissions can be facilitated by not allowing any terminal-to-relay transmissions in some subframes. The RN declares an MBSFN subframe in the DL in order to create a transmission gap. When the UE is informed of MBSFN subframe it processes the first symbols carrying the control information, but ignores the rest of the subframe. The RN switches the RF circuitry for receiving the Un channel from DeNB in that sunbframe.

The set of rules for resource partitioning and multiplexing in the relay physical layer is specified in [9]. The subframes during which DeNB-RN transmission may take place are configured by higher layers. Downlink subframes configured for DeNB-to-RN

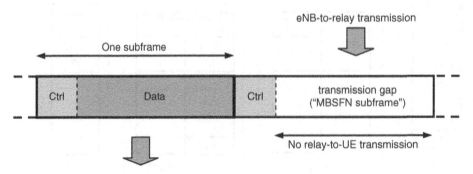

Figure 12.20 Relay Node DL radio frame configuration: normal subframes (left) composed with reference, control signals and user data, and gap in transmission to UE on Uu in declared MBSFN subframes (right). This gap is used for reception of Un by RN from DeNB [8].

transmission shall be configured as MBSFN subframes by the relay node. The set of subframes for downlink or uplink backhaul transmission is semi-statically assigned with the help of a new Relay Physical Downlink Control Channel, R-PDCCH. The R-PDCCH informs the RN about the resource allocation of DL-SCH, and Hybrid-ARQ information related to DL-SCH. It also carries the backhaul uplink scheduling grant.

12.5 Enhanced Physical Downlink Control Channel (E-PDCCH)

3GPP Release 11 introduced a new control channel, the E-PDCCH to support new features of LTE-A like CoMP, DM-RS, beamforming as well as other features to be introduced in later releases. The E-PDCCH does not replace PDCCH, this is a control channel complementary to PDCCH. The E-PDCCH is user-specific, meaning that different UEs can have different E-PDCCH configurations. Each UE can be configured with up to two sets of EPDCCHs.

Design configuration of E-PDCCH is different from PDCCH. Instead of using the first four symbols in a subframe where Downlink Control Information (DCI) is spread over the entire system bandwidth, the E-PDCCH shares the time-frequency grid resources with PDSCH, as shown in Figure 12.21 in a specially configured subframes. Figure 12.21 illustrates the downlink subframe structure in LTE (Release 8) and LTE-A (Release 11), the latter is a subframe with special configuration that contains both PDCCH and E-PDCCH.

The features of e-PDCCH are as follows:

- support for increased control channel capacity,
- support frequency domain ICIC,
- improved spatial reuse of control channel resource,
- support beamforming and/or diversity,
- operate on the new carrier type with Carrier Aggregation,
- coexist on the same carrier as legacy UEs.

The E-PDCCH is configured via RRC signalling. The RRC signalling will indicate to the UE those particular subframes it has to monitor for dedicated E-PDCCH. Also, a signalling message indicates the configuration of the E-PDCCH, specifically, one or two sets of Resource Blocks (RB) pairs and number of RBs per pair. The RB pairs could be of two, four or eight RBs. Similar to PDCCH the E-PDCCH is transmitted using an aggregation of one or several consecutive enhanced control channel elements (ECCEs) where each ECCE consists of multiple enhanced resource-element groups (EREGs). The details of the E-PDCCH format are specified in [3].

12.6 Downlink Multiuser Superposition, MUST

A power-domain non-orthogonal multiple-access (NOMA) scheme has been proposed to LTE Release 13 under a study called downlink MultiUser Superposition Transmission (MUST) [10]. Unlike orthogonal multiple access, NOMA allows multiple users to simultaneously access the same time-frequency resource, by applying superposition coding

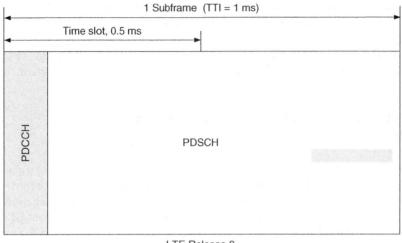

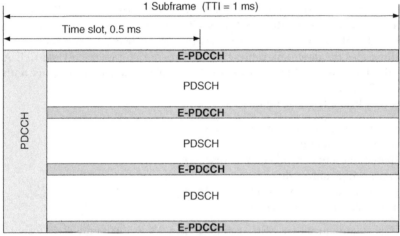

Figure 12.21 Resource allocation for PDCCH and E-PDCCH.

and Successive Interference Cancellation (SIC) method. The objective of the study was to investigate the potential gain of schemes enabling simultaneous transmissions of more than one layer of data for more than one UE within one cell without time, frequency and spatial layer separation.

The MUST is based on coding superposition of different users in the same set of sub-carriers. The target scenario for MUST is the case of transmission in common resource block to two active users with very different channel gains: one is close to the base station and another is located at the cell edge. The channel gain is drastically different due to a substantial difference in the path losses experienced by user at the cell edge compared with the user close to base station. In this case, the base-station transmitter allocates more power for an edge based user and much less power to a signal for the closer user.

The SIC receiving method implies that user close to base station can detect the strongest signal (intended to edge based user) first, then subtracts it from the composite waveform and then decodes its own signal. The edge user may just decode his signal assuming that weak signal transmitted to 'near' user is received at the noise level. More details of NOMA are described in Section 13.5.1.2.

In a practical deployment, the distribution of the users in a given cell is rather arbitrary where a larger difference of distances between the base station and different users can result in different path losses, (i.e. near-far effects). In a typical mobile environment where the path loss exponent is about four, the difference of the path losses could be ~40 dB between a user 5 km away from the BS and another user 500 m away from the BS. The superposition coded-multiplexing signals exploits the large path loss difference between different users in the cell by increasing the number of subchannels or spatial reuse. Therefore, the aggregate data rate of the downlink can be improved at the expense of UE decoding complexity.

12.7 Summary of LTE-A Features

LTEA is deployed as an evolution of LTE Release 8. LTEA is backward compatible with LTE. Key features of LTE-A are:

- Carrier aggregation for aggregated bandwidth up to 640 MHz;
- Relaying functionality as a substitute for a backhaul transmission;
- Advanced multiple antenna transmission technique: Up to $8 \times M8$ DL MIMO, 4×4 UL MIMO, Multiuser MIMO;
- Improved peak data rates up to 1 GHz (Release 10);
- Coordinated Multipoint Transmission supporting macro transmit/receive diversity;
- Downlink MUST improves spectrum efficiency and cell throughput.

References

1 Evolved Universal Terrestrial Radio Access (E-UTRA); User Equipment (UE) radio transmission and reception (Release 14), 3GPP TS 36.101 V14.3.0 (2017–03).
2 Evolved Universal Terrestrial Radio Access (E-UTRA) and Evolved Universal Terrestrial Radio Access Network (E-UTRAN); Overall description; 3GPP TS 36.300 V14.0.0 (2016–09).
3 Evolved Universal Terrestrial Radio Access (E-UTRA); Physical channels and modulation (Release 14), 3GPP TS 36.211 V14.4.0 (2017–09).
4 Evolved Universal Terrestrial Radio Access (E-UTRA); Further advancements for E-UTRA physical layer aspects (Release 9), 3GPP TR 36.814 V9.2.0 (2017–03).
5 Coordinated multipoint operation for LTE physical layer aspects (Release 11), 3GPP TR 36.819 V11.2.0 (2013–09).
6 Study on further enhancements to Coordinated Multipoint (CoMP) Operation for LTE (Release 14), 3GPP TR 36.741 V14.0.0 (2017–03).
7 Evolved Universal Terrestrial Radio Access (E-UTRA); Relay architectures for E-UTRA (LTE-Advanced) (Release 9), 3GPP TR 36.806 V9.0.0 (2010–03).

8 Feasibility study for Further Advancements for E-UTRA (LTE-Advanced) (Release 14), 3GPP TR 36.912 V14.0.0 (2017–03).

9 Physical layer for relaying operation (Release 14); 3GPP TS 36.216 V14.0.0 (2017–03).

10 Study on Downlink Multiuser Superposition Transmission (MUST) for LTE (Release 13); 3GPP TR 36.859 V13.0.0 (2015–12).

13

Further Development for the Fifth Generation

5G is a Fifth Generation of mobile communication technology, anticipated to come to the market sometime after 2020. The driving force for a new generation of technologies is generated by increasing demands from a networked human society. Those demands include more traffic volume, more devices with diverse service requirements, better quality of user experience (QoE) and better affordability by further reducing costs.

The 5G system is expected to be able to provide optimized support for a variety of different services, different traffic loads and different end-user communities. The 5G White paper by the Next Generation Mobile Network Alliance (NGMN) describes a multi-faceted 5G system capable of simultaneously supporting multiple combinations of reliability, latency, throughput, positioning and availability. This is achievable with the introduction of new technologies, both in access and the core parts of the network, which allow flexible, scalable assignment of the network resources.

According to ITU-R studies [1] major usage scenarios for the new 5G system can be largely classified into five categories:

1) *Enhanced Mobile Broadband*: Mobile Broadband addresses the human-centric use cases for access to multimedia content, services and data. This usage scenario covers a range of cases, including wide-area coverage and hotspot, which have different requirements.
2) *Ultra-reliable and low-latency critical communications*: This use case has stringent requirements for capabilities such as throughput, latency and availability. This may cover interactive games, sports, communications to/from drones, robots and emergency communications. Some examples include wireless control of industrial manufacturing or production processes, remote medical surgery, distribution automation in a smart grid, transportation safety and so on, and can be referred as machine types of communications.
3) *Machine Type Communications* (MTC): Two types of MTC can be distinguished: either massive or critical. Massive MTC is characterized by a very large number of connected devices typically transmitting a relatively low volume of non-delay-sensitive data. Devices are required to be low cost and have a very long battery life. Critical MTC refers to applications such as traffic safety/control, control of critical infrastructure and wireless connectivity for industrial processes. Such applications require very high reliability and availability in terms of wireless connectivity, as well as very low latency. While the average volume of data transported to

Introduction to Mobile Network Engineering: GSM, 3G-WCDMA, LTE and the Road to 5G,
First Edition. Alexander Kukushkin.

and from devices may not be large, wide instantaneous bandwidths may be required in order to meet capacity and latency requirements.

A later study by 3GPP [2] added two other 5G use cases:

4) Network Operation that is enhanced with *network slicing*, routing, migration and interworking and energy saving.
5) Enhancement of *Vehicle-to-Everything*: for example, autonomous driving, safety and non-safety aspects associated with vehicles.

The overall objective of 5G is to provide ubiquitous connectivity for any type of device and any kind of application. It is expected that 5G networks will not be based on one specific radio access technology but, instead, 5G accommodates a number of access and connectivity solutions addressing requirements from various usage scenarios. 5G solution will incorporate existing and evolved LTE systems as well as new radio interface technology. The integration of LTE in 5G follows a key interoperability principle of ICT industry. It will allow legacy devices to operate in 5G network within compatible operational bands.

13.1   Overall Operational Requirements for a 5G Network System

The 5G system will support roaming and interworking between home and visited mobile networks by establishing either home or visited network data connectivity and providing subscribed services.

In addition to many new services, the 5G system will support most of the existing EPS services. The existing EPS services may be accessed using the new 5G access technologies even where the EPS specifications might indicate E-UTRA(N) only. The following exceptions will apply:

- CS voice service continuity and/or fallback to GERAN or UTRAN,
- seamless handover between NG-RAN and GERAN,
- seamless handover between NG-RAN and UTRAN, and
- access to a 5G core network via GERAN or UTRAN.

The 5G system will support mobility procedures between a 5G core network and an EPC with minimum impact to the user experience (e.g. QoS, QoE).

13.2   Device Requirements

5G terminals will have a high degree of programmability and over-the-air configurability by the network in terms of access technology and transport protocol. This will enable efficient logical division for different services (slicing) while removing dependency on terminal type. 5G UE will be capable of choosing specific profiles in dynamic way depending on QoS needs, radio node capability and/or radio conditions.

5G UE has to support multiple frequency bands as well as multiple modes (TDD/FDD/mixed). The 5G device will have the capability of concurrent Multi-RAT (multiband) connectivity with aggregation of data flows from different technologies and carriers. Such demanding requirements generate advanced developments in resource, signal processing, signalling and power efficiency of 5G devices.

13.3   Capabilities of 5G

The following parameters are considered to be key capabilities of 5G. The values that follow are targets for research and investigation for IMT-2020 and may be further developed or revised in ITU recommendations.

- The peak data rate for 5G Enhanced Mobile Broadband is expected to reach 10 Gbps and exceed this value in indoor and dense outdoor environments.
- The user experienced data rates are expected as follows: at least 10 Mbps accessible everywhere including sparsely populated areas; 100 Mbps for wide-area coverage, for example in urban and suburban areas and up to 1 Gbps in indoor environment.
- The spectrum efficiency is expected to be three times higher compared to LTE-A for enhanced Mobile Broadband. 5G is expected to support a 10 Mbps/m^2 area traffic capacity in an indoor environment.
- The energy consumption for the 5G radio access network should not be greater than LTE networks deployed today, while delivering the enhanced capabilities.
- Over-the-air latency is to be reduced to 1 ms, capable of supporting services with very low-latency requirements.
- 5G is also expected to enable high mobility up to 500 km/h with acceptable QoS.
- 5G is expected to support a connection density of up to $10^6/_{km^2}$ in massive MTC scenarios.

13.4   Spectrum Consideration

A new 5G air-interface technology development is anticipated to support very high data rate in Enhanced Mobile Broadband scenarios. On the other hand, recently completed 3GPP standard in LTE-A Release 13 [3] for the Narrowband Internet of Things (NB-IoT) aimed at supporting MTC in wide-area applications can also be considered part of 5G technology. The NB-IoT is likely to be deployed in the frequency bands below 2 GHz providing high capacity and deep coverage for a large number of connected devices. The spectrum bandwidth in support of different usage scenarios in 5G (e.g. enhanced mobile broadband, ultra-reliable and low-latency communications and massive machine type communications) would vary. For those scenarios requiring bandwidth from several hundred MHz up to 1 GHz, there would be a need to consider an allocation of the wide-band contiguous spectrum above 6 GHz.

While in 5G some spectrum bands are already included in Release 15 [4, 5], other specific spectrum bands for 5G are yet to be identified by the regulatory bodies. 5G services will require a range of different frequency bands and bandwidths; for instance, MTC services may utilize relatively narrow bandwidth while eMBB require very wide bandwidths for high-capacity usage cases. An overall solution for 5G will include spectrum below 6GHz as well as the spectrum at the higher frequencies in the range of 6–100 GHz. By around 2020, most mobile communications will likely be provided by LTE, therefore,

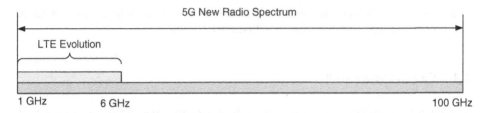

Figure 13.1 Planned frequency spectrum allocation for 5G.

an evolution of LTE to 5G has to support dual connectivity between the LTE operating below 6 GHz and New Radio (NR) technology operating in the range above 6 GHz. The schematic spectrum allocation for further evolution of LTE and 5G New radio is illustrated in Figure 13.1.

The results of a study on radio propagation environment of IMT in the bands between 6 GHz and 100 GH are reported in [6]. The report includes performance simulations results for several different deployment scenarios and describes solutions based on MIMO and beamforming with a large number of antenna elements, which compensate for the increasing propagation loss with frequency.

The practicality of manufacturing commercial transmitters and receivers at millimetre wavelengths is investigated and evidenced by the availability of commercial 60 GHz multi-gigabit wireless systems products and prototyping activities that are already underway at frequencies such as 11, 15, 28, 44, 70 and 80 GHz. Also, the report considers the potential advantage of an integrated backhaul solution using the same spectrum for both access and front-haul/backhaul [6].

13.5   5G Technology Components

Technical and operational characteristics of the 5G system can be supported through advances in technology and techniques. Some technical aspects of the 5G system are discussed in [7] where technology enablers are presented in following groups:

- Technologies to enhance the radio interface
- Technologies to support wide range of emerging services
- Technologies to enhance user experience
- Technologies to improve network energy efficiency
- Terminal technologies
- Network technologies
- Technologies to enhance privacy and security.

The whole spectrum of 5G technology enhancements is out of the scope of this study. We briefly consider a few technological developments related to the radio access part of 5G.

13.5.1   Technologies to Enhance the Radio Interface

The summary of some major developments considered for enhancement in 5G radio interface technology is provided in the following subsections.

13.5.1.1 Advanced Modulation-and-Coding Schemes

The 5G demand for further improvements in spectrum efficiency stimulates a development of the new advanced waveforms and modulation-and-coding schemes, possibly leading to a breakthrough in transceiver design. With a new variety of deployment conditions, different applications may necessitate specific characteristics and performance criteria for transmit waveforms. For example, a sensor network in mass scale MTC drives priority for low complexity, low-cost, low-power consumption devices with relatively robust link budget and low/moderate data rates. In another extreme, interactive real-time video scenario, an indoor cell gives priority to low-latency and the highest possible data rate.

A diverse set of applications anticipated for 5G system ensures a diverse set of performance criteria to waveforms and respective modulation-and-coding schemes. Ideally, future systems may incorporate flexible allocation of time-frequency resources with a broad set of modulation-and-coding schemes that may be supported simultaneously within same bandwidth and transmission time interval. Several options for new waveform formats are being considered by 5G researchers, these include: FBMC (Filter Bank Multicarrier), UFMC (Universal Filtered MultiCarrier), GFDM (Generalized Frequency Division Multiplexing), FTN (Faster-than-Nyquist), Time-Frequency-packed Signalling (TFS), Wave Amplitude Modulation (WAM) and Sparse Code Multiple Access (SCMA). Some examples of 5G waveforms, modulations and coding can be found in the references provided in [7].

At the moment, alternatives to or enhancements of OFDM are still at the research stage of being explored and tested. The new waveform technology is to bring benefit in the following areas [8]:

- Enable steeper spectrum roll-off and bands with difficult constraints
- Reduced peak-to-average power ratio thus improving energy efficiency and tolerating development of cheaper devices, especially M2M)
- Enable framework with low-latency
- Address the M2M specific requirement, especially for low data rates
- Enable synchronous-constrained applications (e.g., CoMP, D2D, MMC)
- High transmission range in higher frequency bands
- Combination with Massive MIMO in higher frequency bands.

So far, there is no consensus on implementation of the new waveform technology. The New Radio platform based on OFDM is modified with new numerology and advanced filtering to stop a roll-off in the OFDM spectrum mask. These new subcarrier spacing enhancements seem to be sufficient for 5G deployment in the frequency band from 6 up to 30 GHz. However, R&D efforts continue in numerous labs anticipated to produce a breakthrough in technology, especially in the millimetre-wavelength range.

13.5.1.2 Non-Orthogonal Multiple Access (NOMA)

With NOMA, multiple users can transmit signals at the same spatial-time-frequency resource during uplink transmission, or the signals of multiple users can be transmitted by eNB at the same spatial-time-frequency resource during downlink transmission.

The NOMA scheme efficiently exploits the channel gain difference between users to achieve high spectral efficiency. At the transmitter site, all individual signals from different users are superimposed into a single waveform. The receiver tries to decode the

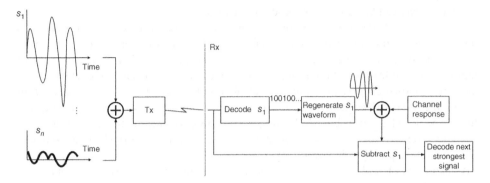

Figure 13.2 Concept of successive interference cancellation.

strongest signal while the other signals are treated as interference. The decoded signal is then regenerated into a waveform and subtracted from the composite received signal. The process is iterated for all desired signals. This method is called Successive Interference Cancellation (SIC), illustrated in Figure 13.2 for n user signals, $s_1 \ldots s_n$, where s_1 is a strongest signal.

NOMA schemes share a common radio resource among multiple users by different means, such as a combination of multiuser power superposition, multiuser space diversity and codebook based multiple access. First, NOMA was studied in LTE-A referred to as a MultiUser Superposition Transmission (MUST). Now it is included in the LTE-A standard [9].

A power-domain NOMA is also called a multiuser power superposition. An example of this version of NOMA is MUST, which is a two-user downlink power-domain NOMA scheme, where a base station serves simultaneously two users at the same OFDMA subcarrier. The concept of power-domain NOMA is as follows.

The base station selects a pair of users with very different channel conditions, UE1 located close to the BS potentially receiving a strong signal and the second user UE2 with poor reception at the cell edge. In the NOMA downlink, more power is allocated to signal s_2 intended for UE2 located further from the BS and the least power to the signal s_1 for UE1 closest to the BS. Both UEs receive the signal that contains inter-user interference due to multiplexing information for both users into the same resource block, as shown in Figure 13.3.

The SIC method is applied to separate each of the multiplexed signals in receiver. The simplified process for a two-user case is shown in Figure 13.4. Being located near

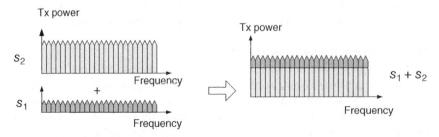

Figure 13.3 Downlink NOMA power allocation.

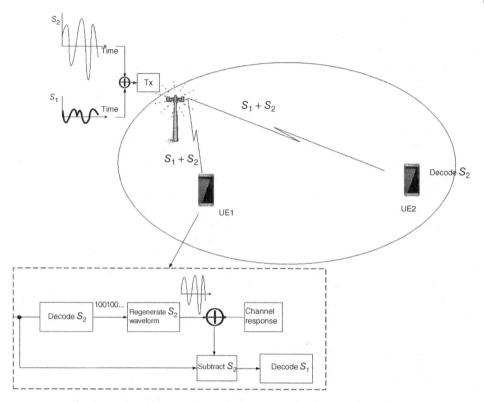

Figure 13.4 Concept of NOMA.

the base station, UE1 receives a very strong s_2 signal in a background of a relatively weak signal s_1. UE1 first decodes s_2 and then regenerates its waveform, as it supposedly received, with the help of estimated channel response. Then UE1 receiver subtracts s_2 from received composite signal $s_1 + s_2$ and decodes s_1. The UE2 receiver can directly decode s_2 without applying the SIC scheme. The s_1 low-power component is likely attenuated down to the noise level due to path loss difference for UE1 and UE2. In the downlink code multiplexing of the signals for multiple users, the SIC receiver decodes the strongest signal first, subtracts it from the received signal and iterates the process until it finds the desired own signal. The performance of SIC depends on the correct cancellation of the interfering signals.

When power-domain NOMA is applied to the large number of users in the cell, the UE with the best channel conditions may have to decode all the other user signals before its own, thus resulting in high processing load and delay. That drawback can be mitigated with a multicarrier NOMA, where the users in a cell are divided into multiple groups. Each group is served by the same orthogonal resource block, and different groups are allocated to different orthogonal resource blocks.

The users within one group are allocated to the same set of subcarriers and intra-group interference is mitigated by using the SIC receiver technology. Different groups of users are allocated to different resource blocks, therefore inter-group interference is avoided. As intended, the number of supported users is greater than the number of available subcarriers. The number of users per subcarrier, nevertheless should not be so large such that overloading will be realized with manageable system complexity. Highest performance for multicarrier NOMA is anticipated when grouping assigns a block of shared subcarriers to the set of users with most different channel conditions.

5G NOMA is still at the research stage. Several NOMA schemes proposed for 5G are discussed in [10]. Special attention is given to NOMA schemes that may overcome the drawback of the SIC power-domain method that hardly works when all users in a group have approximately the same channel conditions. Some new forms of NOMA based on Network-Coded Multiple Access are proposed for implementation. One of such schemes is Lattice Partition Multiple Access (LPMA) [11] based on the orthogonal properties of the lattice code. The users' messages are encoded with lattice codes, multiplied with different prime numbers $p_1, \ldots, p_n$. Before transmission, the module operation $mod(p_1 \cdot p_2 \cdot \ldots \cdot p_n)$ is applied to linear combination of user codes. The UE receiver performs iterative decoding using the modulo operation $mod(p_m)$ starting with strongest signal (assuming s_m). Then the UE has to remove the decoded signal and repeat the procedure until the desired own signal is decoded. The LPMA is claimed to outperform power-domain NOMA in a case when users have similar channel conditions [10].

In order to resolve multiple user signals at the receiver, an advanced interference cancellation technique needs to be implemented. Several interference cancellation techniques are considered for non-linear detector design, see the references in [4].

13.5.1.3 Active Antenna System (AAS)

An AAS integrates active RF components such as power amplifiers and transceivers within an array of antenna elements. The AAS technology will allow deployment of antenna arrays with a large number of antennas enabling a large-scale MIMO systems. The first AASs have been developed for 3G/4G systems, the requirements and basic specifications are documented in 3GPP Releases 13 and 14 [12,13].

The AAS base station radio architecture is represented by three main functional blocks; the *transceiver unit array* (TRXUA), the *radio distribution network* (RDN) and the *antenna array* (AA). The *transceiver units* (TRXU) interface with the baseband processing within the AAS BS.

The TRXUA consists of multiple transmitter units Tx and receiver units Rx. The TRx unit takes the baseband input from the system module in the base station over the OBSAI interface and provides the RF TX outputs at carrier frequency. The transmitted signal of the same band from different *transceiver units* does not appear at a common *TAB connector*. The RF TRX outputs/inputs are distributed to the AA via a RDN. The RDN performs the distribution of the TX outputs into the corresponding antenna paths and antenna elements, and a distribution of RX inputs from antenna paths in the reverse direction. The transmitters (TXs) and receiver units (RXs) can be separated and can have different mapping through the RDN towards the AA.

The transceiver array boundary (TAB) is the point or points at which the TRXUA is connected to the RDN. The point where a TXU or RXU connects with the RDN is equivalent to an 'antenna connector' of a non-AAS BS and is called 'Transceiver Array Boundary connector' (*TAB connector*). The *TAB connector* is defined as conducted requirement reference point. The transmitted signal per carrier from one Transmitter Unit appear at one or more *TAB connector(s)* and the received signals per carrier from one or more *TAB connector(s)* appear at a single RXU.

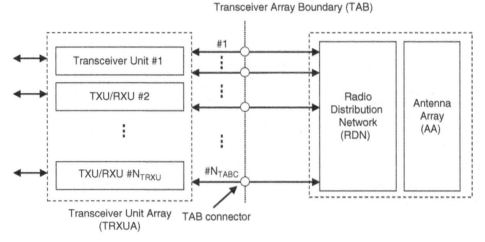

Figure 13.5 General AAS BS radio architecture [12].

Figure 13.5 shows a general AAS BS radio architecture, where N_{TRXU} is the total number of *transceiver units* and N_{TABC} is the total number of *TAB connectors* at the transceiver array boundary.

The transceiver unit array contains an implementation specific number of both transmitter and receiver units. Transmitter units and receiver units may be combined into transceiver units. The transmitter/receiver units have the ability to transmit/receive parallel independent modulated symbol streams.

The composite antenna contains a radio distribution network (RDN) and an antenna array. The RDN is a linear passive network that distributes the RF power generated by the transceiver unit array to the antenna array, and/or distributes the radio signals collected by the antenna array to the transceiver unit array, in an implementation specific way.

Practical composition of active antenna system may include calibration, control and measurement units that split the input signal to respective radios and apply the desired phasing thus forming the antenna beam(s). There could be separate beams for Rx and Tx as well as different beams for different carriers in case of multi-band (multi-RAT) radios and system modules. The schematic composition of an active antenna system with a common control module is shown in Figure 13.6.

As depicted in Figure 13.6, the AAS main blocks may include several radio modules with power amplifiers, cross (X)-polarized antenna array and diplexers, an RET module, internal PSU (Power Supply Unit) with OVP (OverVoltage Protection module) and a Common Module for calibration, measurements and control. The common module is a block in the AAS responsible for beam forming and shaping, synchronization, calibration and control of AAS transceivers. It has interfaces for power supply, to a system module via 2 OBSAI ports.

13.5.1.4 3D Beamforming and Multiuser MIMO (MU-MIMO)

The AAS two-dimensional antennas provide the capability to adjust and control the transmitted beam in the vertical dimension. This approach enables a variety of strategies such as sector-specific elevation beamforming (e.g. adaptive control over the vertical

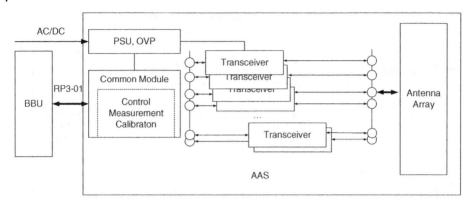

Figure 13.6 Active Antenna System architecture [12].

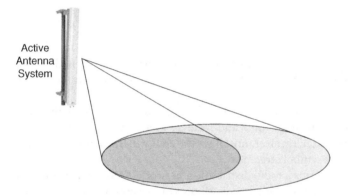

Figure 13.7 Beamforming in the elevation plane with an Active Antenna System.

pattern beamwidth and/or downtilt), advanced sectorization in the vertical domain and user-specific elevation beamforming. Vertical sectorization can improve average system performance through the higher gain of the vertical sector patterns, see Figure 13.7.

Terminal-specific elevation beamforming is a promising method for improving the signal-and-interference-to-noise ratio (SINR) statistics seen by the terminals by pointing the vertical antenna pattern to the direction of the terminal, thus causing less interference to adjacent sectors via steering the transmitted energy in elevation.

13.5.1.5 Massive MIMO

The network system capacity depends on spectral efficiency, system bandwidth and cell density. On the other hand, Shannon's law implies that system capacity is bound by signal-to-interference-and noise ratio (SINR).

Massive MIMO could improve spectrum efficiency or SINR through narrow beamforming. If the base station can realize a high SINR region in wider areas of the cell, then adaptive coding and modulation can provide higher throughput in these areas. Ideally, Massive MIMO should support uniform QoS requirements across the cell area in dense urban, suburban and rural macro cells. Both Massive MIMO and CoMP transmission rely on the availability of reliable Channel Sate Information. The advances in CSI reporting are to be considered in NR network design.

The objective of Massive MIMO is to serve tens of terminals simultaneously using base station antenna arrays with hundreds of antenna elements radiating into a large number of very narrow pencil-like beams. While application of Massive MIMO in the UHF band will improve spectrum efficiency, it is unrealistic to expect improvement in order of magnitude.

The Massive MIMO is best suited in upper band of 5G spectrum, specifically in millimetre wavelengths. Operating at higher frequencies with a reduced array size and increased isolation of inter-cell interference, massive MIMO is expected to be more suitable for small hotspot cells in heterogeneous networks. The small-cell massive MIMO antennas have to be able to flexibly adapt to individual small-cell environments with an optimal number of vertical and horizontal antenna elements and independent transceivers. Deployment of massive MIMO in heterogeneous centralized networks may impose high demands on a front-haul transport capacity.

Management of multiple beams has to be supported with some mechanism of channel estimation and control signalling associated with each antenna beam. As carrier frequencies increase, massive MIMO deployments may become more feasible with small factor antenna arrays. However, millimetre-wave technology rapid deployment may be constrained by cost and power consumption of the RF chips, especially for mobile terminals.

A massive MIMO antenna is made of multiple antenna elements placed in a two-dimensional array. The antenna may consist of hundreds of elements, which implies that the practically possible design is to be based on the AAS concept that combines the antenna, TRX and DA-A/D converters into a single unit. The size of Massive MIMO antenna in the EHF band can be expected in a range of tens of centimetres.

13.5.1.6 Full Duplex Mode

The 5G network is expected to operate in various duplex modes. It includes TDD, FDD with asymmetric and symmetric DL/UL allocation, flexible duplex mode (e.g. unified TDD/FDD frame structure design) and Full Duplex mode.

Flexible and full duplex, as well as schemes that reduce the amount of guard bands and overhead, may improve the overall spectral efficiency. However, given the theoretical limit and technologies, it seems unrealistic to assume order of magnitude improvements.

While flexible duplex could minimize the difference between TDD and FDD design, and reduce the complexity of co-platform implements, full duplex is likely a most revolutionary change in mobile network technology.

Full duplex mode is also called a Simultaneous Transmission and Reception (STR) in the same frequency band with self-interference cancellation. Briefly, full duplex radio is to be able to transmit and receive simultaneously on the same frequency channel. Potential benefits include:

- Doubling capacity
- Potential lower latency full duplex
- Effective adaptation to dynamic DL-UL traffic load with flexible resource allocation
- Integration of backhaul and access for heterogeneous networks in NLOS scenarios
- Full duplex could enable a unified structure of TDD and FDD.

Single channel full duplex is a technology that will bring significant advantages in terms of operation, cost and efficiency to 5G mobile communications systems. Once fully developed, most immediate applications of full duplex can cover backhaul solutions with significantly reduced end-to-end transmission delay, such as:

- self-backhauling,
- millimetre wave backhauling,
- small-cell backhauling,
- relay operation,
- low-latency C-Plane backhaul for CoMP.

The key challenge for full duplex radio is an enormous (over 100 dB) level of interference from transmitter to receiver that needs to be cancelled in order to make a single frequency full duplex radio feasible. The target for interference cancellation level is ~120 dB for a pico cell and ~140 dB for a macro cell [8].

Current full duplex technology is not mature enough for the extreme self-interference cancellation required in mobile network conditions. Most developments in self-interference cancellation area follow a mixed signal design and combine RF analogue cancellation with digital cancellation algorithms operating on the digital baseband IQ samples between the transceiver and the baseband modem. Some tailored solutions relying on combinations of more antenna isolation and advanced interference suppression may be feasible to enable the full duplex for backhauling applications [14, 15]. Some early prototypes are implemented with balun (Balanced to Unbalanced Conversion) techniques or an antenna cancellation technique by means of two transmit antenna separated at half a wavelength [15]. The remaining interference cancellation is performed in the baseband by means of digital cancellation algorithms that are similar to digital pre-distortion (DPD) algorithms widely used in linearization of power amplifiers and other applications.

When fully developed, full duplex will be applied where feasible to resolve issues around FDD (e.g. guard bands) and TDD (e.g. guard time, synchronization). Recommendations from industry suggest that, even if implementation technologies limit the achievable performance by 2020, protocols should be designed to support flexible and full duplex from the beginning if advances in implementation technologies are foreseen [8].

13.5.1.7 Self-Backhauling

The basic concept of self-backhauling is not to use a conventional MW link radio spectrum for the mobile backhaul but, instead, utilize the already licensed mobile radio access spectrum for connection between base stations and core network. The 5G network shall support wireless self-backhaul using NR and E-UTRA with a flexible partitioning of radio resources between access and backhaul functions.

It is anticipated that the 5G network for Enhanced Mobile Broadband will be deployed with a great number of small cells in order to meet demands for high density/high throughput traffic. Deployment of small cells may not always be supported with the fibre connection available for a mobile backhaul. In such circumstances, the reuse of the available access radio spectrum for mobile backhaul can provide a significant cost benefit for deployment and operation. The concept of sharing access spectrum for a backhaul is already implemented in LTE-Advanced Relay Node and will be further developed in 5G.

Wireless self-backhauling is expected to be an integral part of the radio access architecture. Wireless self-backhauling in the radio access network can enable simple deployment and incremental rollout. Network planning and installation efforts can be reduced by leveraging plug and play type features of the Self Organized Network (SON); that is, self-configuration and self-optimization. The requirements for wireless backhauling [16] include the following features:

- multi-hop wireless self-backhauling;
- autonomous adaptation on wireless self-backhaul network topologies to minimize service disruptions;
- topologically redundant connectivity on the wireless self-backhaul.

The requirement for integrated access and backhaul links is quite challenging in the spectrum band below 6 GHz. On the other hand, the large spectrum bandwidth available for NR in the band above 6 GHz makes integrated backhaul practically possible. This may allow easier deployment of a dense network of self-backhauled NR cells in a more integrated manner by building backhaul system design upon many of the control and data channels/procedures defined for providing access to UE. To some extent, LTE-A relay nodes enable such integrated access and backhaul links where relay nodes RN can multiplex access and backhaul links in time, and frequency and space (beamforming).

The operation of the access and backhaul links may be supported on the same (inband) or different (outband) frequencies. Support for outband relays could be important in some specific scenarios. Most of the benefit of wireless self-backhauling is anticipated with inband operation. The cost effective and scalable solution for wireless self-backhauling have not been standardized yet, still being at the R&D stage. Operating NR wireless self-backhauling in a millimetre wave spectrum may present some unique challenges due to requirements for LOS propagation. In some deployment environments, the millimetre wave links may experience relatively long term blockage caused by moving obstacles. Instead of using RRC-Based handover, which may involve core networks, the NG-RAN should be capable of fast switching between and rerouting backhaul between the access nodes, essentially creating the mesh type network.

13.5.2 Technologies to Enhance Network Architectures

Network technologies associated with 5G have to meet requirements for optimal processing of the network node functions and operational efficiency. Major developments are expected in areas of:

- Network virtualization with Software-Defined Networking (SDN);
- Centralized and collaborative system operation based on the Cloud RAN (C-RAN) concept;
- Network Slicing;
- Automation of Operation Support System functions based on advanced Self-Organizing Network technology.

It should be noted, that SDN, SON and C-RAN have been under development for many years. To some extent, these technologies already realized in SW and HW and became operational in current 3G and 4G systems.

13.5.2.1 Software-Defined Network

One of the emerging trends in mobile network architecture is an introduction of SW configurable network nodes that are configurable based on the Software-Defined Network (SDN) architecture and network function virtualization (NFV) for optimal processing of the node functions and improving the operational efficiency of network. The HW for those nodes can be based on ATCA (Advanced Telecommunications Computing Architecture) blades. The ATCA blade is a generic high-speed processor unit that can be SW configured as a switch carrier, processor and so on. Each blade has a standard high-speed Ethernet interface to connect to backplane fabric and other boards. An SDN with virtual functional nodes will provide a novel system with flexibility and enhancements for dynamic, scalable and self-optimization capabilities for data processing and connection with the radio access network.

The Radio Access nodes based on SDR technology have already become reality. Further development in 5G will provide on-demand signal processing and capacity wherever needed. The SDR technology is seamlessly integrated in the 5G network architecture. The 5G core network will evolve on cloud computing and network virtualization to accommodate new applications and services and make a way for a seamless integration with current 3G and 4G networks.

These new capabilities and technologies in the 5G network need to be enhanced to manage the network and associated nodes with new interfaces supporting real-time software updates.

13.5.2.2 Cloud RAN

Future networks will deploy dense networks, which will be more heterogeneous than today. The RAN architecture needs to be very flexible in order to support fully distributed, fully centralized implementations, as well as hybrids of centralized and distributed implementations. A cloud RAN (C-RAN) features centralized and collaborative system operation. The C-RAN network centralizes the baseband and higher-layer processing resources to form a pool so that the processing resources can be managed and allocated dynamically on demand, while the radio units and antenna facilities are deployed in a distributed manner.

The C-RAN architecture can be implemented with a centralized baseband pool and many small cells equipped with radio modules only and are connected via a front-haul interface (CPRI or OBSAI RP3–01). This configuration facilitates joint radio resource management, consolidated traffic steering, effective mobility management of mobility and call re-direction among multiple cells and multiple RATs. The C-RAN with baseband pool enables 'virtual transceiver'; that is, joint transmission and joint processing at the same time to realize the suitable selection and optimal coordination from among a large number of pooled transmitters/receivers. This concept is shown in Figure 14.10 in Chapter 14.

13.5.2.3 Network Slicing

As described in the NGMN white paper [8], diverse new use cases come with great variety of requirements on performance and functionality such as mobility, security, policy control, charging, throughput, latency, reliability and so on. Simultaneous operation of a multitude of service segments and verticals in shared network infrastructure is more effective when different segments are logically isolated from each other into different network slices. Scenarios such as eMBB should not impact on the performance

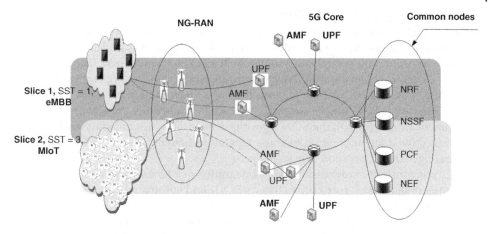

Figure 13.8 5G network slices implemented on a common infrastructure.

of mission critical services. The objective of network slicing is to confine service assurance to a single slice, rather than the whole network.

Figure 13.8 provides a high level illustration of the Network Slicing concept. A network slice is composed of a collection of logical network functions (explained further in Section 13.6.3) that supports the communication service requirements of particular use case(s). Potentially, the operator of the 5G network should be able to dynamically compose an independent set of network functions, possibly from different vendors, into a network slice serving a specific set of applications of market scenarios.

The 5G network shall direct terminals to selected slices based on subscription or terminal type. Network slicing primarily targets a partition of the core network, but it could be possible that 5G RAN may need partitioning of resources for different network slices. The isolation between network slices should ensure that a potential cyber-attack is to be confined to single slice.

A network slice is composed of a collection of 5G network functions and specific RAT settings that support a particular connection type with a specific way of handling the C- and U-planes for specific service. Thus, a 5G network slice can span all domains of the network: software logical modules, specific configurations of the transport network, a dedicated radio configuration, possibly specific RAT, as well as configuration of the 5G device.

Figure 13.8 illustrates the concept of 5G network slices concurrently operated on the same infrastructure. For example, a network slice for eMBB use can be realized by setting a full set of functions distributed across the network. For a network slice supporting a massive sensor network, there is no need for mobility support, therefore only some basic C-plane functions can be configured with contention-based resources for the access.

Certain network slices may need a dedicated infrastructure. In addition, shared infrastructure resources and functions are used between slices as illustrated in Figure 13.9.

In order to realize network slicing, the C- and U-plane functions should be clearly separated, with open interfaces defined between them, in accordance with SDN and C-RAN principles. The objective is to ensure multivendor operation and provisioning of different network functions with forward and backward compatibility.

The 5G Release 15 support for network slicing is described in Section 13.6.3.

13.5.2.4 Self-Organized Network, SON

With full scale LTE deployment and further introduction of 5G system, the number of base stations will expand significantly together with the complexity of a heterogeneous network incorporating several subsystems, such as LTE, 3G and GSM. The concept of the Self-Organizing Network (SON) was introduced to reduce the operating expenditure (OPEX) associated with the management of a huge number of nodes in a multi-vendor network [17]. While the concept and some functionalities are accepted in 3GPP recommendations [18–20], the SON algorithms themselves are not standardized. The SON objective is an automation of some Operation System Support (OSS) functions eliminating manual operation in several areas:

Self-Configuration: automated network integration of a new eNB by auto-connection and auto-configuration, core connectivity (via S1) and automated neighbour site configuration (via X2).

Self-configuration requires an auto-configuration SW agent preloaded in the base station. The base-station agent SW establishes a connection with the SW Repository Server in central Network Management System and downloads the required SW build files. The plan files include the site configuration file, radio and transmission configurations. After the commissioning data is downloaded and activated in the base station, commissioning continues with possible base station tests (antenna line communication tests, Ethernet tests, etc.). Next, a communication exchange involving BS SW agent and commissioner agent in a Central management server follows a standard commissioning procedure resulting in the BS commissioning acceptance and transfer of the control of the base station from commissioner to administration of the operational network.

Self-Optimization: real-time optimization of network parameters based on monitoring performance measurements, fault alarms, notifications and so on.

Self-healing: automated repair that may include fault management and node restart or, for example, power adaptation of neighbour sites to mitigate holes in RF coverage caused by eNB outage. At the base station level, self-healing can be integrated in system maintenance, which includes a supervision system, alarm system, recovery system and diagnostic system.

Self-planning: dynamic re-computation of network plan parameters due to capacity extensions, traffic monitoring or optimization results. The replanned network parameters may include a physical cell identifier, pilot specific settings, neighbour list and so on.

SON Architecture Three major architectural solutions can be proposed for a Self-Organizing Network according to the location of optimization algorithms: Centralized SON, Distributed SON and Hybrid SON.

- *Centralized SON*: In Centralized SON, optimization algorithms are executed in the Centralized Management System. In such solutions, SON functionality resides in a small number of locations at a high level in the architecture. In Centralized SON, all SON functions are located in OAM systems, so it is easy to deploy them. But since different vendors have their own OAM systems, there is low support for optimization cases among different vendors. And it also does not support those simple and quick optimization implement Centralized SON, the existing Itf-N interface needs to be extended.
- *Distributed SON*: In Distributed SON, optimization algorithms are executed in the eNB. In such solutions, SON functionality resides in many locations at a relatively low

level in the architecture. In Distributed SON, all SON functions are relocated in the eNB, so it causes a lot of deployment work. And it is also difficult to support complex optimization schemes, which require the coordination of lots of eNBs. But in Distributed SON, it is easy to support those cases that only concern one or two eNBs and require quick optimization responses. For Distributed SON, the X2 interface needs to be extended.

- *Hybrid SON*: In Hybrid SON, parts of the optimization algorithms are executed in the OAM system, while others are executed in the eNB. In Hybrid SON, simple and quick optimization schemes are implemented in the eNB and complex optimization schemes are implemented in the OAM. So it is very flexible and supports different kinds of optimization cases. And it also supports the optimization between different vendors through the X2 interface. But on the other hand, it costs lots of deployment effort and interface extension work.

SON features have been introduced gradually with the rollout of 4G systems in radio access networks. SON functions have been adapted to the 3G system in order to manage a heterogeneous operator network as a single entity. The introduction of the 5G network triggers further development of the SON to incorporate new technology and features.

13.6 5G System Architecture (Release 15)

13.6.1 General Concepts

The 5G system architecture is defined to support Network Function Virtualization (NFV) and Software-Defined Networking (SDN). The 5G system design envisages structural separation of hardware and software. Key principles of 5G system architecture design are specified in [21]. Those include the following:

- Separate the User-Plane (UP) functions from the Control Plane (CP) function thus ensuring independent scalability, evolution and flexible deployment.
- Modular design enabling efficient Network Slicing.
- Separation of Access Network (AN) and the Core Network (CN).
- The core network is to be designed as a converged core network with a common AN-CN interface integrating different 3GPP and non-3GPP access types.
- The architecture should support concurrent access to local and centralized services.
- Architecture should support roaming with both home routed traffic and local traffic breakout in the visited PLMN.

13.6.2 Architecture Reference Model

The 5G system architecture consists of the set of Network Functions (NF) as illustrated in Figure 13.9.

Authentication Server Function (AUSF). The AUSF supports Authentication Server Function (AUSF)

Core Access and Mobility Management Function (AMF). The Access and Mobility Management function (AMF) includes the following functionality:

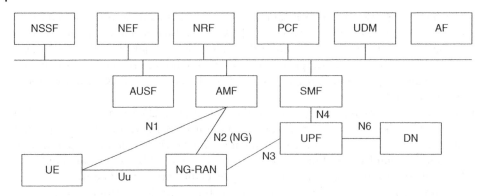

Figure 13.9 5G stand-alone system architecture [21].

- Termination of RAN CP interface (N2).
- Termination of NAS (N1), NAS ciphering and integrity protection.
- Registration management.
- Connection management.
- Reachability management.
- Mobility Management.
- Lawful intercept (for AMF events and interface to the LI System).
- Provide transport for SM messages between the UE and SMF.
- Transparent proxy for routing SM messages.
- Access Authentication and Authorization.
- Security Anchor Function (SEAF). This interacts with the AUSF and the UE, receives the intermediate key that was established as a result of the UE authentication process. In the case of USIM based authentication, the AMF retrieves the security material from the AUSF.
- Security Context Management (SCM). The SCM receives a key from the SEAF that it uses to derive access network specific keys.

Data network (DN), for example operator services, Internet access or third party services.

Network Exposure Function (NEF) provides a means to securely expose the services and capabilities provided by network functions for Application Functions. In addition, it provides a means for the Application Functions to securely provide information to the 3GPP network; for example, Mobility Pattern and communication pattern.

NF Repository Function (NRF) supports the following functionality:

- Service discovery function. Receives a NF Discovery Request from the NF instance and provides the information about the discovered NF instances (be discovered) to the NF instance.
- Maintains the NF profile of the available NF instances and their supported services.

Session Management Function (SMF) includes the following functionality:

- Session establishment, modify and release, including tunnel maintenance between the UPF and AN node.
- UE IP address allocation and management.
- Selection and control of the UP function.
- Configures traffic steering at the UPF to route traffic to the proper destination.
- Termination of interfaces towards policy control functions.

- The control part of policy enforcement and QoS.
- Lawful intercept (for SM events and interface to LI System).
- Charging data collection and support of charging interfaces.
- Control and coordination of charging data collection at UPF.
- Downlink data notification.
- Roaming functionality.

User-plane function (UPF), which includes the following functionality:

- Anchor point for intra-/inter-RAT mobility.
- Packet routing and forwarding.
- Packet inspection and user-plane part of policy rule enforcement.
- Lawful intercept (UP collection).
- Traffic usage reporting.
- Uplink classifier to support routing traffic flows to a data network.
- QoS handling for the user plane; for example, packet filtering, gating, UL/DL rate enforcement.
- Uplink traffic verification (SDF to QoS flow mapping).
- Downlink packet buffering and downlink data notification triggering.

Policy Control function (PCF) supports a unified policy framework to govern network behaviour. The PCF provides policy rules to the Control Plane function(s) to enforce them. It accesses subscription information relevant for policy decisions in a Unified Data Repository (UDR).

Unified Data Management (UDM) supports for the following functionality:

- 3GPP AKA Authentication Credential Processing.
- User Identification Handling.
- Access Authorization.
- Registration/Mobility management.
- Subscription management.
- SMS management.

Application Function (AF) is specified and deployed by the operator. As a trusted function, it is allowed to interact directly with other NFs and with the 3GPP Core Network in order to provide services. Examples of services could be application influence on traffic routing or interaction with the policy framework for policy control [21, 22].

Network Slice Selection Function (NSSF) supports the following functionality:

- Selecting the set of network slice instances serving the UE.
- Determining the allowed Network Slice Selection Assistance Information (NSSAI).
- Determining the AMF Set to be used to serve the UE or, based on configuration, a list of candidate AMF(s), possibly by querying the NRF.

The 5G system architecture identifies the following reference points:

N1:   Reference point between the UE and the AMF.

N2:   Reference point between the New Generation-Radio Access Network (NG-RAN) and the AMF.

N3:   Reference point between the NG-RAN and the UPF.

N4:   Reference point between the SMF and the UPF.

N6:   Reference point between the UPF and a Data Network.

N9:   Reference point between two UPFs.

Also, each Network Function in 5G architecture contains its own service-based interface. Since 5G system architecture allows direct interaction between NFs, two interfaces from both ends are involved in respective dialogue using a common message structure. For instance, an interaction between PCF and SMF enables the PCF to have a dynamic policy and charging control at an SMF by means of:

- Establishment of PDU session by the SMF;
- Request for policy and charging control decision from the SMF to the PCF;
- Provision of policy and charging control decision from the PCF to the SMF;
- Delivery of network events and PDU session parameters from the SMF to the PCF;
- Termination of PDU session by the SMF or the PCF.

13.6.3   Network Slicing Support

13.6.3.1   General Framework
The Network Slice is defined within a PLMN and shall include the Core Network Control Plane and User-Plane Network Functions and, in the serving PLMN, at least one of the following: the NG Radio Access Network function or interworking function to the non-3GPP Access Network.

Network slices may differ for supported features and network function optimizations. The operator may deploy multiple Network Slice instances delivering exactly the same features but for different groups of UEs because they may be dedicated to a customer.

A single UE can simultaneously be served by one or more Network Slices via a 5G access network. A single UE may be served by at most eight Network Slices at a time. The AMF instance serving the UE logically belongs to each of the Network Slice instances serving the UE; that is, this AMF instance is common to the Network Slice instances serving a UE.

The role of various common network functions depicted in Figure 13.9 in support of network slicing is as follows: When the UE sends a Session Management message to establish a PDU session to AMF, the AMF initiates SMF discovery and selection within the selected Network Slice instance. The NRF is used to assist the discovery and selection tasks of the required network functions for the selected Network Slice instance. A PDU session belongs to one and only one specific Network Slice instance per PLMN. Different Network Slice instances do not share a PDU session, though different slices may have slice-specific PDU sessions using the same Data Network Name (DNN).

13.6.3.2   Network Slice Selection Assistance Information (NSSAI)
The NSSAI is a collection of Single Network Slice Selection Assistance Information modules, S-NSSAIs. Each S-NSSAI identifies a Network Slice. The UE subscription may contain multiple S-NSSAIs. One or at most eight of the Subscribed S-NSSAIs can be marked as default S-NSSAI. The NSSAI the UE provides in the Registration Request is verified against the user's subscription data.

A UE can be configured by the Home PLMN with a Configured NSSAI per PLMN. Upon successful completion of a UE's Registration procedure, the UE may obtain from the AMF an Allowed NSSAI for this PLMN, which may include one or more S-NSSAIs. The Allowed NSSAI shall take precedence over the Configured NSSAI for this PLMN.

There can be at most eight S-NSSAIs in the NSSAI sent in signalling messages between the UE and the Network. Each S-NSSAI assists the network in selecting a particular

Table 13.1 Standardized SST values.

Slice/Service type	SST value	Characteristics
eMBB (enhanced Mobile Broadband)	1	Slice suitable for handling of 5G enhanced mobile broadband, useful, but not limited to the general consumer space mobile broadband applications including streaming of High Quality Video, Fast large file transfers etc.
URLLC (ultra-reliable low-latency communications)	2	Supporting ultra-reliable low-latency communications for applications including industrial automation, (remote) control systems.
MIoT (massive IoT)	3	Allowing the support of a large number and high density of IoT devices efficiently and cost effectively.

Network Slice instance. The NSSF uses S-NSSAI for selection the Network Slice instance and the Control plane and User-Plane network functions corresponding to the selected Network Slice serving the UE.

An S-NSSAI is comprised of:

- A Slice/Service type (SST), which refers to the expected Network Slice in terms of features and services;
- A Slice Differentiator (SD), which is optional information that complements the Slice/Service type(s) to differentiate among multiple Network Slices of the same Slice/Service type.

The SST values have to be standardized for establishing global interoperability for slicing in a roaming 5G network. The SSTs standardized in Release 15 are provided in Table 13.1 [21].

13.6.3.3 Selection of a Serving AMF Supporting the Network Slices

After registration with a PLMN the UE is provided with a Temporary User ID (Temp ID) in the RRC connection setup. If the UE for this PLMN has a Configured NSSAI or an Allowed NSSAI, the UE shall provide to the network in RRC and NAS layer a Requested NSSAI containing the S-NSSAI(s) corresponding to the slice(s) to which the UE wishes to register. The NG-RAN selects sn AMF based on a Temp ID provided by UE over RRC. In case a Temp ID is not available, the RAN uses the assistance information provided by the UE at RRC connection establishment (during random access procedure) to select the appropriate AMF instance. If such information is also not available, the RAN routes the UE to a default AMF instance.

The gNB uses the list of supported S-NSSAI(s) previously received in the NG Setup Response message when selecting the AMF with the assistance information. This list may be updated via the AMF Configuration Update message.

The following procedures are then performed in the core network:

- AMF checks whether it can serve all the S-NSSAI(s) from the Requested NSSAI present in the Subscribed S-NSSAIs.
- If this is the case, the AMF remains the serving AMF for the UE. The Allowed NSSAI is then composed of the list of S-NSSAI(s) in the Requested NSSAI permitted based on the Subscribed S-NSSAIs.

- If this is not the case, the AMF queries the NSSF.
- The AMF queries the NSSF with Requested NSSAI, the Subscribed S-NSSAIs, PLMN ID of the subscriber, location information and access technology being used by the UE.
- Based on this information, local configuration and other locally available information including RAN capabilities in the Registration Area, the NSSF does the following:
  1) selects the Network Slice instance(s) to serve the UE;
  2) determines the target AMF Set to be used to serve the UE;
  3) determines the Allowed NSSAI;
  4) based on operator configuration, the NSSF may determine the NRF(s) to be used to select NFs/services within the selected Network Slice instance(s).
- The NSSF returns to the current AMF the Allowed NSSAI and the target AMF Set, or, based on configuration, the list of candidate AMF(s). The NSSF may return the NRF(s) to be used to select NFs/services within the selected Network Slice instance(s).
- Depending on the available information and based on configuration, the AMF may query the NRF with the target AMF Set. The NRF returns a list of candidate AMFs.

The serving AMF shall return to the UE the Allowed NSSAI. It may also indicate to the UE regarding Requested S-NSSAI(s) not included in the Allowed NSSAI, whether the rejection is permanent (e.g. the S-NSSAI is not supported in the PLMN) or temporary (e.g. the S-NSSAI is not currently available in the Registration Area).

Upon successful registration, the UE is provided with a 5G-S-TMSI by the serving AMF. The UE shall include this 5G-S-TMSI in any RRC Connection Establishment during subsequent initial accesses to enable the RAN to route the NAS signalling between the UE and the appropriate AMF.

13.6.3.4   UE Context Handling
Following the initial access, the establishment of the RRC connection and the selection of the correct AMF, the AMF establishes the complete UE context by sending the Initial Context Setup Request message to the NG-RAN over NG-C. The message contains the S-NSSAI as part of the PDU session/s resource description.

The AMF queries the NRF to select an SMF in a Network Slice instance based on S-NSSAI, DNN and other information, for example UE subscription and local operator policies, when the UE triggers the establishment of a PDU session. The selected SMF establishes a PDU session based on S-NSSAI and DNN. Upon successful establishment of the UE context and allocation of PDU resources to the relevant NW slice/s, the NG-RAN responds with the Initial Context Setup Response message. The establishment of a PDU session in a Network Slice to a DN allows data transmission in a Network Slice. A Data Network is associated to an S-NSSAI and a Data Network Name (DNN). NG-RAN confirms the establishment/modification/release of a PDU session associated to a certain NW slice by responding with the PDU Session Setup/Modify/Release Response message over the NG-C interface.

When new PDU sessions need to be established or existing ones modified or released, the 5GC requests the NG-RAN to allocate/release resources relative to the relevant PDU sessions by means of the PDU Session Setup/Modify/Release procedures over NG-C. In the case of network slicing, S-NSSAI information is added per PDU session, so an NG-RAN is enabled to apply policies at the PDU session level according to the SLA represented by the network slice, while still being able to apply (for example) differentiated QoS within the slice.

Further details of the 5G system architecture, interfaces and functionalities can be found in [21].

13.7   New Radio (NR)

NR interface technology will be introduced in 5G in the spectrum band mainly above 6 GHz, while at lower frequencies, the LTE radio interface will be supported in the 5G system. The following are planned capabilities and features of NR in the first Release 15:

- A network can support both LTE and 5G NR, including dual connectivity with which devices have simultaneous connections to LTE and NR.
- Carrier aggregation for multiple NR carriers.
- 5 Gbps peak downlink throughput in initial releases, increasing to 50 Gbps in subsequent versions.
- OFDMA in downlink and uplink, with optional Single-Carrier Frequency Division Multiple Access (SC-FDMA) for uplink.
- Massive MIMO and beamforming.
- Ability to support either FDD or TDD modes for 5G radio bands.
- Numerologies of $2N \times 15$ kHz for subcarrier spacing up to 120 kHz or 240 kHz. This approach supports both narrow radio channels (e.g. 1 MHz), or wide ones (e.g. 400 MHz). Phase 1 is likely to support a maximum of 400 MHz bandwidth with 240 kHz subcarrier spacing.
- Error correction through low-density parity codes (LDPC), which are computationally more efficient than LTE turbo codes at higher data rates.
- Standards-based cloud RAN support, compared with proprietary LTE approaches, which specifies a split between the PDCP and Radio Link Control (RLC) protocol layers.
- Self-contained integrated subframes that combine scheduling, data and acknowledgement.
- Future-proof flexible radio framework that has forward compatibility to future developed services.
- Scalable transmission time intervals with short time intervals for low-latency and longer time intervals for higher spectral efficiency.
- QoS support using a new model.
- Dynamic co-existence with LTE in the same radio channels.

13.7.1   NG-RAN Architecture

The New Generation Radio Access Network (NG-RAN) architecture for NR is based on several principles:

- Logical separation of signalling and data transport interfaces and networks.
- NG-RAN and 5GC functions are fully separated from transport functions. Addressing the scheme used in NG-RAN and 5GC shall not be tied to the addressing schemes of transport functions.

- For each NG-RAN interface, the related Transport Network Layer (TNL) protocol and the functionality are specified. The TNL provides services for user-plane transport, signalling transport.
- Mobility for RRC connection is fully controlled by the NG-RAN.
- One physical network element can implement multiple logical nodes.

Following this principle, the NG-RAN is layered into a Radio Network Layer (RNL) and a Transport Network Layer (TNL). The overall NG-RAN architecture is described in 3GPP TS 38.401 [23]. The NG-RAN architecture, that is the NG-RAN logical nodes and interfaces between them, are defined as part of the RNL. Two types of NG-RAN nodes (base stations) are defined:

- A gNB, providing a NR user plane and control plane protocol terminations towards the UE; or
- A ng-eNB, providing E-UTRA user plane and control plane protocol terminations towards the UE.

The gNBs and ng-eNBs are interconnected with each other by means of the Xn interface. The gNBs and ng-eNBs are also connected via NG interfaces to the 5GC nodes, specifically to the AMF (Access and Mobility Management Function) by means of the NG-C interface and to the UPF (User-Plane Function) by means of the NG-U interface.

The NG-RAN architecture is illustrated in Figure 13.10.

13.7.2 Functional Split

The functional split between access nodes and 5G core is illustrated in Figure 13.11.

The gNB is responsible for Radio Resource Management, encryption and integrity protection of data at air interface, routing of user plane and control plane data to

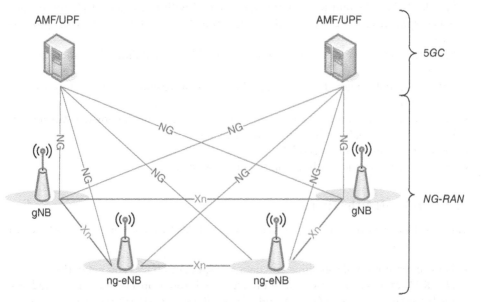

Figure 13.10 Overall architecture [22].

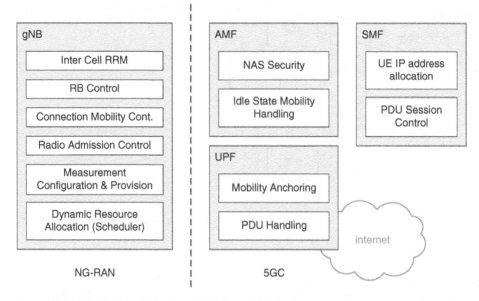

Figure 13.11 Functional split between NG-RAN and 5GC [22].

the UPF and AMF, respectively, connection setup and release, session management, interworking between NR and E-UTRA and network slicing.

The *AMF* main functions are defined in [21]. The AMF is responsible for mobility support and AAA security. The AMF functions include NAS signalling termination and security, inter-core network signalling in support of mobility in the 3GPP access network, Access Authentication and Authorization, Registration Area management, support for intra-system and inter-system mobility, support of network slicing and selection of SMF.

The *UPF* acts as an anchor for intra-/inter-RAT mobility, packet routing and forwarding, packet inspection and the user plane part of policy rule enforcement and QoS handling for the user plane; for example, packet filtering, gating, UL/DL rate enforcement, UL traffic verification (SDF to QoS flow mapping), DL packet buffering and DL data notification triggering.

The *SMF* functions are responsible for session management, the UE IP address allocation and management, selection and control of UP function, the control part of policy enforcement and QoS and DL data notification.

13.7.3   Network Interfaces

13.7.3.1   NG Interface

The NG interface is specified at the boundary between the 5GC and the NG-RAN. Figure 13.11 illustrates the logical division of the NG interface. From the NG perspective, the NG-RAN access point is an NG-RAN node which is either an ng-eNB or a gNB and the 5GC access point is either the control plane AMF logical node or the user plane UPF logical node. Two types of NG interfaces are thus defined at the boundary depending on the 5GC access point the NG-RAN node is connected to: NG-C towards an AMF and NG-U towards an UPF.

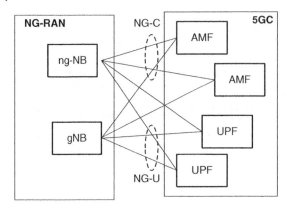

Figure 13.12 NG interface architecture.

The NG-RAN may thus have several NG access points towards the 5GC. The NG interface has the following features:

- it is open;
- it is a point-to-point logical interface between an NG-RAN node and a 5GC node, it is feasible in the absence of a physical direct connection between the NG-RAN and 5GC;
- it supports control plane and user-plane separation;
- it separates Radio Network Layer and Transport Network Layer;
- it is decoupled with the possible NG-RAN deployment variants.

There may be multiple NG-C logical interfaces towards the 5GC from any one NG-RAN node. The selection of the NG-C interface is then determined by the NAS Node Selection. There may be multiple NG-U logical interfaces towards the 5GC from any one NG-RAN node. The selection of the NG-U interface is done within the 5GC and signalled to the NG-RAN Node by the AMF.

The NG interface supports:

- procedures to establish, maintain and release NG-RAN part of PDU sessions;
- procedures to perform intra-RAT handover and inter-RAT handover;
- the separation of each UE on the protocol level for user-specific signalling management;
- the transfer of NAS signalling messages between UE and AMF;
- mechanisms for resource reservation for packet data streams.

The NG user-plane interface (NG-U) is defined between the gNB/eNB and the User-Plane Function (UPF). The transport network layer is built on IP transport and GTP-U is used on top of UDP/IP to carry the user-plane PDUs between the gNB/eNB and the UPF. The NG control plane interface (NG-C) is defined between the gNB/eNB and the AMF. The control plane protocol stack of the NG interface is shown in Figure 13.13. The transport network layer is built on IP transport. For the reliable transport of signalling messages, SCTP is added on top of IP. The application layer signalling protocol is referred to as the NG-AP (NG Application Protocol). The SCTP layer provides guaranteed delivery of application layer messages. In transport, IP layer point-to-point transmission is used to deliver the signalling PDUs.

Figure 13.13 NG interface protocol stack [22].

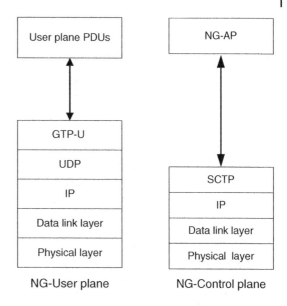

NG-User plane   NG-Control plane

13.7.4 Xn Interface

The Xn User-plane (Xn-U) interface is defined between two gNBs connecting to 5GC, between gNB and ng-eNB connecting to 5GC and between two ng-eNBs connecting to 5GC.

The user-plane protocol stack on the Xn interface is shown in Figure 13.14a. The transport network layer is built on IP transport and GTP-U is used on top of UDP/IP to carry the user-plane PDUs. Xn-U provides non-guaranteed delivery of user-plane PDUs and supports data forwarding and flow control.

The control plane protocol stack of the Xn interface is shown in Figure 13.14b. The transport network layer is built on the SCTP on top of IP. The application layer signalling protocol is referred to as Xn-AP (Xn Application Protocol). The SCTP layer provides the guaranteed delivery of application layer messages. In the transport IP layer, point-to-point transmission is used to deliver the signalling PDUs.

The Xn-C interface manages the UE mobility for connected mode between nodes in the NG-RAN. The Xn-C supports the following functions:

- interface management and error handling (e.g. setup, reset, removal, configuration update);
- connected mode mobility management (handover procedures, sequence number status transfer, UE context retrieval);
- support of RAN paging;
- dual connectivity functions (secondary node addition, reconfiguration, modification, release, etc.).

The user-plane Xn-U supports:

- context transfer from old serving NG-RAN node to new serving NG-RAN node;
- control of user-plane tunnels between the old serving NG-RAN node and new serving NG-RAN node.

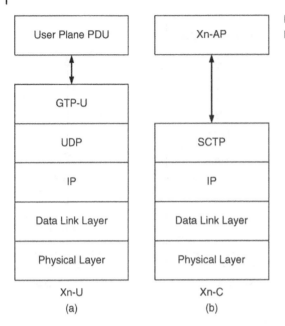

Figure 13.14 Xn protocol stack: (a) user plane and (b) control plane [22].

In a case when gNB connects to the eNB that is still connected to the EPC, the legacy X2 interface will be used for signalling related to UEs connected to the EPC while the new Xn interface will be used for UEs connected to the 5GC.

13.7.5   NG-RAN Distributed Architecture

The NG-RAN supports a distributed base station architecture where a logical gNB may consist of a one major node, gNB-Central Unit (gNB-CU) and several distributed units, gNB-DUs, as shown in Figure 13.15 [23,24]. A gNB-CU and a gNB-DU is connected via F1 logical interface.

In distributed architecture, the only major node gNB-CU is interfacing the core network. Both the NG and Xn-C interfaces for a gNB consisting of a gNB-CU and gNB-DU terminate in the major node gNB-CU. A single unit gNB-DU can be connected to only one central unit, the gNB-CU. The whole set of gNB-CU and connected gNB-DUs is only visible to other base stations nodes, gNBs and the 5GC as a gNB.

All NG-RAN logical nodes and interfaces between them are defined as part of the Radio Network Layer, RNL. For each NG-RAN interface (NG, Xn, F1) the related Transport Network Layer (TNL) protocol and the functionality are specified. The TNL provides services for user-plane transport, signalling transport.

The transport layer address parameter is transported in the radio network application signalling procedures that result in establishment of transport bearer connections. However, those TNL address parameters are transparent in the radio network application protocols.

The formats of the transport layer addresses are further described in 3GPP TS 38.414, 3GPP TS 38.424 and 3GPP TS 38.474. In distributed gNB architecture there is a need for new signalling procedures, such as initial access from distributed unit gNB-DU

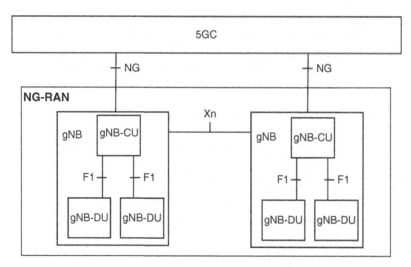

Figure 13.15 NG-RAN gNB architecture [23].

and inter-DU mobility. Distributed gNB architecture supports the mechanism of centralized retransmission of lost PDUs that provides some means to mitigate the link outage between one of the distributed unit and mobile UE.

13.7.5.1 F1 Interface Functions

The F1 interface supports exchange of the traffic data and control information between gNB-DU and gNB-CU. The F1 is a logical point-to-point interface feasible in the absence of a direct physical connection between gNB-CU and gNB-DU. This is an open interface that allows multivendor implementation of the distributed gNB architecture. Some functions of the F1 interface are as follows:

The F1 setup (or update the gNB-DU and gNB-CU configuration) function allows exchange (or update) of theapplication level data needed for the gNB-DU and gNB-CU to interoperate correctly on the F1 interface. The F1 setup is initiated by the gNB-DU.

The error indication function is used by the gNB-DU or gNB-CU to indicate to the gNB-CU or gNB-DU that an error has occurred.

The reset function is used to initialize the peer entity after node setup and after a failure event occurs. This procedure can be used by both the gNB-DU and gNB-CU.

Scheduling of system broadcast information to UE is carried out in the gNB-DU. The gNB-DU is responsible for transmitting the system and paging information according to the scheduling parameters available.

In gNB-CU/gNB-DU distributed architecture, different RRM functions may be located at different locations. The split of RRM functions between gNB-CU and gNB-DU is illustrated in Table 13.2 [23].

13.7.5.2 F1 Protocol Structure

Figure 13.16 shows the protocol structure for F1-C. The Transport Network layer is based on IP transport, with the SCTP for reliable delivery of the F1 Application Part.

Table 13.2 RRM functional split between gNB-CU and gNB-DU.

RM functions	Function description	Location
Radio bearer control	Establishment, maintenance and release of radio bearers	gNB-CU gNB-DU
Radio admission control	Admit or reject the establishment requests for new radio bearers	gNB-CU gNB-DU
Connection mobility control	Management of radio resources in connection with idle or connected mode mobility	gNB-CU
Dynamic resource allocation: packet scheduling	Allocate and de-allocate resources to user and control plane packets	gNB-DU
Cell on/off	Adaptively turn the DL transmission of a cell on and off	gNB-CU gNB-DU
Inter-RAT Radio resource management	Management of radio resources in connection with inter-RAT mobility	gNB-CU

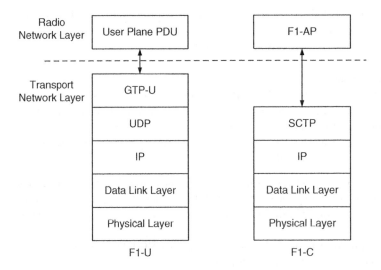

Figure 13.16 Interface protocol structure for F1 [23].

The application layer signalling protocol is referred to as an F1AP (F1 Application Protocol). The user-plane protocol stack is similar but uses a GTP-U on the top of IP.

13.7.6 Radio Protocol Architecture

Radio protocol architecture on the air interface is similar to the predecessor protocol Uu in LTE with separate user and control plane protocol stacks.

13.7.6.1 User Plane

Figure 13.17 shows the Uu radio protocol stack for the user plane, where SDAP, PDCP, RLC and MAC sublayers terminated in gNB on the network side perform the Layer 2

Figure 13.17 User-plane protocol stack [22].

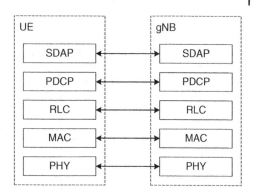

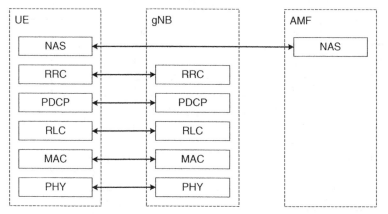

Figure 13.18 Control plane protocol stack [22].

functions illustrated later in Figures 13.19 and 13.20. Figure 13.18 shows the protocol stack for the control plane, where RRC, PDCP, RLC and MAC sublayers are terminated in gNB on the network side perform the functions illustrated in Figures 13.19 and 13.20. The NAS control protocol is terminated in AMF on 5CN side, it performs the functions listed in 3GPP TS 23.501, such as authentication, mobility management and security control.

As can be seen from Figure 13.17, NR introduced a new sublayer Service Data Adaptation Protocol (SDAP) in Layer 2.

The main services and functions of a SDAP sublayer include:

- Mapping between a QoS flow and a data radio bearer;
- Marking QoS flow ID in both DL and UL packets.

The new user-plane protocol layer is applicable for connections to the NG Core. A single protocol entity of the SDAP is configured for each individual PDU session.

Also distinct to NR, a single MAC entity can support one or multiple numerologies and/or TTI durations. MAC use logical channel prioritization and controls which numerology and/or TTI duration a logical channel can use.Layer 2 overall structure for NR is similar to the LTE Layer 2 architecture; most distinctive features and changes selected in the 5G Release 15 standards so far are related to multi-RAT dual connectivity mode and design of the NR physical layer.

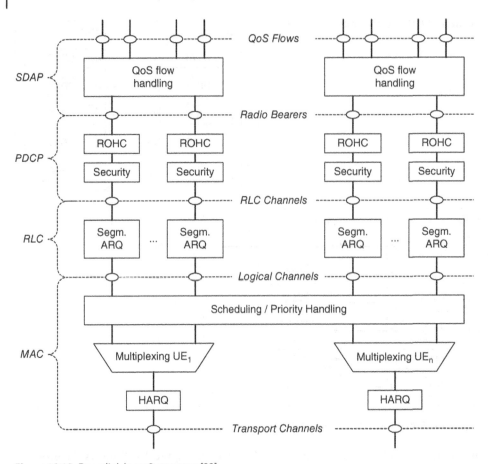

Figure 13.19 Downlink layer 2 structure [22].

13.7.7 NR Physical Channels and Modulation

13.7.7.1 Physical-Layer Design Requirements

5G NR access technology aims to serve use cases with services that may require a range of different bandwidths, such as eMBB, URLLC and MTC. These diverse use case requirements call for a flexible channel arrangement, frame structure and waveforms.

NR supports the paired and unpaired spectrum. The design objectives of NR aim to maximize commonality between the technical solutions, allowing FDD operation on a paired spectrum, different transmission directions in either part of a paired spectrum, TDD operation on an unpaired spectrum where the transmission direction of time resources is not dynamically changed and TDD operation on an unpaired spectrum where the transmission direction of most time resources can be dynamically changing.

DL and UL transmission directions for data can be dynamically assigned on a per-slot basis in the TDM mode. Transmission directions shall include all of the downlink, uplink, sidelink and backhaul link. NR supports at least the semi-statically assigned DL/UL transmission direction as the gNB operation; that is, the assigned DL/UL transmission direction can be signalled to UE by higher-layer signalling. Transmission

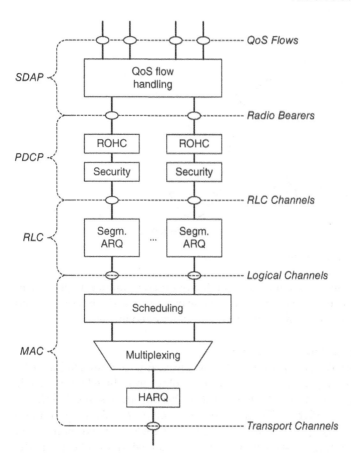

QoS Flows

SDAP

Radio Bearers

PDCP

RLC Channels

RLC

Logical Channels

MAC

Transport Channels

Figure 13.20 Uplink layer 2 structure [22].

in NR is self-contained meaning that demodulation reference signals are mapped onto a given slot and beam making data in a slot is decodable on its own. NR avoids mapping the control channels across the whole system bandwidth.

5G NR will introduce scalable OFDM numerology to support various spectrum bands. The OFDM subcarrier spacing has to increase with channel bandwidth, which increases with carrier frequency in order to keep FFT size and processing complexity manageable. The subcarrier spacing requirements for different bands can be roughly estimated as follows:

Carrier frequency	Subcarrier spacing
Below 6 GHz	15 kHz
6–20 GHz	30–60 kHz
Above 20 GHz	60–480 kHz

Design of 5G NR allows accommodation of mixed numerologies on the same carrier frequency. This may to bring some benefit in support of different services with

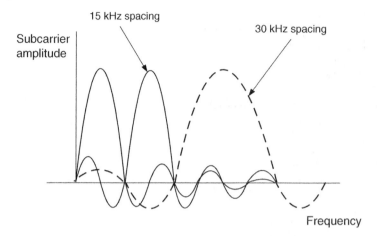

Figure 13.21 Inter-subcarrier interference.

very different latency requirements, however, mixing of different numerologies causes inter-subcarrier interference since energy leaks outside of the subcarrier bandwidth and is picked up by a subcarrier of another numerology, as illustrated in Figure 13.21.

13.7.7.2 Frame Structure and Physical Resources
At least in the first stage of 5G standardization, the 5G physical-layer design follows LTE OFDMA structure and principles yet with a new numerology introduced with scalable subcarrier bandwidth. Specifications 3GPP TS 38.211 [25] introduce some definitions for numerology in 5G NR. Similar to LTE, the basic time unit used in 5G is a minimum sampling interval $T_s = 1/(\Delta f_{max} \cdot N_f) = 0.508$ ns. The $N_f = 4096$ is a maximum size of discrete FFT window and $\Delta f_{max} = 480$ kHz is maximum bandwidth of subcarrier. The size of various fields in the time domain is expressed as a number of time units T_s.

The set of subcarrier bandwidths supported in 5G Release 15 is given by table 4.2–1 in Reference [25]. Multiple OFDM numerologies are supported as given in Table 13.3 where μ and the cyclic prefix for a carrier bandwidth part are given by the higher-layer parameters for the downlink and the uplink.

The length of the cyclic prefix has been proposed to be fixed in terms of number of samples, namely 288 samples in CP [26]. That way, CP processing is uniform across the numerology. Given the different sampling rate for different subcarrier bandwidths, the

Table 13.3 Supported transmission numerologies.

μ	Δf, kHz	Cyclic-prefix type	Symbol duration, μs
0	15	Normal	66.67
1	30	Normal	33.33
2	60	Normal/Extended	16.67
3	120	Normal	8.33
4	240	Normal	4.17
5	480	Normal	2.08

physical length of the CP will vary with subcarrier carrier bandwidth in order to keep a fixed CP overhead in the OFDM symbol. On the other hand, CP physical duration has to cope with a variation in delay spread of the propagation channel being always greater than delay spread.

For instance, at frequencies up to 6 GHz with an FFT size 4096 and subcarrier bandwidth $\Delta f = 15$ kHz sampling interval is $T_s = 16.28$ ns and CP length is 4.69 µs, which is sufficient for non-LOS conditions of system deployment similar to the LTE environment. With millimetre-wave carrier frequencies reliable communications are feasible only within LOS, thus implying a very short cell range. The delay spread in LOS conditions is observed to be much smaller compared with non-LOS and typically is in order of tens of nanoseconds. Assuming the 4096 for FFT window size, 480 kHz subcarrier bandwidth and $T_s = 0.5086$ ns, we obtain the CP length $288 \times T_s = 146.25$ ns.

13.7.8 Frames and Subframes

Downlink and uplink transmissions are organized into frames with $T_f = (\Delta f_{max} N_f / 100) \cdot T_s = 10$ ms duration, consisting of 10 subframes of $T_{sf} = (\Delta f_{max} N_f / 1000) \cdot T_s = 1$ ms duration each. The number of consecutive OFDM symbols per subframe is $N_{symb}^{subframe,\mu} = N_{symb}^{slot} N_{slot}^{subframe,\mu}$. There is one set of frames in the uplink and one set of frames in the downlink on a carrier.

The length of radio frames and subframes in NR is chosen to match the LTE frame structure. Such similarity supports E-UTRA and NR co-existence. For instance, in the case of co-site deployment, cell search and acquisition of the control channels is simplified with aligned frame structure. On the other hand, in order to support multiple numerologies, 5G frame structures also provide flexible substructures for defining slot and symbol transmission timing. Different TTI durations can be defined by the MAC sublayer when using a different number of symbols (e.g. corresponding to a mini-slot, one slot or several slots in one transmission direction).

Which numerologies and/or TTI durations a logical channel of a radio bearer is mapped to can be configured and reconfigured via RRC signalling. The mapping is not visible to RLC, that is the RLC configuration is defined per logical channel with no dependency on numerologies and/or TTI durations and ARQ can operate on any of the numerologies and/or TTI durations the logical channel is configured with. A single MAC entity can support one or multiple numerologies and/or TTI durations. A logical channel prioritization procedure takes into account the mapping of one logical channel to one or more numerologies and/or TTI durations.

New Radio slots and symbols are of flexible lengths and depend on subcarrier spacing. Table 13.4 gives information on relation of slots and frame symbols per subframe depending on subcarrier numerology.

In 5G NR the number of slots and OFDM symbols per subframe is scalable depending on subcarrier bandwidth used. The physical-layer design supports an extended CP. Extended CP will be only one in 60 kHz subcarrier spacing in Release 15. The CP type can be semi-static configured with UE-specific signalling. A UE supporting the extended CP may depend on the UE type/capability.

Just the same as in LTE, Transmission Time Interval (TTI) is determined by subframe duration, for example TTI = 1 ms, but the scheduler works at the slot level, that is minimum scheduling interval is one slot. This allows flexible FDD or TDD implementation

Table 13.4 Frame structure with scalable transmission numerology.

μ	Δf, kHz	N symbol/ slot	N slot/ subframe	N slot/ frame	N sym/ subframe	N sym/ frame
0	15	14	1	10	14	140
1	30	14	2	20	28	280
2	60	14 (12)	4	40	56 (48)	560 (480)
3	120	14	8	80	112	1120
4	240	14	16	160	224	2240
5	480	14	32	320	448	4480

(*)- values with extended cyclic prefix

of NR with the common frame structure. Slot aggregation is supported, that is data transmission can be scheduled to span one or multiple slots depending on subcarrier spacing, as illustrated in Table 13.4.

The 3GPP Working Group is considering the introduction of mini-slot configuration with one or more symbols into the NR specification in the frequency range above 6 GHz. Also, mini-slot structure with two symbols is required for URLLC applications. The Group is also studying the possibility of shortening the LTE subframe (TTI) in order to further reduce latency as is required in some applications.

13.7.9 Physical Resources

13.7.9.1 Resource Grid

For each numerology and carrier, a resource grid of $N_{RB,x}^{max,\mu} N_{sc}^{RB}$ subcarriers and $N_{symb}^{subframe,\mu}$ OFDM symbols is defined, where $N_{RB,x}^{max,\mu}$ is given by Table 13.5 [25] and x is DL or UL for downlink and uplink, respectively. When there is no risk for confusion, the subscript x may be dropped. There is one resource grid per antenna port p, per-subcarrier spacing configuration μ, and per transmission direction (downlink or uplink).

Each element in the resource grid for antenna port p and subcarrier spacing configuration μ is called a resource element and is uniquely identified by the index pair (k, l) where k is the index in the frequency domain and l refers to the symbol position in the time domain.

Table 13.5 Minimum and maximum number of resource blocks.

μ	$N_{RB, DL}^{min, \mu}$	$N_{RB, DL}^{max, \mu}$	$N_{RB, UL}^{min, \mu}$	$N_{RB, UL}^{max, \mu}$
0	24	275	24	275
1	24	275	24	275
2	24	275	24	275
3	24	275	24	275
4	24	138	24	138
5	24	69	24	69

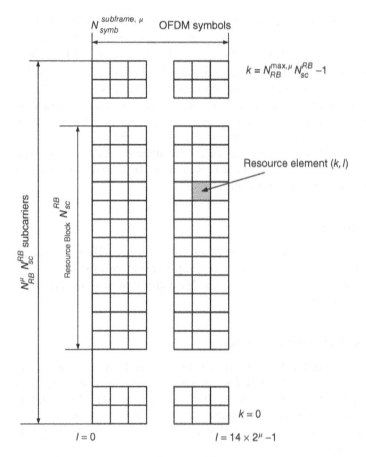

Figure 13.22 Resource grid and resource block [25].

Resource element (k, l) on antenna port p and subcarrier spacing configuration μ is denoted $(k, l)_{p,\mu}$ and corresponds to the complex value $a_{k,l}^{(p,\mu)}$. When there is no risk for confusion, or no particular antenna port or subcarrier spacing is specified, the indices p and μ may be dropped, resulting in $a_{k,l}^{(p)}$ or $a_{k,l}$, see Figure 13.22.

A resource element is classified as 'uplink', 'downlink', 'flexible' or 'reserved'. If a resource element is configured as 'reserved' then neither UL transmission nor DL reception on that resource element is allowed.

13.7.9.2   Resource Blocks

A physical resource block is defined as $N_{sc}^{RB} = 12$ consecutive subcarriers in the frequency domain. Physical resource blocks (PRBs) are numbered from 0 to $N_{RB}^{max,\mu} - 1$ in the frequency domain. The relation between the physical resource block number n_{PRB} in the frequency domain and resource elements (k, l) is given by

$$n_{PRB} = \left\lfloor \frac{k}{N_{sc}^{RB}} \right\rfloor$$

A carrier bandwidth part is a contiguous subset of the physical resource blocks for a given numerology μ_i on a given carrier. The resource blocks in a carrier bandwidth part are numbered from 0 to $N_{RB,x}^{\mu} - 1$ where x is DL or UL for downlink or uplink, respectively. The number of resource blocks in a carrier bandwidth part shall fulfil $N_{RB,x}^{min,\ \mu} \leq N_{RB,x}^{\mu} \leq N_{RB,\ x}^{max,\ \mu}$ where the minimum and maximum values are given by Table 13.5.

A UE can be configured with one or more carrier bandwidth parts in the downlink with a subset of carrier bandwidth parts being active at a given time. The UE is not expected to receive PDSCH or PDCCH outside an active bandwidth part.

A UE can be configured with one or more carrier bandwidth parts in the uplink with a subset of carrier bandwidth parts being active at a given time. The UE shall not transmit PUSCH or PUCCH outside an active bandwidth part.

13.7.10   Carrier Aggregation

Transmissions in multiple cells can be aggregated where up to 15 secondary cells can be used in addition to the primary cell. Unless otherwise noted, the description in this specification applies to each of the up to 16 serving cells. Mechanism of carrier aggregation follows the scheme defined in LTE-A.

13.7.11   Uplink Physical Channels and Signals

On the uplink, a Cyclic-Prefix-OFDM (CP-OFDM)-based waveform is supported in NR. Also, a SC-FDMA (DFT-OFDM) based waveform is supported, complementary to the CP-OFDM waveform at least for an eMBB uplink for up to 40 GHz. A CP-OFDM waveform can be used for single-stream and multi-stream MIMO transmissions, while a DFT-S-OFDM based waveform is limited to single-stream transmissions (targeting for link budget limited cases).

Both CP-OFDM and DFT-S-OFDM based waveforms are mandatory for UEs. The network can decide and communicate to the UE which one of the CP-OFDM and DFT-S-OFDM based waveforms to use. Dynamic switching between transmission methods/schemes is supported.

An uplink physical channel corresponds to a set of resource elements carrying information originating from higher layers. The following uplink physical channels are defined in NR Release 15:

- Physical Uplink Shared Channel, PUSCH
- Physical Uplink Control Channel, PUCCH
- Physical Random Access Channel, PRACH

An uplink physical signal is used by the physical layer but does not carry information originating from higher layers. The following uplink physical signals are defined:

- Demodulation reference signals, DM-RS
- Phase-tracking reference signals, PT-RS
- Sounding reference signal, SRS.

The reference signals DM-RS and SRS are used in a same way as in LTE(A). DM-RS is a user-specific signal that used to estimate radio channel for demodulation. It is confined in the scheduled resource for PUSCH and PUCCH. The SRS is used by a base station in CSI measurements for scheduling and link adaptation.

PT-RS is a newly introduced reference signal in NR. PT-RS is used for compensation of the phase noise in an oscillator. The phase noise impact increases with frequency and, therefore, impact is weighty in the millimetre wave frequency range. For a given UE, the designated PT-RS is confined in the scheduled resource as a baseline. The presence and pattern of PT-RS is RRC configured. Multiple PT-RS densities defined in the time/frequency domain are supported. When present, frequency domain density is associated with dynamic configuration of the scheduled bandwidth.

The signal processing chain for physical channels and signals follows an LTE scheme. This includes scrambling, modulation, layer mapping, precoding, amplitude scaling to conform with the transmit power setting and mapping to the physical time-frequency resource grid.

The NR in Release 15 supports QPSK, 16QAM to 256QAM modulation formats. Additionally, $\pi/2$-BPSK is included in the uplink modulation scheme set for low rate applications that need economically efficient devices in mass deployment, such as mMTC. The physical uplink control channel supports multiple formats with different payloads and numbers of OFDM symbols. A single UE can transmit up to two PUCCHs with different formats [25].

13.7.12 Downlink Physical Channels and Signals

A downlink physical channel corresponds to a set of resource elements carrying information originating from higher layers. The following downlink physical channels are defined:

- Physical Downlink Shared Channel, PDSCH
- Physical Broadcast Channel, PBCH
- Physical Downlink Control Channel, PDCCH.

A downlink physical signal corresponds to a set of resource elements used by the physical layer but does not carry information originating from higher layers.

The following downlink physical signals are defined:

- Demodulation reference signals, DM-RS, for demodulations of PDSCH, PDCCH and PBCCH
- Phase-tracking reference signals, PT-RS, with the same purpose as in uplink
- Channel-state information reference signal, CSI-RS
- Primary synchronization signal, PSS
- Secondary synchronization signal, SSS

In general, the signal processing chain for physical channels and signals is similar to the one developed in LTE. This includes scrambling, modulation, layer mapping, precoding, amplitude scaling to conform with transmit power allocation and mapping to the physical time-frequency resource grid.

To support MIMO transmission, multiple orthogonal DM-RS ports can be scheduled, one per each layer. At the least, the eight orthogonal DL DM-RS ports are supported for SU-MIMO and the maximal 12 orthogonal DL DM-RS ports are supported

Table 13.6 Supported PDCCH aggregation levels.

Aggregation level	Number of CCEs
1	1
2	2
4	4
8	8

for MU-MIMO. With DL DM-RS port multiplexing orthogonality can be achieved by FDM (comb), CDM (Code Division Multiplexing) and TDM.

The number of CSI-RS antenna ports can be independently configured for periodic/semi-persistent CSI reporting and aperiodic CSI reporting. UE can be configured with a CSI-RS resource configuration with up to 32 ports. UE is configured by RRC signalling with one or more CSI-RS resource sets and CSI-RS resources is dynamically allocated from the one or more sets to one or more users. Allocation can be aperiodic (single-shot) and can be on a semi-persistent basis. CSI-RS is mapped in one or multiple symbols. Presence and patterns of PT-RS in a scheduled resource is UE-specific and configured by RRC signalling. In addition to time-frequency synchronization, PSS and SSS are used for a beam training with massive MIMO deployment.

A physical downlink control channel, PDCCH, is structured in a way similar to LTE PDCCH and consists of one or more control-channel elements (CCEs) as indicated in Table 13.6 [25].

CCE consists of six Resource-Element Groups (REGs) where a resource-element group equals one resource block during one OFDM symbol. Resource-element groups within a control-resource set are numbered in increasing order in a time-first manner, starting with 0 for the first OFDM symbol and the lowest-numbered resource block in the control-resource set.

A UE can be configured with multiple control-resource sets. Each control-resource set is associated with one CCE-to-REG mapping only.

13.7.13 SS/PBCH Block

Control information needed for cell search and acquiring basic system information is composed in a self-contained SS/PBCH block. The SS/PBCH consists of 24 contiguous PRB resource blocks with 288 subcarriers in the frequency domain. In the time domain, the SS/PBCH block spans over four OFDM symbols thus making quick acquisition of basic control information for the cell-search and access procedure possible. The resource allocation for the SS/PBCH block is given in Table 13.7 [25].

Here index $v = N_{\text{ID}}^{\text{cell}} \bmod 4$ defines position DM-RS in the frequency domain depending on Physical Cell ID.

A UE can be configured with a set of SS/PBCH block indexes for radio link monitoring in the time domain on a serving cell [27]. The number and OFDM symbol indexes for candidate SS/PBCH blocks depend on carrier frequency and subcarrier bandwidth.

Table 13.7 Resources within an SS/PBCH block for PSS, SSS, PBCH and DM-RS for PBCH.

Channel or signal	OFDM symbol number *l*	Subcarrier number *k*
PSS	0	80, 81, ..., 206
SSS	2	80, 81, ..., 206
PBCH	1, 3	0, 1, ..., 287
DM-RS for PBCH	1, 3	$0 + v, 4 + v, 8 + v, ..., 280 + v, 284 + v$

The overall design of Layer 2 for 5G NR in Release 15 follows the LTE/LTE-A structure. The set of logical channels, mapping to transport channels, functions of MAC, RLC, PDCP, SDAP, RRC sublayers are similar to those in LTE. The details of the Layer 2 design in Release 15 are given in [22].

13.7.14 Coding and Multiplexing

3GPP specifications TS 38.212 [28] describe the mapping of transport to physical channels for 5G NR in Release 15. While still in the development stage, it will detail the scheme of transport block segmentation, channel coding, rate matching and code block concatenations for different uplink and downlink transport channels and control information. The structure and design are similar to LTE format.

Among new features introduced in 5G Release 15 are the new channel coding schemes; namely polar coding for control channels and Low-Density Parity Check (LDPC) coding for transport data channels [29]. Both LDPC and Polar codes outperform turbo codes for high data rates in large block lengths. The polar code for control channel has been selected due to its low complexity and energy efficiency. The LDPC codes allow for high peak throughputs and low-latency. The nature of the LDCP parity check matrix makes possible high coding gains and low coding rates suitable for a use case with high reliability.

13.7.15 NR Dual Connectivity

NR supports a Dual Connectivity (DC) [22] operation whereby a UE in RRC_CONNECTED mode is configured to utilize radio resources provided by two distinct schedulers, located in two gNBs connected via a non-ideal backhaul. As with LTE-A, the objective of Dual Connectivity is to reduce signalling for a mobile UE and improve throughput in a heterogeneous 5G network. Dual Connectivity involves two radio network nodes providing radio resources to a given UE (with active radio bearers), while a single N2 termination point exists for the UE between an AMF and the RAN. gNBs involved in DC may assume two different roles: a gNB may either act as an MgNB (Master gNB) or as an SgNB (Secondary gNB) for connected UE.

NR introduces four radio bearer types for Dual Connectivity: the MCG bearer, MCG split bearer, SCG bearer and SCG split bearer; the latter did not exist in LTE-A Release 12. Those four bearer types are depicted on Figures 13.23 and 13.24.

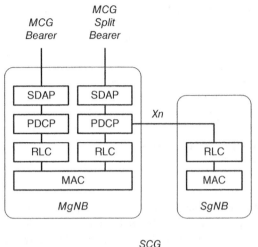

Figure 13.23 MgNB Bearers for Dual Connectivity [22].

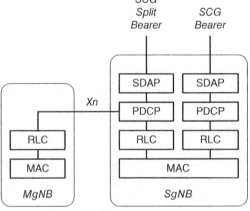

Figure 13.24 SgNB Bearers for Dual Connectivity [22].

The SgNB bearer allows user-plane traffic to be sent over both links similar to a split MgNB bearer. This may reduce the processing load in master node, which normally provides umbrella coverage in HetNet.

13.7.16   E-UTRA and NR Multi-RAT Dual Connectivity

The Multi-RAT Dual Connectivity differs from NR Dual Connectivity where both nodes belong to NR. With Multi-RAT Dual Connectivity (MR-DC), a multiple Rx/Tx UE may be configured to utilize radio resources provided by two distinct schedulers in two different nodes connected via non-ideal backhaul, one providing E-UTRA access and the other one providing NR access. In MR-DC, there is an interface between the MN and the SN for control plane signalling and coordination. For each MR-DC UE, there is also one control plane connection between the MN and a corresponding Core Network entity. The MN and the SN involved in MR-DC for a certain UE control radio resources and are primarily responsible for allocating the radio resources of their cells.

One scheduler is located in the Master Node (MN) and the other in the Secondary Node (SN). The MN and SN are connected via a network interface and at the very least

Figure 13.25 Control plane architecture for EN-DC [30].

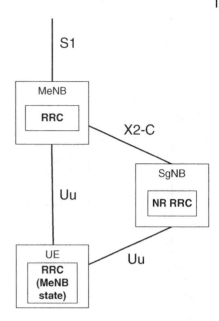

the MN is connected to the core network. Several types of Multi-RAT dual connectivity can be distinguished

1) MR-DC with the EPC, EN-DC. In this case the Master Node is connected to EPC, that is, the 4G core network. In this case a UE is connected to one eNB that acts as a Master Node and one gNB that acts as a SN. The eNB is connected to the EPC and the gNB is connected to the eNB via the X2 interface. Such a connection is named E-UTRA-NR Dual Connectivity (EN-DC) [28]. The control plane architecture for EN-DC is shown in Figure 13.25. In MR-DC with EPC (EN-DC), the involved core network entity is the MME. S1-MME is terminated in Master eNB.

2) MR-DC with the 5GC, NGEN-DC. This is the case where the Master Node is connected to the 5G Core network via the N2 interface. A UE is connected to one ng-eNB that acts as a MN and one gNB that acts as a SN. The ng-eNB is connected to the 5GC and the gNB is connected to the ng-eNB via the Xn interface. This CD mode is named NG-RAN E-UTRA-NR Dual Connectivity (NGEN-DC) [30]. The control plane architecture is shown in Figure 13.26.

3) MR-DC with the 5GC, NE-DC. This is a case similar to NGEN-DC but the gNB and ng-NB swap their roles. A UE is connected to one gNB that acts as a MN and one ng-eNB that acts as a SN. The gNB is connected to 5GC and the ng-eNB is connected to the gNB via the Xn interface. The control plane connectivity is similar to previous case and illustrated in Figure 13.26.

In MR-DC with 5GC (NGEN-DC, NE-DC), the involved core network entity is the AMF. NG-C is terminated in the MN and the MN and the SN are interconnected via Xn-C.

In MR-DC, the UE has a single RRC state, based on the Master Node RRC and a single C-plane connection towards the Core Network. Each radio node has its own RRC entity (E-UTRA version if the node is an eNB or NR version if the node is a gNB), which can generate RRC PDUs to be sent to the UE.

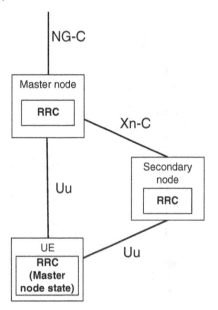

Figure 13.26 Control plane architecture for MR-DC with 5GC [30].

RRC PDUs generated by the SN can be transported via the Master Node to the UE. The MN always sends the initial Secondary Node RRC configuration via a Master Cell Group Signalling Radio Bearer (SRB1), but subsequent reconfigurations may be transported via MN or SN.

In EN-DC and NGEN-DC, at initial connection establishment SRB1 uses E-UTRA PDCP, since the Master Node is an E-UTRA type node (eNB or ng-eNB).

13.7.16.1 Bearer Types for MR-DC Between LTE and NR

In MR-DC, the radio protocol architecture that a particular radio bearer uses depends on how the radio bearer is set up. Four bearer types exist: MCG bearer, MCG split bearer, SCG bearer and SCG split bearer. These four bearer types are depicted in Figure 13.27 for MR-DC with EPC (EN-DC) and in Figure 13.28 for MR-DC with 5GC (NGEN-DC, NE-DC).

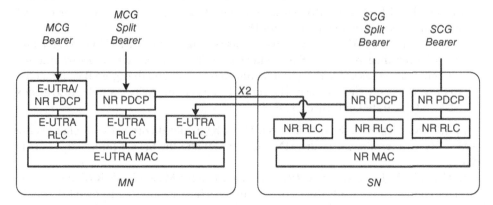

Figure 13.27 Radio Protocol Architecture for MCG, MCG split, SCG and SCG split bearers in MR-DC with EPC (EN-DC) [30].

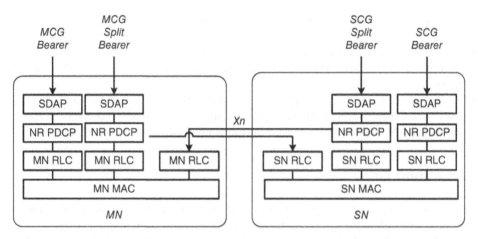

Figure 13.28 Radio Protocol Architecture for MGC, MCG split, SCG and SCG split bearers in MR-DC with 5GC (NGEN-DC, NE-DC) [30].

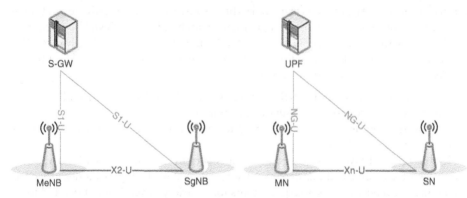

Figure 13.29 U-plane connectivity for EN-DC (left) and MR-DC with 5GC (right) [30].

For EN-DC, the network can configure either E-UTRA PDCP or NR PDCP for MCG bearers while NR PDCP is always used for SCG bearers. For split bearers, NR PDCP is always used and from a UE perspective there is no difference between MCG and SCG split bearers.

In MR-DC with 5GC, NR PDCP is always used for all bearer types.

13.7.16.2 MR-DC User-Plane Connectivity

The U-plane connectivity options between MN, SN and core network depend on bearer options configured. Figure 13.29 illustrates connectivity options to EPC and 5GC. The choice of the bearer option implies:

- For MCG bearers, the user-plane connection to the CN entity is terminated in the MN. The SN is not involved in the transport of user-plane data for this type of bearer(s) over the Uu.
- For MCG split bearers, the user-plane connection to the CN entity is terminated in the MN. PDCP data is transferred between the MN and the SN via MN-SN user-plane

interface. The SN and MN are involved in transmitting data of this bearer type over the Uu.

- For SCG bearers, the SN is directly connected with the CN entity via a user-plane interface. The MN is not involved in the transport of user-plane data for this type of bearer(s) over the Uu.
- For SCG split bearers, the user-plane connection to the CN entity is terminated in the SN. PDCP data is transferred between the SN and the MN via MN-SN plane interface. The SN and MN are involved in transmitting data of this bearer type over the Uu.

13.8 Summary

5G is an emerging technology promising to address an increasing demand for enormous growth in network capacity and make wireless broadband connection a common tool for human and machine-to-machine communication. 5G standardization process has started in Release 14 with a study of the new 5G radio, and continues now with a first phase of specifications in Release 15 that provide both non-standalone and standalone options for the 5G network. 5G is expected to be designed as a mix of further evolution of LTE-A and revolutionary network at higher spectrum bands. The 5G will be integrated with LTE networks, and many 5G features can be implemented as LTE-Advanced Pro extensions prior to full 5G availability. The major up to date features adopted for the first 5G releases can be formulated as follows:

1) Introduction of New Radio (NR) technology and New Generation Radio Access architecture that support full interworking with LTE.
2) New numerology is introduced in the NR physical layer with variable subcarrier bandwidth from 15 to 480 kHz.
3) Flexible physical-layer design allows frame structure configuration with a variable number of slots and symbols capable of supporting flexible duplex modes as well as meeting a diverse set of user requirements for a variety of services anticipated to be provided with 5G.
4) NR supports Multi-RAT dual connectivity when UE can have two LTE and NR radio bearers simultaneously while being connected either to 5G Core or EPC.
5) Both 5G Core and NR supports Network Slicing that allow operators to create an isolated subset of network functions/virtual nodes that run specific functionality needed to efficiently support different needs: such as a fixed IoT service or eMBB smart phones or URLLC. Network Slicing enables the network elements and functions to be configured and reused in each network slice in order to run a specific service. Each slice can have its own network architecture likely built as a Virtual Network Function and network provisioning.

References

1 IMT Vision – Framework and overall objectives of the future development of IMT for 2020 and beyond, ITU-R M.2083–0.
2 Feasibility Study on New Services and Markets Technology Enablers; Stage 1 (Release 14); 3GPP TR 22.891, V14.2.0 (2016–09).

3 Evolved Universal Terrestrial Radio Access (E-UTRA); NB-IOT, 3GPP TS 36.802.
4 New frequency range for NR (3.3–4.2 GHz) (Release 15); 3GPP TR 38.813 V0.0.1 (2017–08).
5 New frequency range for NR (4.4–4.99 GHz) (Release 15); 3GPP TR 38.814 V0.0.1 (2017–08).
6 Technical feasibility of IMT in bands above 6 GHz, Report ITU-R M.2376–0 (07/2015).
7 Future technology trends of terrestrial IMT systems, Report ITU-R M.2320–0 (11/2014).
8 5G White Paper, By NGMN Alliance, 17-February-2015.
9 Study on Downlink Multiuser Superposition Transmission (MUST) for LTE (Release 13); 3GPP TR 36.859 V13.0.0 (2015–12).
10 Ding, Z., Lei, X., Karagiannidis, K. et al., A survey on non-orthogonal multiple access for 5G Networks: Research challenges and future trends, *IEEE Journal on Selected Areas in Communications*, 35(10), Oct. 2017: 2181–2195.
11 Fang, D., Huang, Y.-C., Ding, Z. et al., Lattice partition multiple access: a new method of downlink non-orthogonal multiuser transmissions; *Proceedings of GLOBECOM 2016 – 2016 IEEE Global Communications Conference*. Washington DC, USA, 2016.
12 Radio Frequency (RF) requirement background for Active Antenna System (AAS) Base Station (BS), 3GPP TR 37.842 V13.2.0 (2017–03).
13 Active Antenna System (AAS) Base Station (BS) transmission and reception, (Release 14); 3GPP TS 37.105 V14.0.0 (2017–03).
14 Hong, S., Brand, J., Choi, J. et al., Applications of self-interference cancellation in 5G and beyond, *IEEE Communications Magazine*, February 2014: 114–121.
15 Choi, J., Jain, M., Srinivasan, K. et al., *Achieving single channel, full duplex wireless communication*. Available online at: https://web.stanford.edu/~skatti/pubs/mobicom10-fd.pdf (accessed January 2018). Stanford University, Stanford, CA, USA.
16 Service requirements for next generation new services and markets; Stage 1 (Release 15); 3GPP TS 22.261 V0.1.1 (2016–08).
17 Self-Organizing Networks (SON); Concepts and requirements, 3GPP TS 32.500 V10.0.0 (2010–06).
18 Self-configuration of network elements; Concepts and requirements, 3GPP TS 32.501 V14.0.0 (2017–04).
19 Automatic Neighbour Relation (ANR) management; Concepts and requirements, 3GPP TS 32.511 V14.0.0 (2017–04).
20 Self-Organizing Networks (SON) Policy Network Resource Model (NRM) Integration Reference Point (IRP); 3GPP TS 32.521 V11.1.0 (2012–12).
21 System Architecture for the 5G System; Stage 2, (Release 15); 3GPP TS 23.501 V1.3.0 (2017–09).
22 NR; NR and NG-RAN Overall Description; Stage 2 (Release 15); 3GPP TS 38.300 V1.0.0 (2017–09).
23 NG-RAN; Architecture description (Release 15); 3GPP TS 38.401 V0.2.0 (2017–07).
24 NG-RAN; NG general aspects and principles (Release 15); 3GPP TS 38.410 V0.4.0 (2017–09).
25 NR; Physical channels and modulation (Release 15); 3GPP TS 38.211 V1.0.0 (2017–09).

26 Ali A. Zaidi, A.A., Baldemair, R., Tullberg, H. et al., Waveform and numerology to support 5G services and requirements, *IEEE Communications Magazine*, 54, Nov. 2016: 90–98.

27 NR; Physical-layer procedures for control (Release 15); 3GPP TS 38.213 V1.0.0 (2017–09).

28 NR; Multiplexing and channel coding (Release 15); 3GPP TS 38.212 V1.0.0 (2017–09).

29 Study on New Radio (NR) access technology (Release 14); 3GPP TR 38.912 V14.1.0 (2017–06).

30 Evolved Universal Terrestrial Radio Access (E-UTRA) and NR; Multi-connectivity; Stage 2 (Release 15); 3GPP TS 37.340 V1.0.0 (2017–09).

14

Annex: Base-Station Site Solutions

In earlier days, the GSM base stations used to be housed in large racks installed indoors in air-conditioned rooms. An example of an old site solution is shown in Figure 14.1. Current requirements for base-station design include ensuring easy and cost-effective deployment, scalable modular design, efficient OAM and high reliability in different environmental conditions. All mobile network vendors offer base stations based on Software Designed Radio (SDR) with various options for installing the base-station modules indoor and outdoors, just by using suitable module casing.

The installation requirements for the base-station site normally follow a country's environmental standards for operation of mobile networks. Those typically set constraints for electromagnetic emission shielding, acoustic noise, ingress protection, electrical requirements for grounding, lighting and Over Voltage Protection (OVP). From an operational point-of-view a site has to have an AC or DC power supply and capabilities for backhaul transport connectivity via microwave or fibre to the core network.

Major site constraints may come from power backup requirements due to battery string size and weight. Depending on the requirement for power backup duration, one may need additional racks to house the number of battery strings.

14.1 The Base-Station OBSAI Architecture

The modern base-station architecture is based on the OBSAI standard recently developed by Industry Working Groups. The OBSAI (Open Base-Station Architecture Initiative) [1] defines both open standardized modular base-station structure and digital interfaces between modules. OBSAI provide a complete set of interface specifications to cover the functions of all base-station modules: control, transport, baseband and radio. The standardized functions of each module and defined internal digital interfaces make possible multivendor interoperability and integration of a set of common modules into the base-station structure. The OBSAI Reference Architecture defines functional modules, interfaces between them and requirements for external interfaces.

14.1.1 Functional Modules

There are four main blocks (modules) in defined in an OBSAI base transceiver station (BTS): Radio Frequency (RF) block, baseband block, control and clock block and transport block, as depicted in Figure 14.2.

Introduction to Mobile Network Engineering: GSM, 3G-WCDMA, LTE and the Road to 5G,
First Edition. Alexander Kukushkin.
© 2018 John Wiley & Sons Ltd. Published 2018 by John Wiley & Sons Ltd.

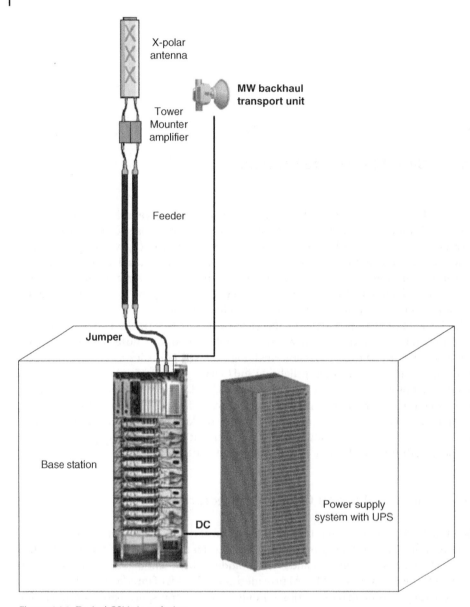

Figure 14.1 Typical GSM site solution.

The *Radio Frequency Block* sends and receives signals to/from portable devices (via the air interface) and converts between digital data and antenna signal. The main functions of RF module are D/A and A/D conversion, up/down conversion, carrier selection, linear power amplification, diversity transmit and receive, RF combining and RF filtering.

The *Baseband Block* performs digital processing of the baseband signal. This includes encoding/decoding, ciphering/deciphering, frequency hopping and spreading/despreading.

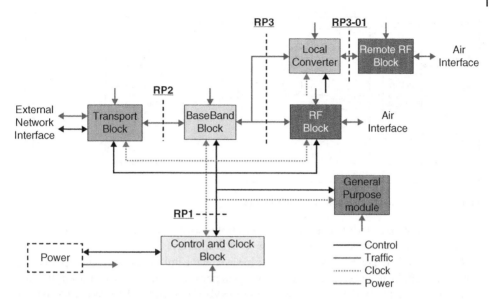

Figure 14.2 OBSAI BTS reference architecture [2].

The *Transport-Block* interfaces to the external network and provides functions such as
    QoS, security functions and synchronization.
Coordination between these three blocks is maintained by the *Control and Clock Block*.

14.1.2 Internal Interfaces

Internal interfaces between the functional blocks are called reference points (RP), shown
in Figure 14.2.

RP1 is the interface that allows communication between the control block and the other
    three blocks. It includes control and clock signals.
RP2 provides a link between the transport and baseband blocks.
RP3 is the interface between baseband block and RF block. RP3–01 is an (alternate)
    interface between the Local Converter and Remote RF block.
RP4 provides the DC power interface between the internal modules and DC power
    sources.

The most important development in OBSAI architecture is development of an RP3
interface that separates the RF and system modules. The motivation for that is the
cost of RF modules and power amplifier, those components account for about 50% of
base-station cost. The open standard for RP3 interface enables multivendor integration
and subsequent cost reduction of RAN architecture.

14.1.3 RP3 Interface

The OBSAI RP3 specification [3] defines the interface between the baseband module
and the RF module, which includes a maximum number of nine pairs of unidirectional
links for every baseband and RF module. Each link represents digitized IQ carrier data.
The sampling rate of the I and Q channels is variable depending on carrier bandwidth

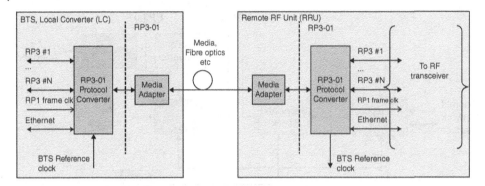

Figure 14.3 Logical structure of RP3–01 interface [3].

with format of up to 16 bits per sample. Those links can be connected with either a mesh or a centralized combiner and distributor (C/D) topology.

RP3 uses a four-layer protocol stack: physical, data-link, transport and application layers. The application layer provides mapping of packets to the payload. The application layer supports LTE, W-CDMA, CDMA2000, GSM/EDGE and 802.16 (WIMAX) packet types, it can also be expanded to accommodate future packet types. The RP3 has a fixed length message of 19 bytes with 16 bytes of data and 3 bytes containing the header.

The transport layer provides end-to-end message routing. The data-link layer frames and synchronizes the messages, while the physical layer sends out the messages on the electrical interfaces that include serializing and coding the data. The RP3 bus clock signal is provided externally via RP1 interface, the RP3 frame is synchronized by 10 ms reference counters.

The RP3–01 is an evolved RP3 interface that realizes a high-speed optical communication link between the Local Converter (LC) module and the remote RF module, called the Remote Radio Head (RRH). The role of the LC is to convert RP3 and RP1 data to the RP3–01 protocol and adapt it to an optical transport media. The RP1 data includes Ethernet and frame clock bursts. The local converter is normally integrated in the Baseband module.

The logical structure of RP3–01 interface is shown in Figure 14.3. The RP3–01 interface supports bi-directional transfer of digitized baseband radio data together with control and air-interface synchronization information. This way the RRH is synchronized to a baseband unit that can be physically separated from RRH at a significant distance.

RP3–01 and RP3 data rates have the same format of integer multiples of 768 Mbps; that is, a 768 Mbps, 1536 Mbps, 3072 Mbps or 6144 Mbps line rate is used.

14.2 Common Public Radio Interface, CPRI

CPRI specifications [4] were created by industry groups in order to standardize the open interface between the radio module and baseband unit. Unlike OBSAI, the CPRI concentrate only on one interface in place of RP3 in OBSAI. The specification covers Layers 1 and 2 of the Open System Interconnect (OSI) stack. Layer 1 supports both an electrical interface, used in traditional base stations, as well as an optical interface for base stations with remote radio equipment. Layer 2 supports flexibility and scalability.

CPRI allows three line bit-rate options. It is mandatory for REC and RE to support at least one of these options, which include 614.4, 1228.8 and 2457.6 Mbps. CPRI does not have a mandatory physical-layer protocol. The protocol used must meet the bit-error-rate (BER) specification, as well as the clock stability and noise requirement. For an optical transceiver, Gigabit Ethernet, 10 Gigabit Ethernet, fibre channel and InfiniBand are recommended.

Both CPRI and OBSAI standards have a link layer to support special requirements in terms of latency and timing synchronization. Each has a physical layer that is based on already existing electrical standards from Ethernet 10 Gigabit Attachment Unit Interface (XAUI) and Gigabit Ethernet. Both also allow different data rates to implement the various market requirements in terms of carriers and sectors. Additionally, the specifications offer a way to commoditize the components of the radio network and complement existing standardization efforts.

14.3 SDR and Multiradio BTS

Another trend in the design of the base stations and mobile handsets is a Software-Defined Radio (SDR) [5, 6]. The SDR is the concept of a wireless device able to change its functionality by switching in different software stacks (or software-like objects) that can be stored locally or downloaded over the air. In practice, it means that the base-station components IF, baseband and bit-stream processing is implemented in general-purpose programmable processors. The resulting software-defined radio (SDR) extends the evolution of programmable hardware, increasing flexibility via increased programmability.

An ideal SDR base station or handset consists of a power supply, an antenna, a multi-band RF converter, a single chip containing ADC/DAC and on-chip general-purpose processor and memory that perform the radio functions, as illustrated in Figure 14.4.

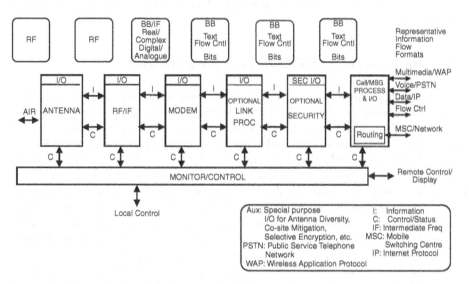

Figure 14.4 Generalized modular SDR architecture [7].

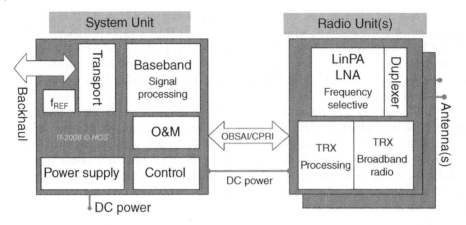

Figure 14.5 OBSAI base-station architecture with separate RF and system modules [8].

While the ideal SDR is still under R&D, the multiradio base station based on the SDR principle with a programmable baseband module is widely available from different mobile network vendors.

The modern base station based on OBSAI and SDR principle consists of one or several base-band modules, also called base-band units (BBU) or system modules (SM) and several RF modules or RRHs (also called Remote Radio Units, RRU), as shown in Figure 14.5. The system module itself can be programmed to support several standards, such as LTE and WCDMA or GSM with appropriate software load. Some vendors can support multiple mobile technology standards (Multi-RAT) in a single module in concurrent mode of operation. Apparently, the common RF module for concurrent operation is possible when all RATs operate in the same RF spectrum band, otherwise different RF modules can be connected to a single system module capable of concurrent processing of Multi-RAT signals to/from RF units. The system module block is scalable in capacity and several system blocks can be chained to perform as a single high-capacity processor.

14.4 Site Solution with OBSAI Type Base Stations

The OBSAI architecture allows various options for base-station site solutions including feederless installation as well as distributed site architecture. All base-station vendors produce OBSAI modules suitable for conventional indoor installation in cabinets as well as for outdoor installation on walls, poles and floors. Both system module and RF module has a closing at IP65 protection level and can withstand a wide range of temperatures for reliable operation. With a feederless site, the RF or RRH module is installed very close to the antenna, thus allowing short jumper cables for connecting the RF module to the antenna. The baseband system module is then connected to remote RF module via RP3–01 interface with an optical single or multimode cable as schematically shown in Figure 14.6. The optical physical layer can be implemented by means of Small Form factor Pluggable (SFP) multimode optical transceivers shown in Figure 14.7.

The DC feed to the feederless site in Figure 14.6 is provided either by the vendor power system or third party products. The input for System and RF Modules is a standard

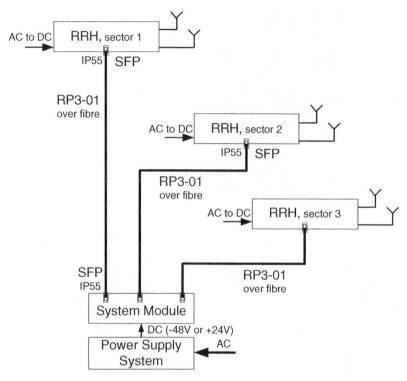

Figure 14.6 Feederless site solution example with OBSAI architecture.

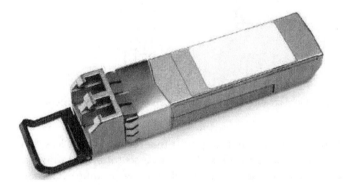

Figure 14.7 SFP, a small form factor pluggable optical transceiver.

−48 V or +24 V DC, depending on vendor. There are two basic alternatives to supply power to the feederless site and its modules:

1) DC feed from System Module to RF Module, Figure 14.5a. However, in this case, distance separation between RF and system module is limited to losses in DC cabling. The longer wire requires a larger cross-section to contain path loss.
2) DC feed to RF Modules from the vendor or third party AC/DC power supply system.

The RRHs in Figure 14.6 are connected to the sector antenna with short jumpers. An example of a compact RF solution from the Nokia Flexi RF module is shown in

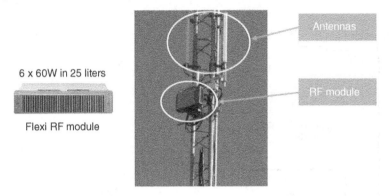

6 x 60W in 25 liters

Flexi RF module

Figure 14.8 Compact Nokia RF module site solutions [9].

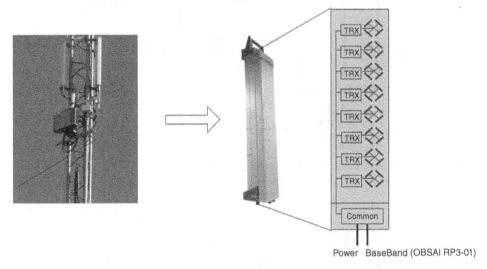

Power   BaseBand (OBSAI RP3-01)

Figure 14.9 Integration of RF modules into an antenna array forming an Active Antenna System (AAS) [9].

Figure 14.8. This module delivers 6 × 60 W output power in a form factor of less than 25 litres. That single module can provide MIMO2x2 for a three-sector base station.

With an active antenna system (AAS), the on top installation has become even more compact, as shown in Figure 14.9. In this case the RF module is integrated with the phased antenna array. The AAS module has an OBSAI RP3–01 interface to the system module and two SFP connectors; the second SFP port is used for chaining to another AAS module to achieve RF coverage for another sector.

14.4.1   C-RAN Site Solutions

With C-RAN, the System Modules (BBUs) can be kept in a central location (possibly indoor) together with backhaul transport units. The RRHs can be placed remotely in suitable outdoor locations, such as masts or rooftop spaces for optimal RF coverage, and connected via a RP03–01 or CPRI front-haul interface over optical fibre cables between

Radio   Front-haul   Baseband pooling   Backhaul   Aggregation/Core

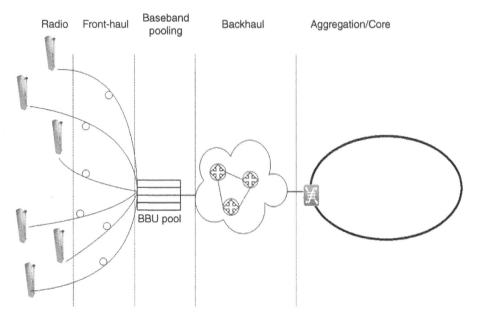

Figure 14.10 Centralized (Cloud) RAN architecture.

system module and RRH. The schematic connection diagram is shown in Figure 14.10. Development of SDR enables introduction of multi-RAT baseband pooling across multiple carriers and frequency bands.

References

1 Open Base Station Architecture Initiative. Website, available at: www.obsai.com.

2 OBSAI, Open Base Station Architecture Initiative BTS System Reference Document Version 2.0. Available online at: http://www.obsai.com/specs/OBSAI_System_Spec_V2.0.pdf (accessed January 2018), 2006.

3 OBSAI, Reference Point 3 Specification Version 4.2. Available online at: http://www.obsai.com/specs/RP3_Specification_V4.2.pdf, 2010 (accessed January 2018).

4 CPRI, Common Public Radio Interface: Website. Available at: http://www.cpri.info/.

5 Tuttlebee, W.H.W. (ed.) *Software Defined Radio: Baseband Technologies for 3G Handsets and Basestations*. John Wiley & Sons, Inc., 2004.

6 Mitola, J. *Software Radio Architecture: Object-Oriented Approaches to Wireless Systems Engineering*, John Wiley & Sons, Inc., 2000.

7 SDR Forum, *Software Defined Radio Forum Base Station Working Group Base Station System Structure Document No. SDRF-01-P-0006-V2.0.0*. Available online at: http://www.wirelessinnovation.org/assets/work_products/Reports/sdrf-01-p-0006-v2_0_0_basestation_systems.pdf (accessed January 2018), 2002.

8 Reconfigurable Radio Systems (RRS), ETSI TR 102 681 V1.1.1 (2009–06).

9 Nokia, *Multi-antenna Optimization in LTE Extended Coverage, Enhanced Data Rates and Higher Capacity with Existing Macro Sites*, Available online at: https://tools.ext.nokia.com/asset/200187 (accessed January 2018), 2015

Index

Introduction to Mobile Network Engineering: GSM, 3G-WCDMA, LTE and the Road to 5G,
First Edition. Alexander Kukushkin.
© 2018 John Wiley & Sons Ltd. Published 2018 by John Wiley & Sons Ltd.